CPC UCW
LLYFRGELL
LIBRARY
ABERYSTWYTH

AF616159

ENZYME ENGINEERING

Volume 4

A Continuation Order Plan is available for this series. A continuation order will bring delivery of each new volume immediately upon publication. Volumes are billed only upon actual shipment. For further information please contact the publisher.

ENZYME ENGINEERING
Volume 4

Edited by

Georges B. Broun
Département de Génie Biologique
Université de Technologie
Compiègne, France

Georg Manecke
Institut für Organische Chemie
Freie Universität Berlin
Berlin, Federal Republic of Germany

and

Lemuel B. Wingard, Jr.
Department of Pharmacology
School of Medicine
University of Pittsburgh
Pittsburgh, Pennsylvania, U.S.A.

PLENUM PRESS • NEW YORK AND LONDON

The Library of Congress cataloged the second volume of this title as follows:

Engineering Foundation Conference on Enzyme Engineering, 2d, Henniker, N. H., 1973.
Enzyme engineering.

Called volume 2 in continuation of a volume with the same title published in 1972, which contains the papers of the 1st Engineering Foundation Conference on Enzyme Engineering.
1. Enzymes — Industrial applications — Congresses. I. Pye, E. Kendall, ed. II. Wingard, L., ed. III. Title. [DNLM: 1. Biomedical engineering — Congresses. 2. Enzymes — Congresses. W3 EN696]
TP248.E5E53 1973 660'.63 74-13768
ISBN 0-306-35282-6 (v. 2)

ACKNOWLEDGMENTS

Permission has been received from the various publishers for reproduction of the following figures.

p. 126-127, Fig. 1, 2 *Biotechnol. Bioeng.* (John Wiley & Sons, Inc.)

p. 145, Fig. 1 *Eur. J. Biochem.* (Springer-Verlag)

p. 295-297, Fig. 1, 2 *Prog. Biophys. Molec. Biol.* (Pergamon Press)

p. 456, Fig. 1 *FEBS Lett.* (Elsevier/North Holland)

This volume contains most of the papers presented at the Fourth Engineering Foundation Conference on Enzyme Engineering, cosponsored by the DECHEMA (Deutsche Gesellschaft für chemisches Apparatewesen e. V), and held at Bad Neuenahr, Federal Republic of Germany, September 25—30, 1977

Library of Congress Catalog Card Number 74-13768

ISBN 0-306-40021-9

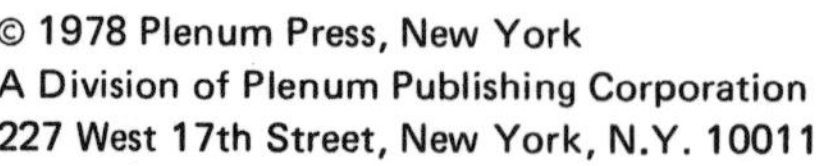

A Division of Plenum Publishing Corporation
227 West 17th Street, New York, N.Y. 10011

Printed in the United States of America

ORGANIZATION OF THE CONFERENCE

COSPONSORS:

The Engineering Foundation
345 East 47th Street
New York, NY 10017
U.S.A.

DECHEMA (Deutsche Gesellschaft fur chemisches Apparatewesen e. V.)
Frankfurt (Main)
F. R. Germany

EXECUTIVE COMMITTEE:

Georg Manecke.	Executive Chairman
Georges B. Broun	Program Chairman
Ichiro Chibata	Member
Howard H. Weetall.	Member
Lemuel B. Wingard, Jr. . .	Permanent Member
Sanford S. Cole.	Conference Director
Klaus Buchholz	Conference Secretary
Dino Dinelli	Honorary Guest

ADVISORY BOARD:

R. Axen
I.V. Berezin
R. Cavanna
T.M.S. Chang
C.K. Colton
D. Dinelli
P. Dunnill
L. Goldstein
W.E. Hornby
D. Jaworek
T. Keleti
J.O. Konecny
M.D. Lilly
M.A. Mitz
K. Mosbach
M.H. Nielson
E.K. Pye
H. Samejima
G. Schmidt-Kastner
B.J. Schnyder
P.V. Sundaram
S. Suzuki
D. Thomas
W.R. Vieth
M. Wilchek
O. Zaborsky

ORGANIZATION OF THE CONFERENCE (CONT'D)

PROGRAM COMMITTEE:

G.B. Broun, Chm.
I.V. Berezin
C.K. Colton
D. Dinelli
P. Dunnill
L. Goldstein
D. Jaworek
J.O. Konecny
K. Mosbach
E.K. Pye
H. Samejima
S. Suzuki

FINANCIAL SUPPORT:

Federal Ministry for Research and Technology
Bonn, F. R. Germany
Engineering Foundation
New York, U.S.A.
DECHEMA
Frankfurt (Main), F. R. Germany
AS Astra Ferment
Sodertalje, Sweden
Alfa-laval AB
Tumba, Sweden
BASF AG
Ludwigshafen, F. R. Germany
Bayer AG
Leverkusen, F. R. Germany
Beecham Pharmaceuticals
Worthing, England, U.K.
C. H. Boehringer Sohn.
Ingelheim, F. R. Germany
Boehringer Mannheim GmbH
Mannheim, F. R. Germany
Givaudan Forschungsgesellschaft AG
Dubendorf, Switzerland
Henkel & Cie GmbH
Dusseldorf, F. R. Germany
Hoechst AG
Frankfurt (Main), F. R. Germany
Hoffmann-La Roche AG
Grenzach, F. R. Germany
ICI Imperial Chemical Industries
London, England, U.K.
Institut Choay
Paris, France

ORGANIZATION OF THE CONFERENCE (CONT'D)

FINANCIAL SUPPORT (CONT'D)

E. Merck
Darmstadt, F. R. Germany
Miles Kali-Chemie GmbH & Co. KG
Hannover, F. R. Germany
Miles Laboratories Ltd.
Slough, England, U.K.
NOVO Industri A/S
Bagsvaerd, Denmark
Pharmacia Fine Chemicals AB
Uppsala, Sweden
Rhone-Poulenc SA, Centre Recherche
Vitry, France
Roehm GmbH
Darmstadt, F. R. Germany
Schering AG
Berlin, F. R. Germany
The Swedish Sugar Company
Malmo, Sweden

NEXT ENZYME ENGINEERING CONFERENCE

SUMMER 1979: Enzyme Engineering V tentatively scheduled for U.S.A.

EXECUTIVE COMMITTEE:

Howard H. Weetall.	Executive Chairman
Garfield Royer	Program Chairman
Ichiro Chibata	Member
Peter Dunnill.	Member
Lemuel B. Wingard, Jr.	Permanent Member
Sanford S. Cole.	Conference Director

Preface

The unique catalytic properties of enzymes and the numerous techniques for immobilization of enzymes and cells continue to maintain a high degree of practical and scientific interest in this area called Enzyme Engineering.

This fourth International Enzyme Engineering Conference was the first to be held outside of the United States. Europe was chosen as the site primarily to enable greater participation by investigators from that continent. The Engineering Foundation of New York, which was the principal sponsor of the first three conferences, was most fortunate in having the DECHEMA (Deutsche Gesellschaft fur chemisches Apparatewesen e. V.) of Frankfurt (Main), F. R. Germany as the cosponsor for this fourth conference. The success of the conference also was due in large part to the generous financial support, especially by the government of the Federal Republic of Germany, as well as by European enzyme and chemical companies.

The fourth conference, held September 25-30, 1977 at Bad Neuenahr, Federal Republic of Germany was certainly successful, with 240 participants from 23 countries, representing many academic disciplines and occupational specialties. At this conference special emphasis was placed on the immobilization of whole cells and organelles, medical applications of immobilized enzymes and organelles, and the industrial status and future for immobilized biological materials.

This volume contains most of the papers presented at the fourth conference. The names of the session cochairmen and committee members are included in appreciation of their efforts in making this 1977 conference a success. The preparation of this volume was carried out by all three editors, with the detailed editing, proofing, and assembling of the final copy done in Pittsburgh. The editors are indebted to Ms. Hall of the Department of Pharmacology of the University of Pittsburgh for her excellent job in retyping all of the edited manuscripts in a form suitable for direct reproduction by the publisher. The authors were most understanding in that many of the papers had to be shortened considerably in order to keep the size of the volume within prescribed limits; and the need for quick publication allowed for only the most minimal editor-author interaction.

Special thanks are due to Dr. K. Buchholz and Mrs. J. Gramberg of the DECHEMA and to Dr. Sandford S. Cole and the staff of the Engineering Foundation Conferences office for making this conference possible.

Georges B. Broun
Georg Manecke
Lemuel B. Wingard, Jr.

March 1978

Contents

Contributed Papers:

Contributed Papers:

Contributed Papers:

KEYNOTE PAPERS

GENETIC ENGINEERING

Alain Rambach

Institut Pasteur
Paris, France

I have the great pleasure to speak to you today about genetic engineering. If I were a scientist working in mechanics, people would expect me to find out the rules of mechanics; and if I were a mechanical engineer, people would expect me to know how to use those rules to design or repair mechanical devices. As a geneticist, I am asked to find out the mystery of heredity, the message that is coded in the genes, and the rules that govern the expression of the chromosomes. And as a genetic engineer, I am asked to use those rules to design or repair chromosomes or genes.

People will ask me to be able to draw on a blackboard a new biological structure and later to be able to construct this new structure. They will, for instance, expect me to design a hybrid vegetable species that will produce potatoes in the ground together with tomatoes on top of the ground. Or they will ask me to create sheep, having their wool already colored in blue, red, or green. Or, they will ask me to destroy the genes which are responsible for cancer in humans, or to build a bacterium that will produce a human hormone.

However, practical engineering does not necessarily follow quickly after scientific ideas are presented. For example, in the field of transport a pioneer scientist of the 13th century thought that a globe made of thin copper and filled with air might fly. This was scientific imagination. But it took five hundred years for the

Jacques & Joseph Montgolfier brothers to build their balloons that actually floated across Paris in 1783. Only then, had a scientific idea been turned into practical engineering.

I wish to describe the present state of genetic engineering and to show how the 50 year old science of molecular genetics only recently yielded a striking new sort of engineering. In other words I want to tell what extraordinary fantasies the genetic engineer should be able to do in the future and what fantasies he will never succeed in doing. I believe that one can predict what is feasible and what is not feasible in a given field. The feasibility for the earlier examples is as follows: a vegetable species that will produce potatoes and tomatoes (no); sheep having their wool colored in blue, red, or green (no); the destruction of genes responsible for cancer in humans (no); and a bacterium that will produce a human hormone (yes).

Let us substantiate these evaluations by a rapid survey of the science of genetics. In a species, like the human, all individuals are basically the same generation through generation. It is because their basic structure is encoded in the chromosomes which are transmitted to offsprings, expressed, and repeated in the billions of cells which constitute the organism. The chromosomes carry the hereditary information, a little like a magnetic tape. Then the task of the genetic engineer appears to be nice and easy: he will manipulate the chromosomes, put them back in a cell, let the cell multiply, and see the newly designed organism expressed. But to know, really, how to manipulate the chromosomes he must understand better their content. The central element is the DNA which is like a string of about 1,000,000,000 letters. The DNA contains units, or genes, of about 1000 letters each. A gene can be copied into messenger RNA that can be translated into proteins, such as enzymes.

With that knowledge, up to 5 years ago it was possible mostly to destroy the DNA for instance by irradiation. If the irradiation was minimal, it would generate random mutations. This approach has been used to improve industrial organisms and to move some genes like the *galactose* or the *lactose* genes of *Escherichia coli* by transduction onto a bacteriophage. But those techniques were very slow. It was 10 or 20 years after their genetic characterization that the *gal* and *lac* genes were moved. So little was done because one cannot extract the chromosome intact and put it back in the cell while retaining biological activity. Let me remind you that the chromosome contains about 10^9 letters. Thus one had to rely on the techniques of *in vivo* manipulations with intact cells.

What is new today? First the availability of the restriction endonucleases, which cut purified DNA at specific sites, yielding DNA fragments that have cohesive ends and can be joined (1). The second discovery is that of DNA vectors. These are small extrachromosomal DNA molecules which can replicate autonomously, which can be purified and opened at one site, and which are small enough to remain biologically active. The first vectors to be constructed were bacterial plasmids (2) and bacteriophages found in several laboratories, including ours (3-5). The DNA is cut, the vector also is cut, and both are joined (this technique is called *in vitro* DNA recombination); the hybrid DNA is put back into the cell and replicated by the millions. The latter takes less than 10 hours in the case of a bacterium. The DNA fragment propagated in the hybrid is "cloned." This allows the preparation of large quantity of pure genes. However, the real breakthrough comes from the simple fact that the DNA that is inserted in a given cell can originate from any species. For example, it is not a problem to clone elephant genes in a bacterium. The species barrier does not exist anymore. The first cloning of foreign DNA in *E. coli* was from *Staphylococcus* DNA in 1973 (5). Soon after mouse DNA, frog DNA, fruit fly DNA and others were used. It came as a shock to the scientific community. Those experiments were done either with purified genes or with mixtures of DNA fragments ("shot gun" experiments).

A second technique has been developed, where it is not a fragment of chromosomal DNA but the DNA copy of a messenger RNA that is cloned. Researchers have obtained fantastic chimeric structures by this technique, such as a bacterium containing the rabbit hemoglobin gene or a bacterium containing the gene of a hormone. A third technique, the ultimate of genetic engineering, consists in writing a DNA sequence corresponding to a given protein. The DNA sequence is then synthesized and cloned.

What can come from cloning? Preliminary studies have been done on the expression of foreign genes cloned in a cell, mostly in *E. coli*. *Staphilococcus* genes and *Bacillus subtilis* genes have been shown to be expressed in *E. coli*. Genes from yeast and, as we showed last year, genes from a higher eukaryote, the fruit fly

Drosophila, have been expressed in *E. coli*. The most extraordinary achievement today seems to be, if the rumor is true, the actual expression of a cloned human hormone gene in *E. coli*.

My own belief is that generally we shall be able to get expression of a cloned gene in the new host, either directly after the transfer of the foreign gene or with a few tricks to turn on the gene. Would this mean that bacteria will be the cheap factories of the future for all sorts of products? I think so; and this opinion is shared by some research institutes and by some established industries who decided to work in the field. There has even been a company created to work specifically on the applications of genetic engineering.

However, I would like to point out several possible problems. In some cases, already demonstrated, a gene may be fragmented on the DNA and not be simply at one locus on the DNA. One could also imagine that proteins expressed in a new host could be degraded there. The proteins may not have the normal configurations or they may not be processed as they should be. However the latter are only hypothetical problems.

A few possible projects might be to synthesize pure and safe vaccines by preparing a pure polypeptide of the pathogenic factor. Another could be to construct plant species resistant to specific herbicides or of new nutritional value. And finally one could invent new enzymes or new proteins.

In conclusion I would like to stress that there is a strong argument going on about the safety of DNA recombinant experiments. Anyone planning such work should be cautious in the design of their experiments.

REFERENCES

1. JACKSON, D.A., SYMONS, R.H., & BERG, P. *Proc. Nat. Acad. Sci. USA 69*:2904, 1972.
2. COHEN, S.N., CHANG, A.C.Y., BOYER, H.W., & HELLING, R.B. *Proc. Nat. Acad. Sci. USA 70*:3240, 1973.
3. MURRAY, N.E. & MURRAY, K. *Nature 251*:476, 1974.
4. THOMAS, M., CAMERON, J.R., & DAVIS, R.W. *Proc. Nat. Acad. Sci. USA 71*:4579, 1974.
5. RAMBACH, A. & TIOLLAIS, P. *Proc. Nat. Acad. Sci. USA 71*:3927, 1974.

THE PHOTOSYNTHETIC APPARATUS

L. O. Krampitz

Department of Microbiology
Case Western Reserve University
Cleveland, Ohio, USA

The purpose of this paper is to review briefly what is known about the oxygen-evolving type of photosynthesis, its efficiency, and the possibility of employing the photosynthetic apparatus as a means of accomplishing the biophotolysis of water to hydrogen and oxygen.

The capture of radiation energy of the visible spectrum for reducing carbon dioxide to complex organic material occurs in two biological forms: (a) the oxidation of water with the evolution of oxygen, which occurs in green plants and the various algae, and (b) the oxidation of inorganic substances such as ferrous ions, reduced forms of sulfur, and organic molecules to reduce carbon dioxide with no evolution of oxygen, which occurs in a variety of bacteria and other microscopic life. The limitation of space will permit only a discussion of the oxygen-evolving type, as found in green plants and the various algae.

The generalized equation for photosynthesis is:

$$2H_2A + CO_2 \xrightarrow{hv} 2A + CH_2O + H_2O$$

where H_2A represents the reductant, A the oxidized product, and CH_2O, cell material on the level of oxidation of carbohydrate. For oxygen-evolving photosynthesis the equation is:

$$2H_2O + CO_2 \xrightarrow{hv} O_2 + CH_2O + H_2O$$

The function therefore of this photosynthetic apparatus is to convert radiant energy between the range of 400 and 700 nm to chemical energy, which reduces carbon dioxide to cell material. Chlorophyl and other light harvesting pigments absorb solar wavelengths between 400 and 700 nm. This represents about half the total solar emission. Due to the high content of these pigments almost total absorption between 400 and 700 nm is approached.

Research in photosynthesis is divided into two main groups: mechanism of electron transfer from the oxidation of water to the reduction of CO_2 and mechanism of fixation of CO_2 to carbohydrate and other cellular materials. Both mechanisms will be discussed.

Fig. 1 portrays the mechanism of the oxidation of water with the evolution of oxygen. The ordinate is the oxidation-reduction potential in volts at pH 7.0. Nature has provided for two photosystems: II and I, operating in series to place electrons from the oxidation-of water to a negative potential of considerable reducing power. At the positive end of the scale is the average potential of the water-oxygen couple, +0.8V. The light harvesting pigments, chlorophyl, carotenoids and others with the necessary enzymatic components of photosystem II form a strong oxidant Z^+, which oxidizes water evolving oxygen. The photochemically expelled electron reduces a quinone substance whose oxidation-reduction potential is approximately 0.0 volts. A connecting link between photosystem II and photosystem I provides for a mechanism of dark oxidation. The electron moves to a positive potential of approximately +0.4 volt. The pathway being from plastoquinone to cytochrome B559, to cytochrome f, and to plastocyanin. The energy of this oxidation is sufficient to generate an evergy righ phosphate ester bond (∿ 10 kcal) which the plant uses for synthetic processes.

At this point a light harvesting component P700 of photosystem I drives an electron to a negative potential of approximately -0.7 volts. The electron deficient P700 is reduced by the electron originating from water through the interconnecting link between the two photosystems. The nature of the electron acceptor at approximately -0.7 volt is not known and is represented by X in Fig. 1. The electron moves next in a dark reaction to a relatively

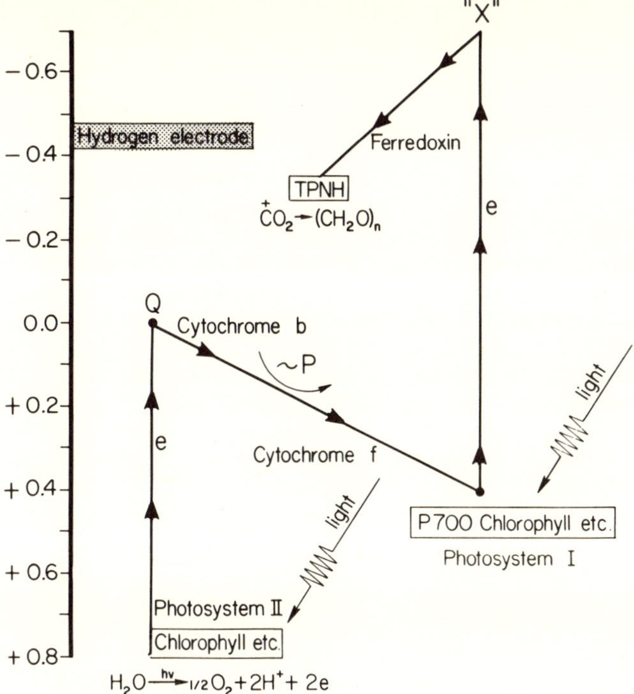

Fig. 1. Oxidation reduction potentials for the sequential transfer of electrons from water to TPNH in photosynthesis.

small molecular weight non-heme iron-sulfur containing protein, ferredoxin, whose oxidation-reduction potential is -0.42 volt, a value almost identical to the hydrogen electrode. The reduced ferredoxin reduces a well known coenzyme, triphosphopyridine nucleotide (TPN) which is required for carbohydrate biosynthesis. The oxidation-reduction potential of the nucleotide is -0.34 volts. By well known biochemical reactions the reduction of carbon dioxide by reduced TPN occurs.

The reducing capacity of these photosystems could be used for oxidizing water to obtain reduced substances for energy purposes other than for biomass in the form of accumulated plant material. The two photosystems operating in series could oxidize water, evolve oxygen, and reduce ferredoxin. This last compound could be used further.

Ferredoxin, which was discovered in an anaerobic bacterium *Clostridium pasteuranium* has the ability to fix elemental nitrogen; and when cultured on carbohydrate, it forms large quantities of hydrogen. The organism contains a hydrogenase which catalyzes the reduction of protons by electrons if the latter are of the potential of the hydrogen electrode. It was found that reduced ferredoxin was the electron precursor for the reduction of protons as catalyzed by the enzyme hydrogenase. Ferredoxin is a one-electron carrier also present in the photosynthetic apparatus of oxygen evolving photosynthesis, albeit of a smaller molecular weight than that found in the bacterium. The two ferredoxins are somewhat interchangeable in the two biochemical systems. The question arises then, can the electrons of the reduced ferredoxin formed in the photosynthetic apparatus be funneled into a hydrogenase system for the reduction of protons to elemental hydrogen. Our laboratory has been working on this problem; and subsequent to the discussion of the pathway of fixation of carbon dioxide, some of our results relating to the mechanisms of hydrogen formation by the biophotolysis of water will be presented.

PATHWAY OF CO_2 FIXATION OR REDUCTION

The primary pathway of CO_2 reduction in most plants is that discovered by Calvin *et al.* (4); however other pathways have been found. Upon short period exposure of plant leaf preparations to light and $^{14}CO_2$ the latter was fixed most extensively into the carboxyl group of 3-phospho-glyceric acid (3-PGA). This fixation of $^{14}CO_2$ was dependent upon light. It will be seen later that the enzymatic carboxylation of the reduced carbon moiety *per se* is a dark reaction if that moiety is supplied to the enzyme with CO_2 in the absence of light. If one were to assume that the fixation occurred by $^{14}CO_2$ fixation with a two-carbon moiety, the latter would have to be ethylene glycol monophosphate, which was known not to exist in photosynthetic metabolism. It was shown that 1,5-ribulose diphosphate was carboxylated, forming 2 moles of 3-PGA. The enzyme has been crystallized and many of its properties determined including that the fixation reaction occurred in the dark.

The next step is a phosphorylation of 3-PGA to 1-3, diphosphoglyceric acid by a kinase employing two moles of adenosine triphosphate (ATP), the latter being generated by oxidative reactions in photosystems II and I. By means of the enzyme, 3-phosphoglyceraldehyde dehydrogenase, 1,3-diphosphoglyceric acid is reduced by the TPNH formed by the two photosystems to 3-phosphoglyceraldehyde. Triose isomerase establishes an equilibrium between 3-phosphoglyceraldehyde and dihydroxyacetone phosphate. These latter two trioses are condensed by aldolase to form 1,6-fructose diphosphate, a hexose. Having formed one mole of hexose from the fixation of one mole of CO_2 and reduction by TPNH formed by photosystems I and II (neglecting phosphorylation by ATP) the question arises how is 1,5-ribulose diphosphate reformed in order that fixation of CO_2 may continue. This is done by a series of reactions occurring in a cyclic manner such that 6 moles of CO_2 are fixed to form a hexose and at the same time regenerate 6 moles of 1,5-ribulose diphosphate in order to initiate the cycle again. The needed 18 moles of ATP are formed by cyclic oxidative phosphorylations which occur in photosystems I and II. The 12 needed TPNH are formed by the photochemically driven reactions of photosystems I and II.

The Calvin cycle was thought for years to be the only mechanism for CO_2 fixation; but evidence with sugar cane, maize, and similar grasses showed that another pathway or pathways were present (5). Under certain conditions preparations from these plant tissues fixed $^{14}CO_2$ more extensively into C_4 acids, such as malic and aspartic, than into the C_3 acid 3-PGA. The fixation of CO_2 in the C_4 plants occurs mainly by reaction with phosphoenolpyruvate to produce oxalacetate and Pi. The oxaloacetate can be reduced to malate by photosystems I and II, and aminated to aspartate or decarboxylated to pyruvate. It is thought that on account of anatomical structures in the C_4 plants the decarboxylation of oxalacetate occurs in the portion of the plant where the resulting CO_2 can be fixed by 1,5-ribulose diphosphate carboxylase of the Calvin cycle. Therefore, the so-called C_4 plants also possess the C_3 or Calvin cycle, type of fixation. Inasmuch as the C_4 plants possess both types of CO_2 fixation, they are more efficient photosynthetically.

Let us now consider the efficiency of the photosynthetic system. If we assume a quantum efficiency of one for each of the two photosystems operating in series, two photons are required to move one electron across a thermodynamical voltage barrier of approximately 1.3 V. The efficiency EFF is 1.3ev/2/hv. Where hv is the energy of one photon expressed in electron volts, i.e., Planck's constant times the frequency of the radiation. At 680 nm the energy of one photon is 1.83 ev; and since two photons are involved per electron in photosystems I and II the above expression becomes 35%. The radiation energy impinging upon the earth's surface between 400 and 700 nm is approximately 43% of the total radiation. Therefore, of the total solar energy available to the photosynthetic apparatus there is 15% (0.43 X 35) conversion efficiency. This theoretical figure is high since in all probability photons are not captured with a quantum efficiency of one; and within photosystems I and II there are cyclic oxidations and reductions. Electrons at the negative potential formed by the two systems reduce oxidants formed by the photosynthetic apparatus thus lowering the overall efficiency.

The theoretical efficiency of carbohydrate biosynthesis can also be calculated. One mole each of CO_2 and H_2O are converted to carbohydrate and O_2. Four moles of electrons are required for the reduction of one mole of CO_2 to one mole of CH_2O. Since each mole of electrons require two einsteins of photons in the two photosystems, the total photon requirement is 8 einsteins. CH_2O represents one-sixth of a glucose moiety. The free energy stored in this reaction is approximately 114 kcal/mole of CO_2 reduced. One mole of photons at 575 nm contains 49.7 kcal, therefore the theoretical efficiency is 114/8/49.7 or 28.6%. Since the ratio of photosynthetically active radiation to total solar energy is 0.43 one must reduce the above theoretical efficiency by 0.43 x 28.6 = 12.1%. Again this efficiency is high owing to respiratory functions of the plant, which oxidize the carbohydrate back to CO_2 and water.

BIOPHOTOLYSIS OF WATER TO HYDROGEN

For the photosynthetic system we have empolyed a blue-green alga, *Anacystis nidulans*. This organism grows photosynthetically on CO_2, nitrate, phosphate, and an array of trace metals. Upon lyophilization, the harvested cells retained photosystems I and II but could no longer reduce CO_2 to carbohydrate. These lyophilized cells, in contrast to freshly harvested cells, were permeable to a variety of electron acceptors including TPN. When suspended in Tris buffer, pH 7.6 at 35°C, these lyophilized cells reduced TPN and evolved oxygen upon illumination with 150,000 lux of white light. 4.8 μmoles of TPNH were formed in 30 min. In the presence of monuron, an inhibitor of photosystem II, there was no evolution of oxygen nor reduction of TPN, indicating that the electrons originated from the oxidation of water. In addition if the reaction mixture was not illuminated, there was no evolution of oxygen nor reduction of TPN.

Inasmuch as the photosynthetic apparatus of this algal preparation could place the electrons from water at a potential near the hydrogen electrode, the possibility exists that by proper catalysis protons could be reduced to form hydrogen. Certain bacteria have a hydrogenase and form hydrogen under growth conditions. Several-years ago Peck and Gest (6) devised a convenient assay for hydrogenase, using pH 6.5 phosphate buffer, sodium dithionite, methyl viologen, and the hydrogenase under an atmosphere of argon. The source of the hydrogenase was from *Clostridium kluyveri*. Methyl viologen is an oxidation reduction dye, whose potential is at the hydrogen electrode. Sodium dithionite is the reductant, which chemically reduces the methyl viologen and serves as an electron precursor for the hydrogenase. It will be observed that there was a rapid evolution of hydrogen of 27.6 μmoles/hr. The reaction must be carried out under anaerobic conditions since the reduced methyl viologen is autoxidizable and the hydrogenase is labile to effects of oxygen.

It has been known for several years (7-10) that methyl viologen will serve as an electron acceptor in the photosynthetic apparatus, providing the oxygen which is formed by photosystem II is efficiently removed. A mixture of 600 μg chlorophyl, 150 μmoles Hepes buffer at pH 7.4, 25 μmoles methyl viologen, and 6.0 μ moles ammonium chloride in a total volume of 6.0 ml were incubated 30 min with constant evacuation and flushing with argon to remove oxygen; and 9.1 μmoles of reduced methyl viologen were formed. When hydrogenase from *E. coli* was added anaerobically, 3.2 μmoles of hydrogen were formed. Control experiments without illumination or inhibited with monuron formed no reduced methyl viologen.

In order to circumvent the cumbersome technique of removing the oxygen, which reoxidizes the reduced methyl viologen, we employed the photosynthetically reduced TPN as the precursor of electrons for the reduction of protons by the hydrogenase. Unfortunately the TPN^{+}-TPNH couple has an oxidation-reduction potential of -0.34 v, which is about 0.1 V more positive than the hydrogen electrode; so that this thermodynamic barrier must be overcome. To accomplish this we employed a second photosynthetic system found in the non-sulfur purple bacterium, *Rhodopseudomonas capsulata*. This organism grew profusely photosynthetically under anaerobic conditions with malic acid as its carbon source, producing large amounts of hydrogen. The organism does not possess photosystem II; therefore no oxygen was evolved when grown photosynthetically on malate. The mechanism by which malate forms hydrogen is not thoroughly understood; however, the organism has a malic dehydrogenase which carries out the reversible reaction between malate and TPN to form oxalacetate and TPNH. Freshly harvested cells of the organism were not permeable to TPNH, however lyophilized cells were. The hydrogenase activity of the latter was destroyed; but the malic dehydrogenase activity was intact. Therefore, we employed a mixture of lyophilized cells and freshly harvested cells. The lyophilized cells carried out the reaction between TPNH oxalacetate, and lyophilized *R. capsulata* to form TPN^+ and malate. The freshly harvested cells formed hydrogen photosynthetically from the malate, which also was oxidized to oxalacetate. Fig. 2 summarizes the events for the formation of hydrogen from water by the two photosynthetic organisms, the algae and the bacteria. In the top compartment is the formation of TPNH from water by the algae. The TPNH enters the lower right compartment, reducing oxalacetate to malate by the lyophilized *R capsulata*. The oxidized TPN returns to the upper algae compartment. The malate enters the lower left compartment where it forms hydrogen by the fresh *R. capsulata*. Diffusion of the components, indicated by arrows, permits the continuous formation of hydrogen.

In an experiment with the combined cells 90.0 µg chlorophyl/ml in lyophilized algae, 10.0 mg/ml freshly

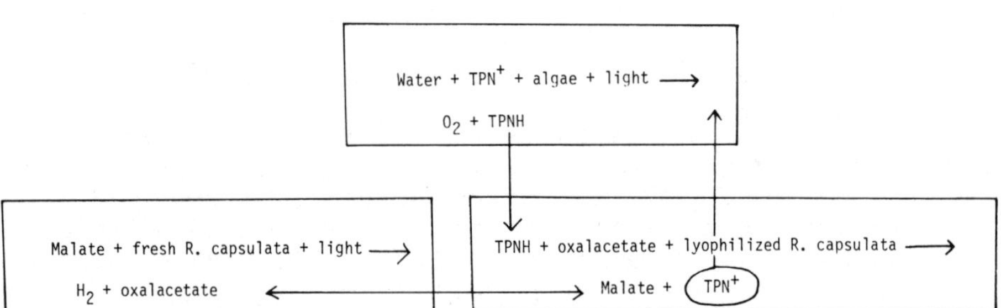

Fig. 2. Schematic for hydrogen formation using algae and bacteria. See text for description.

harvested *R. capsulata*, 4.0 mg/ml lyophilized *R. capsulata*, 1.0 μ mole/ml oxalacetate, 1.0 μmole/ml TPN, and 25.0 μmole/ml Tris buffer pH 7.6 were mixed in 5.0 ml total volume and incubated at 30°C under 150,000 lux white light. 6.2 μmoles of hydrogen were obtained in 30 min. No hydrogen was obtained without illumination or when the system was poisoned with monuron, indicating the electrons for the reduction of protons originated from water.

While the experiments cited above demonstrate that the biophotolysis of water with the formation of hydrogen is possible, the experimental design is much too complicated to be of practical use. In my opinion in order for the biophotolysis of water with the formation of hydrogen to be practical, photosystems I and II must form a reduced product which is stable; and at the same time the product should be a precursor of electrons for the reduction of protons to hydrogen in a dark reaction (11).

There exists in some bacteria an enzyme complex which catalyses a sequence of reactions. Formate is cleaved to CO_2 and hydrogen. This reaction is rapid and freely reversible with an equilibrium of one. It has been shown that the total reaction consists of at least two reactions. The second reaction consists of a formic dehydrogenase and an unknown electron carrier X, the products being 2Xe and CO_2. The third reaction is the oxidation of 2Xe and the reduction of protons to hydrogen and X, as catalyzed by hydrogenase. It has been shown that for the reduction of CO_2, i.e., the reverse reaction represented in reaction two, a reduced artificial electron carrier such as reduced methyl viologen can substitute for 2Xe, thus dispensing with the need for hydrogenase. The following interpretation can be made: if the formic dehydrogenase component of equation two were present in the algae, photosystems I and II would form 2Xe and the reduction of CO_2 would occur, forming formate photosynthetically as a reduced substance. Formate being a stable substance could be used as the bacterial substrate for the formation of CO_2 and hydrogen in a dark reaction, i.e., in a fermentation tank. Unfortunately the algae or any other known photosynthetic apparatti do not reduce CO_2 to formate.

We are attempting to insert the formic dehydrogenase gene for reaction two from *E. coli* into the blue-green alga *Anacystis nidulans*. In order to accomplish this the location of the gene on the bacterial chromosome must first be worked out. By techniques of microbial genetics we have found the formic dehydrogenase gene to be at 72 min on the 90 min circular chromosome of *E. coli*. An attempt is now being made to insert the gene into the algae. The methodology is standard, either by transduction, which employs an algal virus, or transformation, which is a procedure of making the algae competent for accepting plasmid deoxyribonucleic acid (DNA), a vehicle used for inserting foreign markers. A.V. Morey in our laboratory has found that of the blue-green algae examined they possess plasmid DNA; consequently we are hopeful of inserting the formic dehydrogenase gene into the algae.

ACKNOWLEDGMENT

Supported in part by a grant from Ethyl Corporation.

REFERENCES

1. MORTENSON, L.E., VALENTINE, R.C., & CARRAHAN, J.E. *Biochem. Biophys. Res. Commun.* 7:448, 1962.
2. SAN PIETRO, A. in "Light and Life," (W.D. McElroy and B. Glass, eds.), Johns Hopkins Press, Baltimore, 1961, p. 631.
3. TAGAWA, K. & ARNON, D.I. *Nature 195*:537, 1962.
4. BENSON, A.A. & CALVIN, M. *Science 105*:648, 1947.
5. HATCH, M.D. & SLACK, C.R. *Biochem. J. 101*:103, 1966.
6. PECK, H.D. & GEST, H.J. *Bact. 71*:70, 1956.
7. KOK, B., RURAINSKI, H.J. & OWENS, O.V.H. *Biochim. Biophys. Acta 109*:347, 1965.
8. ZWEIG, G. & AVRON, M. *Biochem. Biophys. Res. Commun. 19*:397, 1965.
9. BLACK, C.C. Jr. *Biochim. Biophys. Acta 120*:332, 1966.
10. YOCUM, C.F. & SAN PIETRO, A. *Biochem. Biophys. Res. Commun. 36*:614, 1969.
11. BENEMAN, J.R., BERENSON, J.A., KAPLAN, N.O. & KAMEN, M.D. *Proc. Nat. Acad. Sci. USA 70*:2317, 1973.

Session I
PRODUCTION AND EXTRACTION OF MICROBIAL ENZYMES

Chairmen: H. Samejima and F. Wagner

INCREASING THE PRODUCTION OF ENZYMES VIA FERMENTATION

R.J. Clark III, M.L. King and J.L. Gainer

Department of Chemical Engineering
University of Virginia
Charlottesville, Virginia, USA

In many industrial fermentation processes the ultimate limitation on the rate of reaction is the rate at which nutrients can be supplied to the microbial cells (1). Supplying these materials to the microorganisms through the media in which they are growing is basically a problem in diffusional mass transfer. Most research has centered on ways to minimize diffusional resistances to mass transfer. However, methods designed to improve the molecular diffusivities of the nutrients in the media to the growing cells have been ignored, for the most part.

Several models for the dependence of microbial growth on substrates have been suggested. Monod (2) assumed that the rate of growth was limited by the availability of a single substrate. In Monod's model, it is assumed that population density does not affect the growth rate; and the model says nothing about the multiplication rate. In a batch culture the model predicts a phase of exponential growth followed by a stationary phase when the substrate is exhausted. In the case in which the limiting substrate is a dissolved gas, the Monod equations can be modified for mass transfer. One of the equations which results, for a chemostat, is:

$$\frac{d\bar{c}}{dt} = D(c_f - \bar{c}) - \frac{1}{\gamma}\left[\frac{U_m X \bar{c}}{K_m + c}\right] + K_L a(c^* - \bar{c}) \qquad \text{(Eq. 1)}$$

where: $\bar{c}$ = dissolved gas concentration in bulk liquid
c_f= dissolved gas concentration in feed liquid
c^*= gas concentration that would be in equilibrium with gas phase
t = time
D = dilution rate
U_m= maximum specific growth rate
X = cell mass per unit volume
K_m= Michaelis constant
K_La=volumetric mass transfer coefficient

For oxygen uptake this equation can be rewritten as:

$$\frac{d\bar{c}}{dt} = D\ (c_f - \bar{c}) + K_La\ (c^* - \bar{c}) - RX \qquad \text{(Eq. 2)}$$

where: R = specific oxygen uptake rate per unit mass of fermenting biomass.

In Eq. 2 there is no term which accounts specifically for the transport across the boundary layer of liquid surrounding the microbes. Any changes in this resistance would show up as changes in R, provided that this mass transport were the rate-controlling factor in the oxygen transport from the bulk liquid to the microbe. In such a case it should be possible to increase the specific oxygen uptake rate by increasing the oxygen diffusion through the liquid. It has been shown that carotenoids enhance oxygen transport in solutions (3). The use of such compounds could then increase the metabolic rate of the microbes. If such microbes produce exoenzymes, this could result in increased enzyme production rates.

EXPERIMENTAL

Crocetin, a carotenoid, was added to several fermentations of *Bacillus subtilis*. Tests were done in a chemostat containing 500 ml of nutrient broth medium to determine if crocetin increased the specific respiration rate of *B. subtilis*. The dissolved oxygen concentrations were measured with a sterilizable galvanic membrane electrode. The study was conducted in the manner suggested by Bandyopadhyay and Humphrey (4).

The effect of crocetin on enzyme production also was studied, using 1-ℓ erlenmeyer flasks. These were stoppered and contained inlets for a heater, an air line and a sampling tube. Nutrient broth at 8 g/l was used for the medium. Assays were done for protease and amylase activities.

RESULTS AND DISCUSSION

The effect of crocetin on the respiration rate of *B. subtilis* is shown in Table 1. The introduction of crocetin substantially increased the respiration rate at the highest concentration used. This effect presumably was due to a change in oxygen diffusivity; however, a metabolic effect of the crocetin could not be excluded, based on these data.

The production of proteases and amylases also were studied. These data are listed in Tables 2 and 3.

TABLE 1

EFFECT OF CROCETIN ON THE RESPIRATION RATE OF *BACILLUS SUBTILIS*

Cell Concentration (cells/ml X 10^6)	Crocetin Conc. (mg/liter)	Respiration Rate* (mg O_2/min/cell X 10^{10})
3.44	0	1.24
3.44	4	1.41
3.44	8	1.48
3.30	16	2.10

TABLE 2

INCREASE OF PROTEASE ACTIVITY WITH CROCETIN

Crocetin Concentration (mg/liter)	Percentage Increase in Protease Activity*
0	0
12.5	40
25.0	92
50.0	112
62.5	90

*30°C

TABLE 3

EFFECT OF CROCETIN ON AMYLASE ACTIVITY

Crocetin Concentration (mg/liter)	Percentage Increase in Amylase Activity*
0	0
12.5	0
25.0	0
37.5	230

*30°C

Since the production of enzymes was increased in the bacterial fermentation using crocetin, it was decided to see if the same thing would happen in the case of fungi. Since celluloses are produced by the fungus *Trichoderma viride*, the effect of crocetin on extracellular enzyme production was studied in this organism. *T. viride* was grown in shake-flask cultures, using Solka Floc as the cellulose substrate, and the production of C_1 enzymes was determined with a standard cellulase assay. The effect of crocetin on these cultures is shown in Table 4 below.

TABLE 4

INCREASE IN CELLULASE ACTIVITY WITH CROCETIN

Crocetin Concentration (mg/liter)	Percentage Increase in Cellulase (C_1) Activity*
0	0
12.5	45
25.0	35
37.5	25
50.0	25
62.5	0

*Mandels Media, 30°C.

As can be seen, the crocetin again caused an increase in cellulase activity, either by increasing production of the enzymes or by enhancing the activity of the enzyme, *per se*. In all cases studied, crocetin caused an apparent increase in enzyme activity. This may be important in the commercial preparation of enzymes, and should be studied in further detail.

REFERENCES

1. HARRISON, D.E.F. *J. Appl. Chem. Biotech* *22*:417, 1972.
2. MONOD, J. *Ann. Inst. Pasteur* *79*:390, 1950.
3. GAINER, J.L. & CHISOLM, B.M. in "Chemical Engineering Applications in Medicine," in press.
4. BANDYOPADHYAY, B. & HUMPHREY, A.E. *Biotechnol. Bioeng.* *13*:583, 1971.

CULTURAL CONDITIONS FOR THE MICROBIAL PRODUCTION OF β-TYROSINASE AND TRYPTOPHANASE

H. Yamada*, H. Kumagai**, H. Enei***, H. Nakazawa***, and K. Mitsugi***

Departments of Agricultural Chemistry* and Food Science and Technology**
Kyoto University, Kyoto and
Central Research Laboratories,***
Ajinomoto Co., Inc., Kawasaki, Japan

β-Tyrosinase (tyrosine phenol-lyase: EC 4.1.99.2) and tryptophanse (tryptophan indole-lyase: EC 4.1.99.1) are enzymes which respectively catalyze the degradation of L-tyrosine and L-tryptophan, and require pyridoxal 5'-phosphate (PLP) as a cofactor. Crystalline preparations of these enzymes were prepared in our laboratories from *Escherichia intermedia* and *Proteus rettgeri*, and their properties were established in some detail. The crystalline enzymes were shown to catalyze a variety of α,β-elimination (Eq. 1), β-replacement (Eq. 2), and the reverse of α,β-elimination reactions (Eq. 3) (1-3).

$$\text{L-RCH}_2\text{CHNH}_2\text{COOH} + \text{H}_2\text{O} \longrightarrow \text{RH} + \text{CH}_3\text{COCOOH} + \text{NH}_3 \qquad \text{(Eq. 1)}$$

$$\text{L-RCH}_2\text{CHNH}_2\text{COOH} + \text{R'H} \longrightarrow \text{L-R'CH}_2\text{CHNH}_2\text{COOH} + \text{RH} \qquad \text{(Eq. 2)}$$

$$\text{R'H} + \text{CH}_3\text{COCOOH} + \text{NH}_3 \longrightarrow \text{L-R'CH}_2\text{CHNH}_2\text{COOH} + \text{H}_2\text{O} \qquad \text{(Eq. 3)}$$

For β-tyrosinase R can be phenolyl, -OH, -SH, -Cl and R' phenolyl. For tryptophanase R can be indolyl, -OH, -SH, -Cl and R' indolyl.

We proved that these enzymes catalyze the synthesis of L-tyrosine, L-tryptophan and their related amino acids in significantly high yields. Enzymatic processes for producing these amino acids have been developed using bacterial cells with high activities of β-tyrosinase and tryptophanase (4-7).

CULTURE CONDITIONS FOR THE PRODUCTION OF β-TYROSINASE

To select microorganisms which produce β-tyrosinase at high levels, the activity of this enzyme in microorganisms was investigated. The enzyme activity occurred widely in a variety of bacteria, most of which belonged to the *Enterobacteriaceae*, *Escherichia*, *Proteus* and *Erwinia* (8). A strain of *Erwinia herbicola* (ATCC 21434), which showed the highest activity, was selected for the present investigation.

Erwinia herbicola produced β-tyrosinase at a markedly high level, when it was grown at 28°C for 28 hours in a medium containing 0.2% L-tyrosine, 0.2% K_2HPO_4, 0.1% $MgSO_4.7H_2O$, 2 ppm Fe^{2+} ($FeSO_4 \cdot 7H_2O$), 0.01% pyridoxine-HCl, 0.6% glycerol, 0.5% succinic acid, 0.1% DL-methionine, 0.2% DL-alanine, 0.05% glycine, 0.1% L-phenylalanine and 12 ml of hydrolyzed soybean protein in 100 ml of tap water, with the pH controlled at 7.5 throughout cultivation.

β-Tyrosinase is an inducible enzyme; and the addition of L-tyrosine to the medium is essential for formation of the enzyme (9). However, when large amounts of L-tyrosine were added, both enzyme formation and cell growth were repressed by the phenol liberated from L-tyrosine. L- and D-Phenylalanines did not induce this enzyme by themsleves; but they showed a strong synergistic effect on the induction of enzyme by L-tyrosine (Table 1) (10). L-Phenylalanine has been shown to be a competitive inhibitor of β-tyrosinase (K_i=2.0x10^{-3}M) (11) and maintained the concentration of L-tyrosine in the medium during the cultivation, as shown in Fig. 1.

Under the culture conditions described above, β-tyrosinase was efficiently accumulated in the cells of *Erwinia herbicola* and made up about 10% of the total soluble cellular protein.

CULTURE CONDITIONS FOR THE PRODUCTION OF TRYPTOPHANASE

The tryptophanase activity occurred widely in bacteria belonging to *Enterobacteriaceae*. A strain of *Proteus rettgeri* (AJ 2770) was selected for the investigation

TABLE 1

EFFECT OF SYNERGISTIC MATERIALS ON THE INDUCTION OF β-TYROSINASE BY L-TYROSINE

Synergistic material*	Enzyme activity** (mg/ml)
-	1.25
L-Phenylalanine	6.25
D-Phenylalanine	4.93
Phenylpyruvic acid	6.00
D-Tyrosine	0.80
L-Phenylglycine	0.95
m-Fluoro-L-phenylalanine	1.00
p-Methoxy-L-phenylalanine	1.05

*The materials were added to the medium at the concentration of 0.1%.

**Enxyme activity was determined by measuring the amount of 3,4-dihydroxyphenyl-L-alanine (L-dopa) synthesized under the conditions described in Fig. 1.

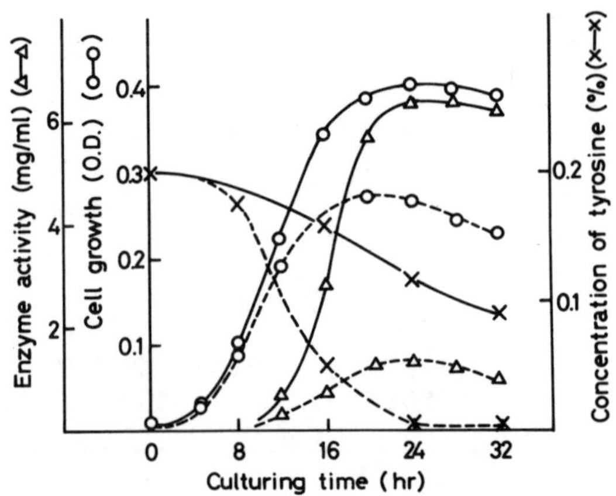

Fig. 1. Time course of the formation of β-tyrosinase in the presence (——) or absence (- - -) of L-phenylalanine. L-Phenylalanine was added to the medium at the concentration of 0.1%. The growth was determined by measuring the optical density at 562 nm of the culture broth after it was diluted 26 times. The enzyme activity was determined by measuring the amount of L-dopa synthesized in a reaction mixture containing 200 mg of DL-serine, 100 mg of pyrocatechol, 50 mg of ammonium acetate, 20 mg of sodium sulfite, 10 mg of EDTA; and the cells were harvested from 10 ml of the culture broth in a total volume of 10 ml, pH 8.0. The mixture was incubated at 22°C for 1 hour.

(12). *Proteus rettgeri* efficiently produced tryptophanase, when it was grown at 28°C for 16 hours in a medium containing 0.6% tryprophan, 4% Sorpol W-200 (polyoxyethylene alkyl phenol ether), 1% soybean protein hydrolyzate, 6% corn steep liquor, 0.3% yeast extract, 0.3% succinic acid, 0.06% L-cystine, 0.06% L-arginine-HCl, 0.03% DL-methionine, 0.03% L-proline, 0.3% KH_2PO_4 and 0.1% $MgSO_4 \cdot 7H_2O$ in tap water, with the pH adjusted to 7.0.

Addition of L-tryptophan, as inducer, to the culture medium was essential for enzyme formation. However, when large amounts of L-tryptophan were added, the enzyme formation was inhibited by the indole liberated from L-tryptophan. In fact, formation of the enzyme was enhanced by adding Sorpol W-200 which removed indole from the medium. Fig. 2 shows the effects of L-tryptophan and Sorpol W-200 on the formation of tryptophanase. A similar effect as that with Sorpol W-200 was observed when various nonionic detergents were added to the medium (Table 2).

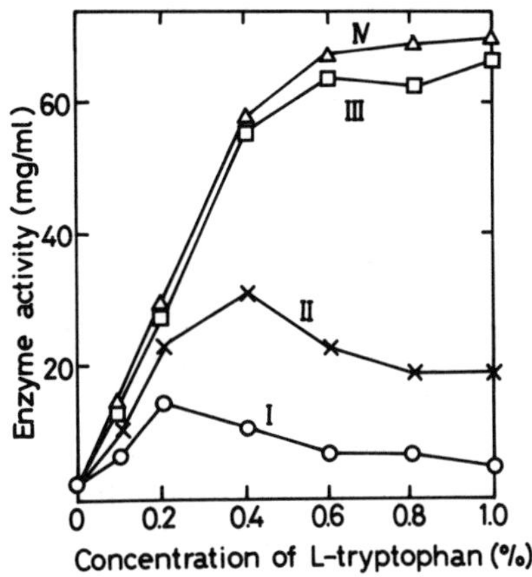

Fig. 2. Effect of L-tryptophan and sorpol on the formation of tryptophanase. Sorpol was added to the basal medium at the concentrations of 0 (I), 1 (II), 3 (III) and 5% (IV). The enzyme activity was determined by measuring the amount of L-tryptophan synthesized in a reaction mixture containing 200 mg of sodium pyruvate, 200 mg of ammonium acetate, 150 mg of indole, 0.5 mg of PLP, 5 mg of sodium sulfite; and the cells were harvested from 5 ml of the culture broth in a total volume of 5 ml. After the pH was adjusted to 8.8, the mixture was incubated at 37°C for 12 hours.

TABLE 2

EFFECT OF NONIONIC DETERGENTS ON THE FORMATION OF TRYPTOPHANASE

Nonionic detergent*	Enzyme activity** (mg/ml)
None	7
Sorpol W-200: P alkyl phenol ether***	83
Liponox NCK: P alkyl phenol ether	62
Liponox LCR: P lauryl alcohol ether	15
Triton XN-100: P alkyl aryl ether	62
Pegnol 1500: P alkyl ether	78
Nonal 208: P alkyl aryl ether	70
Tween 85: P sorbitan trioleate	59
Nissan Nonion NS-230: P alkyl phenol ether	80
Nissan Nonion ST-221: sorbitan monostearate	81
Nissan Nonion LP-20: Sorbitan monolaurate	20
Nissan Staform: Fatty ethanol amide	12
Sorbon T-20: P sorbitan monolaurate	35
Sorbon S-20: Sorbitan monolaurate	12
Tohol N-270: Fatty alkylol amide	30

*Nonionic detergents were added to the medium at the concentration of 5%

**Enzyme activity was determined by measuring the amount of L-tryptophan synthesized under the reaction conditions described in Fig. 2.

***P is polyoxyethylene

Under the culture conditions described above, tryptophanase was accumulated in growing cells of *Proteus rettgeri* as about 6% of the total cellular protein.

SUMMARY

Culture conditions for the microbial production of β-tyrosinase and tryptophanase were described. The enzymes were efficiently accumulated in growing cells of the bacteria by adding the inducers and the competitive inhibitor or the detergent to the media. Thus, the cells would be directly used as the enzymes for the enzymic preparation of L-tyrosine, L-tryptophan and their related amino acids.

REFERENCES

1. YAMADA, H. & KUMAGAI, H. *Adv. Appl. Microbiol.* *19*: 249, 1975.
2. NEWTON, W.A. & SNELL, E.E. *Proc. Natl. Acad. Sci. USA* *51*:382, 1964.
3. YOSHIDA, H., KUMAGAI, H. & YAMADA, H. *Amino Acid Nucleic Acid* *31*:113, 1975.
4. ENEI, H., MATSUI, H., OKUMURA, S. & YAMADA, H. *Biochem. Biophys. Res. Commun.* *43*:1345, 1971.
5. ENEI, H., NAKAZAWA, H., MATSUI, H., OKUMURA, S. & YAMADA, H. *FEBS Lett.* *21*:39, 1972.
6. NAKAZAWA, H., ENEI, H., OKUMURA, S., YOSHIDA, H. & YAMADA, H. *FEBS Lett.* *25*:43, 1972.
7. YAMADA, H., YOSHIDA, H., NAKAZAWA, H. & KUMAGAI, H. *Acta Vitaminol. Enzymol.* *29*:248, 1975.
8. ENEI, H., MATSUI, H., YAMASHITA, K, OKUMURA, S. & YAMADA, H. *Agric. Biol. Chem.* *36*:1861, 1972.
9. KUMAGAI, H., MATSUI, H. & YAMADA, H. *Agric, Biol. Chem.* *34*:1259, 1970.
10. ENEI, H., YAMASHITA, K., OKUMURA, S. & YAMADA, H. *Agric. Biol. Chem.* *37*:485, 1973.
11. KUMAGAI, H., KASHIMA, N., YAMADA, H., ENEI, H. & OKUMURA, S. *Agric. Biol. Chem.* *36*:472, 1972.
12. NAKAZAWA, H., ENEI, H., OKUMURA, S. & YAMADA, H. *Agric. Biol. Chem.* *36*:2523, 1972.

BATCH CULTURE EXPERIMENTS IN THE OPTIMIZATION OF CONTINUOUS FERMENTATION OF AN INTRACELLULAR ENZYME

Georg Skot

Novo Industri A/S
Bagsvaerd, Denmark

The two most important research and development tasks in the fermentation industry are the selection of high yielding mutants and the optimization of medium formulation. Nearly all significant improvements in productivity are the results of investigations in these two areas.

Optimization of continuous fermentation processes present special problems. The optimal values of important operating parameters, such as dilution rate, cell mass and aeration rate depend on the medium composition and on the mutant. A new mutant or a new medium can, therefore, not be tested under a single set of experimental conditions. In addition to this difficulty, continuous culture experiments are often afflicted by practical problems, such as clogging of nutrient feed tubes, drift of oxygen electrodes, and degeneration of the microorganisms. Due to the expense and time involved in continuous culture experiments, we have sought a more effective way of optimizing continuous enzyme fermentation processes.

The case I present here is the optimization of the production of an intracellular enzyme, glucose isomerase, by *Bacillus coagulans*.

METHOD

Batch experiments were used to test new substrates and new mutants. The enzyme yield under continuous culture conditions was calculated on the basis of the results of the batch experiments. The batch experiments were carried out in a 10-l Biotec fermentor at 50°C and pH 6.8 with variable agitation, an aeration rate of 0.5 vol/vol/min, and a 48-hr cycle. A metabolizable sugar or organic acid carbon source was fed through a peristaltic pump with the feed rate determined by a pH-control unit. In this way the carbon consumption rate was easy to measure and proved to be constant during the oxygen-limited phase of the batch process. Growth limitation was due simultaneously to oxygen and the carbon source. The oxygen transfer coefficient also was found to be constant in the present case.

For a batch fermentation the kinetics of growth of the cell mass concentration x can be described in terms of the substrate consumption rate ds/dt, the yield constant Y_G, the maintenance requirement M, or several constants k_1, k_2 and k_3. The method of experimental determination of Y_G and M is shown in Fig. 1. Integration of the growth rate equation of Fig. 1 gives a fourth constant k_4.

The constant k_2 was determined on the basis of measurements of the culture volume and the supply rate for the carbon source. Knowing k_2, k_1 was obtained from

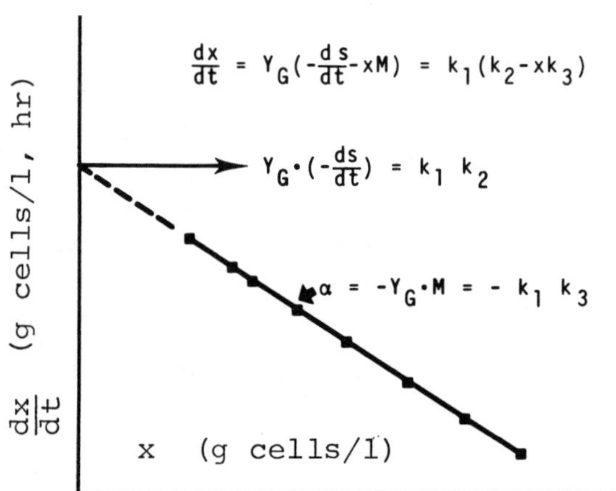

Fig. 1. Graphical determination of Y_G and M. See text for definition of symbols.

the intercept and k_3 from the slope of Fig. 1. The integration constant k_4 was calculated using the integrated equation and cell mass data during the constant growth rate period. Thus, all of the parameters for the oxygen-limited phase of the fermentation were obtained.

Since glucose isomerase is an intracellular enzyme, the rate of enzyme formation E depends on both the specific enzyme activity A_x (defined as the enzyme activity A in units/ml divided by x and the cell mass concentration x. The following model was used:

$$E = \frac{dx}{dt} \times A_x \qquad \text{(Eq. 1)}$$

For continuous culture the steady-state condition in a single-stage chemostat may be expressed as

$$U = D = \frac{1}{x}\frac{dx}{dt} \qquad \text{(Eq. 2)}$$

Where U is the specific growth rate, the expression for enzyme formation therefore can be rearranged to

$$E = UxA_x \qquad \text{(Eq. 3)}$$

This is the function which has to be maximized in a continuous enzyme fermentation.

RESULTS AND DISCUSSION

Fig. 2 shows the pattern of a typical batch experiment where cell mass concentration, oxygen tension, and glucose isomerase activity were recorded. As you see, the fermentation was oxygen-limited during a major part of the batch growth period. The increase in oxygen tension at the end of the fermentation indicated that the growth factors in the medium were limiting. In Fig. 3 dA/dx or simply dA_x is shown as a function of the reciprocal of the specific growth rate. The curve shows that the specific productivity varied strongly with the specific growth rate. In area I, where the growth rate was higher than approximately 0.12 hr^{-1}, the specific productivity was proportional to the reciprocal of the specific-growth rate. In area II, for specific growth rates between 0.06 and 0.12 hr,$^{-1}$ the specific productivity was nearly independent of the growth rate. Specific growth rates below

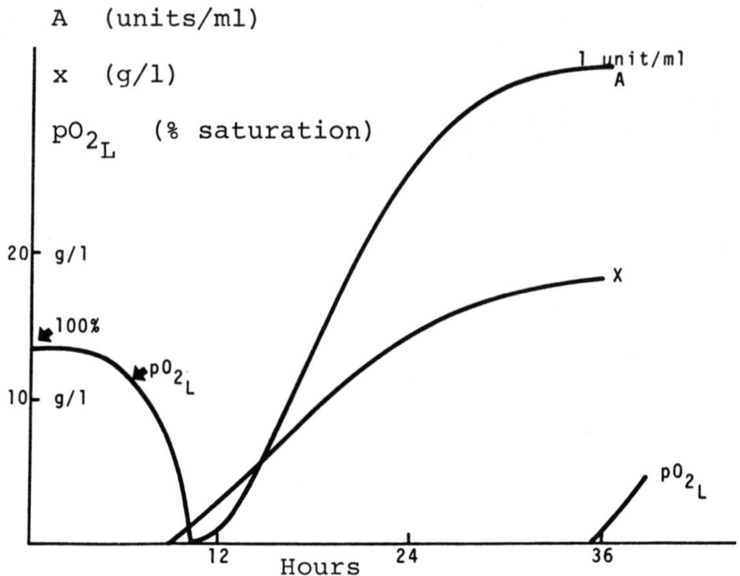

Fig. 2. Batch fermentation results.

0.06 formed a third area where the enzyme productivity decreased for decreasing growth rate.

In the previous batch experiment, a maximum cell mass concentration of 18 g/l was obtained. By using the same medium and previously determined values of Y_G and M, we designed a batch experiment where we selected an aeration rate and subsequently a carbon source consumption rate which yielded a cell mass concentration less than 18 g/l. In this experiment, the oxygen was rate limiting throughout the fermentation. The results (Fig. 4) differ from the curve from the previous batch experiments (Fig. 3) in that the productivity did not decrease with decreasing specific growth rate, but was constant for all values of U less than 0.15 hr^{-1}. The difference may be explained by assuming that the specific enzyme activity was a function of the substrate concentration. A possible explanation for this dependency may have been that certain components affected the enzyme formation rate. With increasing cell concentration the substrate concentration decreased, and the enzyme formation was repressed earlier than growth.

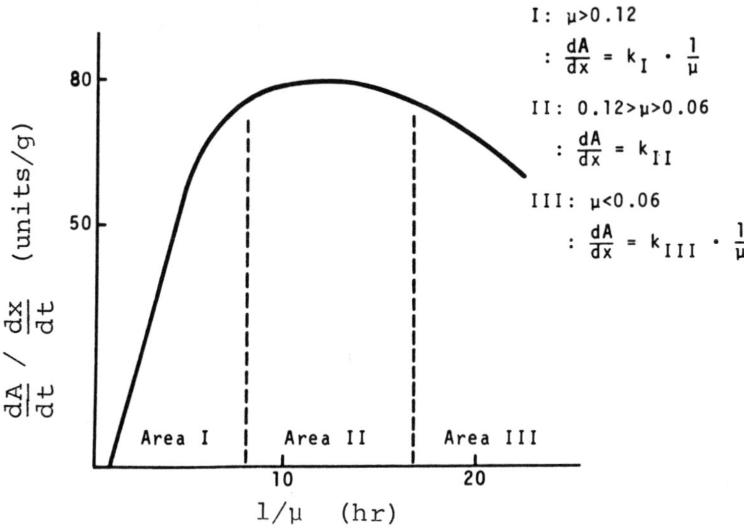

Fig. 3. dA_x as a function of the reciprocal of the specific growth rate.

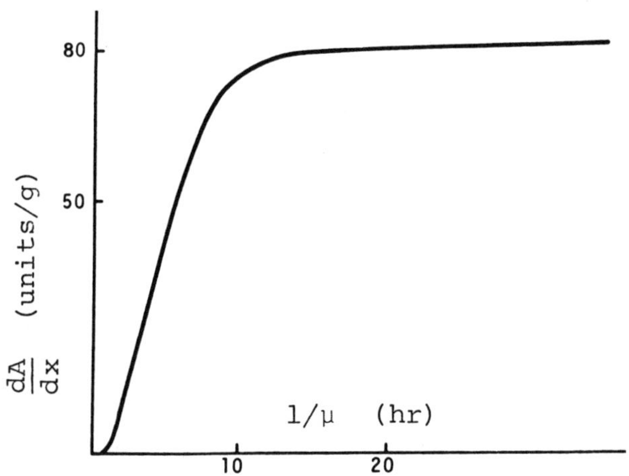

Fig. 4. Fig. 3 type plot at different fermentor operating conditions.

In order to develop the conditions to maximize the rate of enzyme formation (Eq. 3), the product of cell concentration and specific enzyme activity was plotted as a function of cell mass (Fig. 5). The enzyme activity reached a maximum at a cell density of 12-14 g/l. The corresponding dilution rate was obtained from a plot of the product of the specific growth rate and the specific enzyme activity versus the reciprocal specific growth rate. In the range from U_{max} to U = 0.12, the productivity per g cells was constant. Below the value 0.12 the productivity decreased with decreasing U. From the point of view of economics the optimal specific growth rate, i.e. the optimal dilution rate in a continuous fermentation, was around 0.12 hr^{-1}.

In a single batch experiment we have determined some basic growth and production parameters which may be used to design a continuous fermentation process utilizing the same strain and the same medium. The enzyme formation has been described as a function of the substrate concentration and of the specific growth rate. The optimal conditions for a continuous fermentation thus can be calculated. We can, furthermore, use the results to calculate the carbon source consumption rate and hence the aeration rate, which will yield the required 12-14 g cell/l at the optimal dilution rate in a continuous culture.

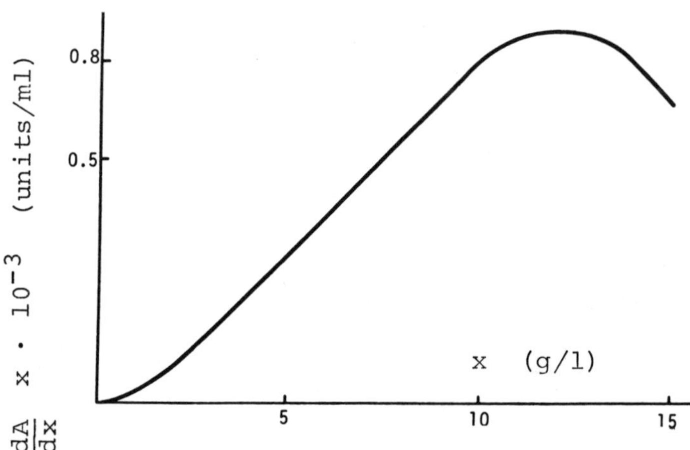

Fig. 5. Specific enzyme activity times cell concentration versus x.

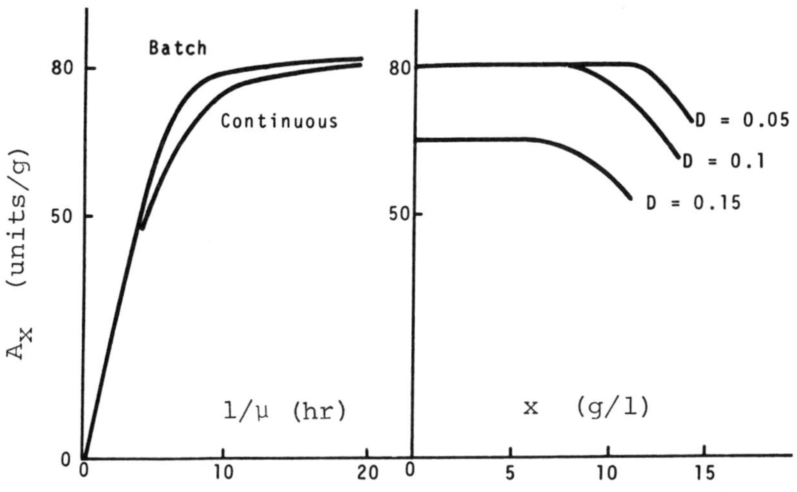

Fig. 6. Continuous fermentation results.

In order to test whether the theory which was developed here holds in a practical case, we performed an actual continuous culture experiment at the same conditions as in the batch experiments. Fig. 6 shows the results of three months experiments in a 350 ml chemostat. There was fairly good agreement between the batch and continuous culture data on the specific activity as a function of dilution rate. Continuous fermentation results for the specific activity as a function of cell mass concentration showed a curve of much the same shape as for the batch fermentation. It was interesting to note from the continuous fermentation experiments that the degree to repression of the enzyme synthesis caused by depletion of the medium was dependent on the specific growth rate.

The case of glucose isomerase production which has been discussed here shows how fairly accurate results can be obtained from a batch fermentation in 48 hours. To generate the same data in continuous culture experiments would last more than 2000 hours.

PROPERTIES OF HEAT STABLE ENZYMES OF EXTREME THERMOPHILES

Tairo Oshima

Mitsubishi-Kasei Institute of Life Sciences
Machida, Tokyo, Japan

(Not presented at conference due to last minute inability of author to attend.)

Thermophilic organisms have become the subject of intense research interest in the recent decade. Biochemists are interested in molecular mechanisms of thermophilly and of biological adaptation to high temperatures (1). Thermophiles also are useful materials for the study of such biochemical reactions as protein biosynthesis (2), nucleic acid synthesis (3), and oxidative phosphorylation (4) in which components from mesophilic sources are so labile that the elucidation of reaction properties is often difficult. In addition, thermophiles are fascinating organisms for enzyme engineering since they are resistant to heat and other chemical denaturants. Also, insight into the molecular basis of unusual stabilities of these enzymes will make it possible to find out a new, effective way of chemical modification for making highly stable enzymes from unstable biocatalysts of mesophilic organisms.

With special attention to the possibilities of application in enzyme engineering, enzymes of an extremely thermophilic bacteria, *Thermus thermophilus*, isolated from a Japanese hot spa (5) have been studied. From results obtained so far the properties common to thermophile enzymes can be listed as follows (6): (a) stable to heat without exception, usually in the absence of specific protective cofactors, (b) stable also to other chemical denaturants, (c) other properties are generally similar to those of the corresponding enzyme of mesophilic organisms, (d) amino acid composition and secondary structure contents of a thermophile enzyme are also similar to those of the mesophile counterpart, and (e) the changes in chemical structure which make the protein thermostable seem to be relatively subtle.

STABILITY

We so far have studied the biochemical properties of more than a dozen enzymes from *T. thermophilus* (5), an extreme thermophile having a maximum growth temperature of over 80°C. We have found that the enzymes are resistant to heat without exception. In most cases, these enzymes are usually stable in a highly purified state, in the absence of polycationic metal ions or specific ligands bound to these enzymes, and in a crude extract. Thus, the unusual stability is based on the intrinsic intramolecular interactions of the polypeptide chain (6). The thermal stability of an enzyme often correlates positively to the upper limit of the growth temperature of the organism from which the enzyme is extracted. An example is illustrated in Fig. 1. Glyceraldehyde-3-

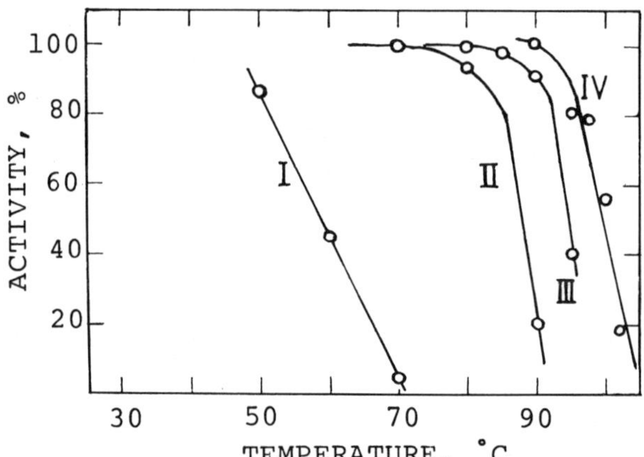

Fig. 1. Thermal inactivation profile of glyceraldehyde-3-phosphate dehydrogenase. I: from non-thermophilic organism (pig); II: from moderate thermophile (*B. stearothermophilus*); III: from *T. thermophilus* (extreme thermophile); IV: from *T. aquaticus* (extreme thermophile). The enzyme was heated for 5 min (except IV where 10 min) at the indicated temperature; then the remaining activity was assayed at 30°C (7, 8, 14).

phosphate dehydrogenases of thermophiles are much more stable than that of pig muscle. Among the thermophile enzymes, those from extreme thermophiles (7,8), such as *T. aquaticus* and *T. thermophilus* which are capable of growing at over 75°C, are more tolerant than those of moderately thermophilic bacteria, such as *Bacillus stearothermophilus*. Similarly, phosphoglycerate kinase (9) from an extreme thermophile, *T. thermophilus* is more stable than that of a moderate thermophile as listed in Table 1. These facts indicate that extreme thermophiles may be more useful than moderate ones as sources of enzymes for industrial applications.

TABLE 1

COMPARISON OF ENZYMATIC PROPERTIES OF PHOSPHOGLYCERATE KINASES OF VARIOUS SOURCES (9, 10, 17)

Item	Mesophilic		Thermophilic	
	Rabbit	Yeast	*B. stearothermophilus*	*T. thermophilus*
Crystals	Needles & rods	Needles & rods		Needles or plates
MW	47000	47000	42000	43000
Subunit	1	1	1	1
pI	7.0	7.2	4.9	5.0
Opt. pH	6-9.2		5.3-8.5	5.5-8.0
Km(ATP)(mM)	0.42	0.48	2.9	0.28
Km(3PG)(mM)	1.4	1.3	2.2	1.79
Substrate specificity*	ATP(100) ITP(77) GTP(56)	ATP(100) ITP(70) GTP(55)	ATP(100) ITP(42) GTP(27)	ATP(100) ITP(0) GTP(0)
Heat Stability (min)**	55°(15)	62°(15)	70°(10)	90°(10)

*Relative activity when ATP is replaced by other purine nucleotides.
**Conditions for 50% loss of enzyme activity, time in ( ).

Thermophile enzymes are, generally speaking, resistant to chemical denaturants such as urea, guanidine, detergents, organic solvents and high salt concentrations. They are often more resistant to acid and alkaline than the corresponding mesophile enzymes. Some enzymes of thermophiles are activated by the addition of such denaturants. For instance, *T. thermophilus* glyceraldehyde-3-phosphate dehydrogenase is activated up to five fold by the addition of ethanol (8). As far as the catalytic activity at room temperature, the enzyme is active in the presence of up to 40% ethanol.

OTHER CATALYTIC PROPERTIES

Except for unusual stability to heat and chemical agents, other catalytic and molecular properties are generally similar to those of the mesophile counterpart. Tables 1 and 2 summarize two examples. In both cases,

TABLE 2

COMPARISON OF ENZYMATIC PROPERTIES OF GLYCERALDEHYDE-3-PHOSPHATE DEHYDROGENASES FROM MESOPHILIC AND THERMOPHILIC SOURCE (7,8,14-16)

Item	Mesophilic		Thermophilic	
	Rabbit	*E. coli*	*B. stearothermophilus*	*T. thermophilus*
MW	140,000	144,000	144,000	130,000
Subunit structure	α_4	α_4	α_4	α_4
Coenzyme	NAD	NAD	NAD	NAD
Km(GAP)	10^{-5}	--	10^{-5}	3×10^{-4}
Km(NAD)	10^{-5}	--	10^{-5}	10^{-5}
Activity (U/mg enzyme)	164(30°)	40(30°)	50-100(25°)	55*(30°)
Active site	Cys	Cys	Cys	Cys
Am_2SO_4 ppt (%)**	>50	65-93	72-90	65-75
Activity after 80°, 10 min	0	0	40%	>90%

*Under optimum conditions.
**Saturation of ammonium sulfate necessary for collection of enzyme protein by precipitation.

no significant difference was observed between the thermophile and mesophile enzymes in respect to molecular weight, subunit structure, Km values, coenzyme requirement, essential amino acid residues for the catalytic action, and pH optimum. Although the molecular weights of the thermophile phosphoglycerate kinases were slightly smaller and their specificities to nucleotide triphosphate were narrower than those of the counterparts from mesophilic sources, these differences were not remarkable. The isoelectric points of thermophile phosphoglycerate kinases were significantly lowered compared with those of yeast and animal enzymes, but their implications in the unusual stability were open to question.

AMINO ACID COMPOSITION

It has been known that at the gross level, the amino acid compositions of thermophile proteins are similar to those of the corresponding proteins from mesophilic organisms (11). Although often Glx, Arg, and Leu are increased and Asx, Lys, Ser, and Cys are decreased in thermophile proteins, these differences are small and varied (1). If the amino acid composition is used to predict structural parameters such as average hydrophobicity or number of α-helix breakers, there appears no significant difference in these predicted parameters of thermophilic and non-thermophilic enzymes. Comparison of CD spectra suggest that there is no significant difference in the secondary structure of thermophile enzymes and the corresponding proteins of non-thermophilic organisms.

THERMODYNAMIC PROPERTIES

We have studied the thermodynamic properties of some proteins from *T. thermophilus* (12). In the case of thermophile phosphoglycerate kinase, the ΔG values for the denaturation in the presence of suitable concentrations of guanidine hydrochloride did not depend on temperature: that is, $\partial \Delta G/\partial T = - \Delta S$ was nearly zero. On the other hand, for the denaturation of yeast phosphoglycerate kinase under analogous conditions, the ΔG values greatly depended on temperature with a sharp maximum at around 28°C, which is the temperature of maximum stability defined by Brandts (13). We are now trying to find out the relation between changes in thermodynamic properties and chemical structure, especially amino acid replacements which affect the change of entropy of thermal denaturation. This line of study is expected to identify the structural elements which enable the thermophile enzyme to be resistant to heat and chemical agents.

REFERENCES

1. ZUBER, H., "Enzymes and Proteins from Thermophilic Microorganisms," Experientia Supplementum 26, Birkhauser Verlag, Zurich, 1976.
2. FRIEDMAN, S.M. *Bacteriol. Rev. 32*:27, 1968.
3. TSUJI, S., SUZUKI, K. & IMAHORI, K. *Nature 261*:725, 1976.
4. YOSHIDA, M., OKAMOTO, H., SONE, N., HIRATA, H. & KAGAWA, Y. *Proc. Nat. Acad. Sci. U.S.A. 74*:936, 1977.
5. OSHIMA, T. & IMAHORI, K. *Intern. J. Syst. Bacteriol. 24*:102, 1974.
6. OSHIMA, T. *Seikagaku 48*:895, 1976.
7. HARRIS, J.I. & WATERS, M. in "The Enzymes," vol. 13 (Boyer, P.D., ed.), Academic Press, New York, 1976, p. 1.
8. FUJITA, S.C., OSHIMA, T. & IMAHORI, K. *Eur. J. Biochem. 64*:57, 1976.
9. NOJIMA, H., OSHIMA, T. & NODA, H. in preparation.
10. SUZUKI, K. & IMAHORI, K. *J. Biochem. 76*:771, 1974.
11. SINGLETON, R. JR. & AMELUNXEN, R.E. *Bacteriol. Rev. 37*:320, 1973.
12. NOJIMA, H., IKAI, A., OSHIMA, T. & NODA, H. *J. Mol. Biol. 116*:429, 1977.
13. BRANDTS, J.F. *J. Am. Chem. Soc. 86*:4291, 1964.
14. SUZUKI, K. & HARRIS, J.I. *FEBS Lett. 13*:217, 1971.
15. VELICK, S.F. & FURFINE, C. in "The Enzymes, vol. 7 (P.D. Boyer, H. Lardy, & K. Myrback, eds.) Academic Press, New York, 1963, p. 243.
16. D'ALESSIO, G. & JOSSE, J. *J. Biol. Chem. 246*:4326, 1971.
17. SCOPES, R.K. in "The Enzymes," vol. 8 (P.D. Boyer, ed.) Academic Press, New York, 1973, p. 335.
18. BIGELOW, C.C. *J. Theor. Biol. 16*:187, 1967.

AQUEOUS TWO-PHASE SYSTEMS FOR THE LARGE-SCALE PURIFICATION OF ENZYMES

M.-R. Kula, A. Buckmann, H. Hustedt, K.H. Kroner, and M. Morr

Gesellschaft fur Biotechnologische Forschung mbH
Braunschweig-Stockheim, Federal Republic Germany

For the production of enzymes the need exists for more efficient isolation and purification techniques. The commercial isolation of intracellular enzymes still resembles normal laboratory procedures, only made larger. For the first steps in isolation, where large quantities and volumes are involved, new processes are particularly needed with the potential for easy scale up and continuous processing under conditions which favor the stability of the biological materials. The removal of cell debris from crude extracts, for example, is still a major technical problem (1). Centrifuges with the necessary capacity have low g-forces and are often inefficient. Filtration as an alternative process is handicaped by the colloidal nature of the suspended materials and by the high viscosity of the extracts, which make filtrations slow and cumbersome. Therefore, we have been searching for a better procedure for the clarification of crude extracts and have investigated a liquid-liquid separation technique for large scale work (2, 3).

The aqueous two-phase systems we tried were described mainly by Albertsson (4). Aqueous two- or multiphase systems are formed as a result of the so-called incompatibility of certain polymers. If a solution of a highly water miscible polymer, e.g. polyethylene glycol, is mixed with the solution of a second highly water miscible polymer, e.g. dextran, one may obtain two phases depending on the concentrations of the polymers. Albertsson determined phase diagrams for a number of systems. A partition coefficient was defined as K = C top/C bottom for such aqueous two-phase systems.

It is advantageous that partition is independent of the absolute concentration over a fairly wide range. Any substance will prefer the phase where maximum numbers of interactions are possible or where a minimum in the energy content of the system is reached. Partition in general is described by the Bronstedt equation $\ln K = \lambda M/kT$, where M stands for the molecular weight of the substance, k for the Boltzmann constant, and T for the absolute temperature. λ is a factor comprising all the characteristics of the system. It is obvious that for large values of M, small deviation in λ will have a pronounced effect on the partition. Therefore, partition has been successfully employed working with very large entities, e.g. separation of different cell types (5), enrighment of phages (6), and purification of DNA (7). Such large macromolecules show a one sided partition ($K>100$; $K<0.01$), while proteins normally have partition coefficients between 0.1 and 10 and small molecules like buffer ions and salts around 1.0.

In such aqueous two-phase systems the following parameters influence the partition: polymers composing the two-phase system, average molecular weight of the polymers, their concentration, kind of ions present, ionic strength, pH, and temperature. All of these parameters do not operate independently of one another; and at present no general theoretical basis is available to analyze and describe such complex systems. Favorable conditions have to be found by experiments. For the successful application of such systems for the clarification of crude extracts, the following conditions should be met: (a) the desired protein and cell debris and cells should partition in different phases; (b) the partition coefficient for the protein and the volume ratio of the phase system should be such that the desired product can be separated in high yield; preferentially in a single step; (c) favorable two-phase systems should be separable employing a commercial separator, taking advantage of the highly advanced technology of this area of chemical engineering; and (d) the removal of the polymers constituting the phase system should be easily performed.

EXPERIMENTAL RESULTS

This approach is demonstrated here with the enzyme pullulanase, produced by *Klebsiella pneumoniae* and a variety of other microorganism. It splits glucose α-1,6-bounds neighboring glucose α-1,4-bounds to yield oligosaccharides (8). Pullulanase debranches starch and is therefore of commercial interest for starch based processes.

Experiments indicated that *K. pneumoniae* cells partitioned entirely into the bottom phase of a polyethylene glycol-dextran system. The partition coefficient for pullulanase was found initially as 0.3 but was increased by exploring different conditions for the two-phase system. We derived the following composition for a suitable phase system: 9% PEG 4000, 1.75% Dextran T 500, 300 mM phosphate buffer pH 7.8 (9).

Before the partition step pullulanase was solubilized from *K. pneumoniae* by treatment with cholate (10). This left the majority of the cells intact. The cells influenced the phase system, as evidenced in the drastic changes in volume ratio with increasing cell concentration, from 8.8 without cells to 1.8 with 25% cells in the system (weight moist cells/weight). The partition coefficient for pullulanase also decreased with increasing cell concentration. The volume ratio as well as the partition coefficient would limit the yield that could be obtained for a single partition step. The yield was calculated as follows:

$$\text{yield (\%)} = \frac{100}{1 - \dfrac{V_b}{V_t \quad K}} \qquad \text{(Eq. 1)}$$

where V_b and V_t are the volumes of the bottom and top phases, respectively. Fig. 1 shows the observed yield for pullulanase as a function of the cell concentration. For a 25% cell suspension still approximately 90% of the enzyme is separated in one step.

The viscosity of the cell suspension also increased quite steeply above 25% cell concentration and influenced the throughput of the separator. The throughput Q can be described in terms of d, the average diameter of particle or droplet; ΔP, difference in density between phases; η, dynamic viscosity of liquid; g, gravitational acceleration; w, angular velocity; Φ, tangens of angle of the disc stack; Z, number of disc in the stack; r_1, outer radius of disc stack; and r_2, inner radius of disc stack. The expression is:

$$Q = \frac{d^2 \ (\Delta P) \ g \ 2 \ \pi \ w^2 \ \Phi \ Z \ (r_1{}^3 - r_2{}^3)}{18 \ \eta \qquad 3g} \qquad \text{(Eq. 2)}$$

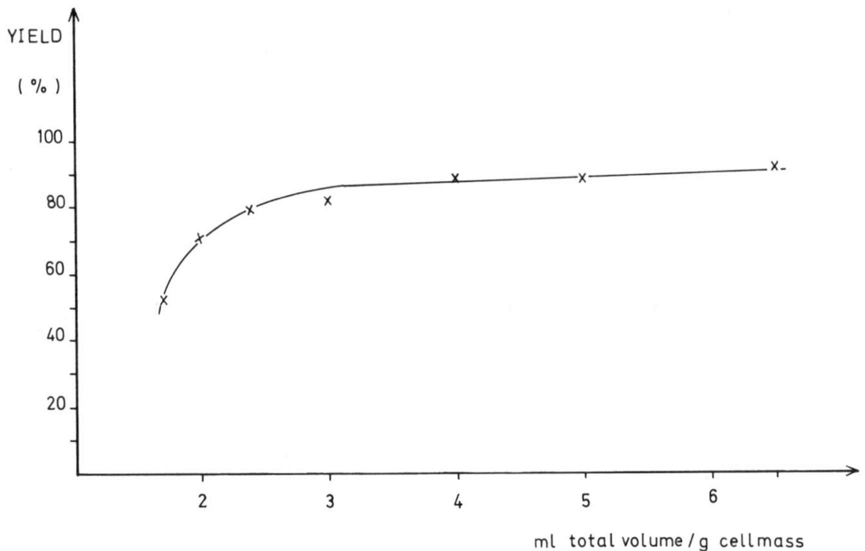

Fig. 1. Dependence of the activity yield on the proportion total volume/cellmass.

To test the performance of the commercial unit in the separation of an aqueous two-phase system, we used the smallest available separator, an open disk stack Gyrotester B from Alfa-Laval. The correct positioning of the interphase and the optimal flow rate for the separation of an aqueous two-phase system were determined by experiments. Both parameters were dependent on the physical properties of the medium processed. With the Gyrotester B the position of the interphase was influenced by the length of the regulating screw and not by a gravity disk. The optimal length was 13.5 mm. The top phase could be obtained in almost 100% purity. The optimal flow rate was around 200 ml/minute. From these values a residence time of 120 seconds was calculated. The loss of top phase at a flowrate of 200 ml per minute was only 2%. The short residence time helped to keep the temperature relatively constant during the separation. This was important since deviations in the temperature could have led to homogeneity in the phase system, thereby abolishing the separation.

Table 1 summarizes a number of experiments. In the larger volume test with *K. pneumoniae* the yield of a 200 l fermentation was processed in a two-phase system and the enzyme obtained in 88% yield. 90% yield was calculated using laboratory data. Thus, a scale up by a factor of 5000 was accurately described in this experiment. Since the enzyme was solubilized from the intact cells by treatment with cholate, the isolation of pullulanase could be thought of as a special case. However the test with *S. carlebergens* demonstrated that the aqueous two-phase system was generally applicable for the separation. α-Glucosidase is a rather labile enzyme. It can be separated from the cell debris of *Saccharomyces carlsbergensis* after breaking the yeast by freezing and thawing. The partition coefficient for α-glucosidase so far could be raised only to 1.5; the yield of 74% for a single separation step compared well with 78%, as calculated.

For the removal of the phase components, e.g. polyethylene glycol or dextran, a number of methods are available, e.g. ultrafiltration or adsorption of the desired enzymes on ion exchange resins. Furthermore new two-phase systems can be generated by adding salt to polyethylene glycol rich top phases and shifting the desired product to the salt phase (9).

Several experiments reported in the literature show that the selectivity of the extraction can be enhanced by employing affinity ligands anchored to one of the polymers constituting the phase systems, a concept comparable to affinity chromatography (11-15). Also, hydrophobic ligands have been used to exploit hydrophobic interactions for the specific extraction of certain proteins (16-18). It can be envisaged that by a series of partition steps fast and efficient processes for large scale enzyme isolations could be developed.

TABLE 1

SUMMARY OF EXPERIMENTS

Variable	Phase System alone	*Klebsiella pneumoniae*		*Saccharomyces carlsbergens*
cell mass (kg)	-	1.30	6.65	0.83
total volume (ℓ)	-	4.20	22.25	2.5
PEG 4000 (%w/w)	9	9	9	9
Dextran T500 (%w/w)	2	2	1.75	2
V_t/V_b	9	1.8	1.8	2.6
P_t/P_b	0.95	0.90	0.90	0.91
total activity (U)	-	41×10^3	186×10^3	229×10^3
activity in top phase (U)	-	37.3×10^3	165×10^3	170×10^3
activity yield (%)	-	91	88	74

ACKNOWLEDGMENTS

Part of this work was supported by the biotechnology program of the Bundesministerium fur Forschung und Technologie der Bundesrepublik Deutschland (Projekt BCT 17). We like to thank K. Wolter and W. Stach for excellent technical assistance.

REFERENCES

1. NAEHER, G. & THUM, W. in "Industrial Aspects of Biochemistry," vol. 1 (B. Spencer, ed.), FEBS 1974, p. 47.
2. KULA, M.-R. & KRONER, K.H. *Abstr. 5 Inter. Fermen. Symp. Berlin* 3.14, 1976.
3. KULA, M.-R., KRONER, K.H., DUREKOVIC, A. & STACH, W. German Offenlegungsschrift 2,616,584; 1977.
4. ALBERTSSON, P.A. "Partition of Cell Particles and Macromolecules," 2nd ed., Wiley, New York, 1971.
5. WALTER, H., ERIKSSON, G., TAUBE, O. & ALBERTSSON, P.A. *Exptl. Cell Res. 64*:486, 1970.
6. ALBERTSSON, P.A. *Methods in Virology 2*:303, 1967.
7. PATTERSON, J.B. & STAFFORD, D.W. *Biochemistry 9*:1278, 1970.
8. EISELE, B., RASCHED, I.R. & WALLENFELS, K. *Eur. J. Biochem. 26*:62, 1972.
9. HUSTEDT, H., KRONER, K.H. & KULA, M.-R. this volume.
10. BENDER, H. *Arch. Mikrobiol. 71*:331, 1970.
11. TAKERKART, G., SEGARD, E. & MONSIGNY, M. *FEBS Lett. 42*:218, 1974.
12. FLANAGAN, S.D. & BARONDES, S.H. *J. Biol. Chem. 250*:1484, 1975.
13. FLANAGAN, S.D., TAYLOR, P. & BARONDES, S.H. *Nature 254*:441, 1975.
14. HUBERT, P., DELLACHERIE, E., NEEL, J. & BAULIEU, E. E. *FEBS Lett. 65*:169, 1976).
15. BUCKMANN, A.F., MORR, M. & KULA, M.-R., this volume.
16. SHANBHAG, V.P. & JOHANSSON, G. *Biochem. Biophys. Res. Comm. 61*:1141, 1974).
17. WESTRIN, H., ALBERTSSON, P.A. & JOHANSSON, G. *Biochem. Biophys. Acta 436*:696, 1976.
18. WALTER, H. & KROB, E. *FEBS Lett. 61*:290, 1976.

PROCEDURE FOR THE SIMULTANEOUS LARGE-SCALE ISOLATION OF PULLULANASE AND 1,4-α-GLUCAN PHOSPHORYLASE INVOLVING LIQUID-LIQUID SEPARATIONS

H. Hustedt, K. H. Kroner and M.-R. Kula

Gesellschaft fur Biotechnologische Forschung mbH.
Braunschweig-Stockheim, Federal Republic Germany

In enzyme isolation procedures liquid-liquid partition can be employed using aqueous two-phase systems, based on two hydrophylic polymers like polyethyleneglycol (PEG) and dextran or on a suitable polymer and certain salts (PEG and phosphate) (1). We have used such a two-phase systems for the simultaneous isolation of pullulanase (2) and 1,4-α-glucan phosphorylase (3) from *Klebsiella pneumoniae*.

K. pneumoniae cells were stirred for one hour in 1% sodium cholate containing 0.4 M sodium phosphate pH 7.5. This solubilized pullulanase from the outer membrane (2), while the main part of the cells remained intact. PEG 4000 and dextran T 500 were added. After equilibration the system (9% PEG, 2% dextran, 25% cells, 0.35 M sodium phosphate pH 7.5) was separated using a Gyrotester B. The pullulanase partitioned into the top phase, whereas cells and cell debris were in the bottom phase. The top phase was diafiltrated with 0.02 M sodium phosphate pH 7.0 using a hollow fiber cartridge with a cut off of 100,000. Then the pullulanase was precipitated by addition of N-cetyl-N,N,N-trimethyl ammonium bromide (CTAB). The precipitate, which was collected easily by low-speed centrifugation, was dissolved in 0.4 M NaCl. Traces of CTAB were removed by a Dowex ion exchanger.

The total yield of pullulanase was 70%, at an enrichment factor of about 8 and purity of about 80%. No α-amylase activity was present.

The highly viscous dextran bottom phase was diluted 1:2 with 0.02 M potassium phosphate, and passed through a Manton-Gaulin Homogenizer at 600 atm. PEG 1550, potassium phosphate, and potassium chloride were added and the bottom phase, containing the cell debris, was discarded. To the top phase, which contained the phosphorylase, PEG 1550, potassium phosphate, and potassium chloride were added. In this system the phosphorylase partitioned into the bottom phase, which was then diafiltrated on a hollow fiber cartridge (cut off 50,000) with 0.02 M sodium phosphate pH 7.8, containing 0.02 M NaCl. From this solution the phosphorylase was isolated by adsorption to carboxymethyl-Sephadex (3). The total yield of phosphorylase was 55% at a purity of about 90% and a purification factor around 35. It was nearly free of glycosyl transferase activity.

The procedure demonstrates that liquid-liquid partition can be a useful method for the large-scale purification of enzymes.

REFERENCES

1. ALBERTSSON, P.A. "Partition of Cell Particles and Makromolekules," Wiley, New York, 1971.
2. BENDER, H. *Arch. Microbiol.* *71*:331, 1970.
3. LINDER, D., KURZ, K., BENDER, H. & WALLENFELS, K. *Eur. J. Biochem.* *70*:291, 1976.

PURIFICATION AND IMMOBILIZATION OF A FUNGAL β-GALACTOSIDASE (LACTASE)

J. L. Leuba, F. Widmer and D. Magnolato

Nestle Products Technical Assistance Co. Ltd.
Research Department
La Tour-de-Peilz Switzerland

PART A: Resolution of three *Aspergillus niger* β-galactosidase isoenzymes.

Several glycosidases secreted by the fungus *Aspergillus niger* have recently been purified and characterized (1-5). These enzymes were found to be glycoproteins; and their pH optima ranged from 3.5 to 5.0. We now report the purification of the *A. niger* β-galactosidase from a commercial preparation. The occurrance of this enzyme in multiple molecular forms has been suggested by earlier work (6). The aim of this study was to achieve separation and characterization of these isoenzymes.

The commercial β-galactosidase preparation was purchased from the RAPIDASE Company (59-Seclin, France). The purification steps were as follows: 1. Molecular sieving on Sepharose 6B; 2. Anion-exchange on DEAE-Sepharose CL-6B; 3. Hydrophobic chromatography on Octyl-Sepharose CL-4B; 4. Cation-exchange on CM-Sepharose CL-6B. Step 3 ensured clear-cut separation of three isoenzymes, which accounted for 83%, 8% and 9% of the total β-galactosidase activity, respectively. They were glycoproteins with an estimated molecular weight of 125,000 (carbohydrate content: ∿20%), isoelectric pH of about 4.6, and pH optima between 2.5 and 4.0. They were heat-stable up to about 60°C; and their K_m values for lactose were found to be 85 mM, 105 mM and 125 mM, respectively.

The present method of purification differs from those previously published by using a hydrophobic chromatography step, for isoenzymes resolution. The relative proportions of the isoenzymes might depend on the nutritional conditions imposed on the fungus.

PART B: Immobilization of *Aspergillus niger* β-galactosidase on chitosan.

Immobilization of β-galactosidases from different sources (bacteria, fungi) has been extensively studied. The natural polysaccharide chitosan (partially deacetylated chitin from arthropod shells) is particularly suitable for enzyme fixation, e.g. protease immobilization (7). We now describe the use of chitosan as an insoluble support for either crude or purified *A. niger* β-galactosidase preparations.

Rigid textured chitosan particles were activated with an aqueous glutaraldehyde solution as cross-linking agent (2.5% or 5%), and were mixed with enzyme solutions. The most active immobilized systems were obtained with *crude* enzyme preparations and 2.5% glutaraldehyde solutions. The following properties were observed: 1. The immobilized enzyme hydrolyzed lactose in acidic wheys, U.F. permeates and pure lactose solutions at comparable rates; 2. The pH activity profiles were not altered (pH optimum about 3.5); 3. $NaBH_4$-reduction of the β-galactosidase-chitosan conjugates increased their stability on repetitive use, and at temperatures up to 60°C; 4. The insoluble preparations were stable on storage (maximum loss of activity after 2 months at 4°C: <3%); 5. Stability on continuous use: after 1 month use with acidic whey (75% lactose hydrolysis), a β-galactosidase-chitosan column was found to retain 80% of its initial activity.

The efficiency of our immobilized β-galactosidase system can be shown to be generally greater than those of other preparations (e.g. immobilization on azoglass, alkylaldehyde glass, collagen, etc.). The stability of the chitosan-enzyme conjugates should allow their use for lactose hydrolysis on an industrial scale. Chitosan is readily prepared from an abundant natural polymer, and its cost should therefore not be prohibitive.

REFERENCES

1. RUDICK, M.J. & ELBEIN, A.D. *J. Biol. Chem.* *248*:6506, 1973.
2. RUDICK, M.J. & ELBEIN, A.D. *Arch. Biochem. Biophys.* *161*:281, 1974.
3. RUDICK, M.J. & ELBEIN, A.D. *J. Bacteriol.* *124*:534, 1975.
4. ELBEIN, A.D., ADYA, S. & LEE, Y.C. *J. Biol. Chem.* *252*:2026, 1977.
5. ADYA, S. & ELBEIN, A.D. *J. Bacteriol.* *129*:850, 1977.
6. LEE, Y.C. & WACEK, V. *Arch Biochem. Biophys.* *138*: 264, 1970.
7. LEUBA, J.L. British Patent 1,476,741: 1977; German Patent DAS 2,522,484: 1977.

A XYLANASE FROM *Schizophyllum commune*

M.G. Paice,* L. Jurasek,* M.R. Carpenter,**
and L.B. Smillie**

Pulp and Paper Research Institute of Canada,*
Pointe Claire, P.Q., Canada and
Department of Biochemistry
University of Alberta, Edmonton,**
Alberta, Canada

Having previously described (1) the production and isolation of cellulases and xylanases from a white rot fungus, *Schizophyllum commune*, we now report some details of the structure of one of the xylanases.

The major xylanase enzyme (Xylanase A) present in the culture filtrate of *Schizophyllum commune* was purified from an ethanol precipitate by simple chromatographic procedures. DEAE-Sephadex A50 separated Xylanase A from other protein components when used with the pH gradient of 9 to 5. Chromatography of the Xylanase A fraction on Sephadex G50 produced an electrophoretically pure enzyme in a yield of 10 mg per litre of culture filtrate.

Using the soluble portion of larchwood xylan as substrate, the pure enzyme (concentration 1.6×10^{-6} μmole/ml) hydrolysed 2.8×10^{5} β1-4 xylosidic bonds per minute at 30°C and pH 5 (optimum). The major hydrolysis product, as determined by gas chromatography of the trimethylsilyl derivatives, is xylobiose (∿70%) with xylose as the major monosaccharide. Xylanase A has no measurable activity against carboxymethyl cellulose. It has a molecular weight of 31,000, determined from calibrated SDS-urea-polyacrylamide gels. The enzyme is rich in acidic and aromatic amino acid residues, features shared by other xylanase enzymes isolated from *Stereum sanguinolentum* (2) and *Trametes hirsuta* (3), but unlike them it is not a glycoprotein. Some polysaccharide associated with the purified enzyme could be completely removed by SDS urea treatment at 100°C with 20% retention of enzymic activity.

The amino-terminal sequence of Xylanase A (Fig 1), determined with a Beckman Sequencer, Model 890B, shows no homology with any known protein sequence deposited with the Atlas of Protein Sequence and Structure (Washington, D.C.). However, surprisingly little sequence data about polysaccharide hydrolyzing enzymes is available. A homology with lysozyme is suspected but more sequence information is required to test the hypothesis.

1 2 3 4 5 6 7 8 9
NH_2–Ser–Gly–Thr–Pro–Ser–Ser–Thr–Gly–Thr–

10 11 12 13 14 15 16 17 18 19
Asp–Gly–Gly–Tyr–Tyr–Tyr–Ser–Trp–Trp–Thr–

20 21 22 23 24 25 26 27
Asp–Gly–Ala–Gly–Asp–Ala–Thr–Tyr–

Fig. 1. Amino Terminal Sequence of Xylanase A.

The development of enzyme-like catalysts has already shown some promise. A synthetic polymer catalyst resembling aryl sulfatase produced a 10^2 fold rate enhancement compared to the natural enzyme (4). A polymer catalyst that hydrolyses model substances of cellulose shows Michaelis-Menten type catalytic behaviour (5).

As expected, owing to the high molecular weight of the substrate, much of the activity of xylanases is lost after immobilization (6). A synthetic xylan-hydrolyzing catalyst could combine the stability of the immobilized xylanase with the activity of the free enzyme. The apparent commercial interest (7) in the non-cariogenic sweetener xylitol (obtained by hydrogenating xylose) may provide the incentive necessary to bring this idea to fruition.

REFERENCES

1. PAICE, M.G. & JURASEK, L. "TAPPI Forest Biology Wood Chemistry Conference Proceedings," Madison, Wisconsin, 1977, p. 113.
2. ERIKSSON, K.E. & PETTERSSON, B. *Int. Biodetn. Bull.* 7 (3):115, 1971.
3. TOMAN, R. *ACS/CIC 2nd Joint Conference*, Montreal, 1977.
4. KIEFER, H.C., CONGDON, W.I., SCARPA, I.S. & KLOTZ, I.M. *Proc. Nat. Acad. Sci. USA* 69:2155, 1972.
5. ARAI, K. & OGIWARA, Y. *Makromolekulare Chemie* 176: 2871, 1975.
6. PULS, J., SINNER, M., & DIETRICHS, H.H. *Trans. Technical Section, Canadian Pulp Paper Asso.* 3:TR64, 1977.
7. MOORE, K.K. *Forest Product Development* 11 (4):66, 1977.

Session II

INDUSTRIAL APPLICATIONS OF ENZYMES: ENGINEERING ASPECTS

Chairmen: C. Colton and V. Jensen

IMMOBILIZED LACTASE FOR WHEY HYDROLYSIS: STABILITY AND OPERATING STRATEGY

W.H. Pitcher, Jr.

Corning Glass Works
Corning, New York, USA

A number of previous studies have described the development of immobilized lactase for the hydrolysis of the lactose in cheese whey (1-7). From cost estimates (1,6,7), it is apparent that one of the most critical factors in determining the cost of hydrolysis is the stability of the immobilized enzyme (IME) in actual use. Although some data are available concerning immobilized lactase stability as a function of feed composition (7), they are not extensive. Particularly in the area of whole whey or whey still containing protein and salts, information is limited (1,7). A second factor, reactor operating strategy, has received even less attention.

The purpose of the studies described here was to characterize the performance of one specific type of immobilized lactase, β-galactosidase from *A. niger* immobilized on porous silica via the silane-glutaraldehyde method, with various feeds and to determine reasonable options for reactor operating strategy.

Basic characterization of immobilized lactase and its performance has been reported by Ford and Pitcher (7). The rate of lactose hydrolysis to glucose and galactose is inhibited by one of the products, galactose. The experimental values for the Michaelis K_m and inhibition K_i constants are 0.0528 and 0.0054 M, respectively. The integrated rate equation is given by Eq. 1:

$$E = 16.667\ F \left[\frac{(K_i - K_m)}{K_i} \quad (So - S) + \left(\frac{K_m S_o}{K_i} + K_m \right) \ln\ (S_o/S) \right]$$

In this expression E is in units of enzyme activity, defined as the amount of enzyme necessary to convert one μmole/min of lactose to glucose and galactose at $S >> K_m$ and at the specified temperature. This equation applies to a plug flow reactor with no product in the feed. S_o is lactose concentration in the feed; S is lactose concentration in the product; and F is the flow rate in ml/hr.

The optimum pH is about 3.5; the activation energy is 12 kcal/gmole. The effect of temperature on half-life also can be described by an Arrhenius-type plot with a deactivation energy of 40.6 kcal/gmole.

METHODS AND MATERIALS

The lactase enzyme (Wallerstein LP, β-galactosidase from *Aspergillus niger*) was immobilized on porous silica bodies by the aqueous silane - glutaraldehyde method (8). The porous silica bodies 30/45 mesh₀(U.S. Standard Sieve), had an average pore diameter of 370A.

Feed or substrate materials included Kraft edible lactose, acid whey (Crowley Food, Inc.), and deproteinized acid whey (ultrafiltrate permeate) (Crowley). The lactose feed solutions contained <200 ppm zepharin-chloride, a quaternary amine for prevention of microbial growth.

The lactose levels in the whey feeds were determined by the Lane and Eynon (9) procedure for reducing sugars. Glucose concentrations were measured using Worthington Biochemical Corp. Glucostat reagent.

Feed solutions were passed continuously down through 1.5 cm diameter columns containing packed beds of 5 to 16 g of immobilized lactase derivative. Columns were water-jacketed to maintain the desired temperature to within ±0.5°C. The ultrafiltered whey deashing was accomplished using Rohm and Haas Amberlite IRA-93 and Diamond Shamrock Duolite C-25D ion exchange resins. Other experimental details are included in the discussion of results.

RESULTS AND DISCUSSION

Enzyme stability data at 50°C for various modified whey and lactose feeds are listed in Table 1. Half-lives were calculated by a linear regression of the logarithm of activity using Eq. 1 versus operating time. Similar data for whole whey feeds are included in Table 2.

Several researchers have observed greatly decreased apparent enzyme half-lives for whole whey versus lactose feeds. Little has been reported concerning the reasons for this difference or whether the problem can be alleviated. Activity lost during operation with whole whey has been partially restored by cleaning and treatment of the immobilized lactase with an iodoform reagent (10).

From the data in Tables 1 and 2, it is apparent that either deashed whey, still containing protein, or deproteinized (ultrafiltered) whey, containing salts, gave significantly longer half-lives than whole whey. Addition of sodium chloride or protein (Crowley Food Mil-Pro 50A, 50% whey protein) to lactose each gave relatively long half-lives. Evidently the interaction between salts and protein was primarily responsible for the decreased apparent enzyme activity. In fact, a visible coating was formed around each immobilized enzyme particle. This coating was readily stained with Napthol Blue Black (Amido Black 10B, Sigma) indicating its proteinaceous character. Removal of this coating by ultrasonic cleaning restored most of the original activity. Daily ultrasonic cleaning resulted in a 37-day projected half-life versus the 2 to 6 days observed otherwise for whole whey.

Since hydrolysis of the whey protein with pepsin did not increase half-lives, it seemed likely that amino acids or peptide chains may also have contributed to coating the carrier. Back washing the carrier bed with water or phosphate buffer did appear to cause some removal of this protein coating. However, half-lives still remained low. The fluidized bed approach, in spite of the continuous upward flow which might potentially have swept away clogging and coating feed constituents, resulted in no significant improvement over fixed-bed reactors.

Higher pH, 4.7 versus 4.0, resulted in increased apparent half-life perhaps because of effects on the salt interaction with protein. Evidently the protein coating rate was retarded at reduced temperatures.

TABLE 1

HALF-LIVES AT 50°C

Feed	pH	Initial Activity (U/g)	Half-Life (days) Mean	95% Confidence Limits Lower	95% Confidence Limits Upper	Days Operated
5% Lactose	3.5	575±61*	126±40*			70-92
5% Lactose	4.0	539	125	96	182	64
5% Lactose, 0.5% NaCl	4.0	534	79	61	110	49
Acid Whey, ultrafiltered, ion-exchanged	3.5	540	57	45	78	54
Acid Whey, ultrafiltered, conc.	4.0	436	54	44	71	58
by RO, 0.206-0.264M	4.0	312	43	35	55	43
De-ashed Whey, 0.130M	3.5	466	30	23	41	32
5% Lactose, 0.9% Mil-Pro 50-A	4.6	296±40	no significant activity loss			16

*Average of 4 runs, ± values indicate standard deviations.

TABLE 2

HALF-LIVES WITH WHOLE WHEY FEED

Description	pH	Temp. (°C)	Half-Life (days) Mean	95% Confidence Limits Lower	95% Confidence Limits Upper	Days Operated
Reduced Temperature	3.5	40	38	30	49	44
Control	3.5	50	5.6	4.5	7.7	6
Ultrasonic Cleaning	3.5	50	37*	--	--	8
Pepsin-treated Whey	4.0	50	2	--	--	4
Fluidized Bed 120/200 Mesh Carrier	4.0	50	3	--	--	3
Higher pH	4.7	50	16.8	9.9	54.4	21

*Corrected for IME weight loss during operation and cleaning.

It appears that protein removal from whey, followed successively by deashing and hydrolysis would be the most logical scheme for producing products of commercial interest.

One immobilized enzyme topic that has received little attention is reactor operating strategy. This covers the choice of the size and number of reactors, temperature policy, and length of operating time for a given load of immobilized enzymes. The most common strategy used commercially is to maintain an approximately constant production rate by means of multiple reactors operated isothermally with staggered reactor start-up times (11). The number of reactors required to maintain the production rate within given tolerance levels is given by:

$$R_p = \exp\ (-(H/N)\ \ln 2) \qquad \text{(Eq. 2)}$$

where R_p = ratio of low to high production rate, H = number of half-lives of immobilized enzyme utilization, and N = number of reactors. An example is shown in Fig. 1, where N = 7, H = 2 and R_p = 0.82. This strategy is the optimal one if the percentage decrease in half-life with temperature is greater than the percentage increase in activity. The minimum allowable temperature is then chosen as the operating temperature.

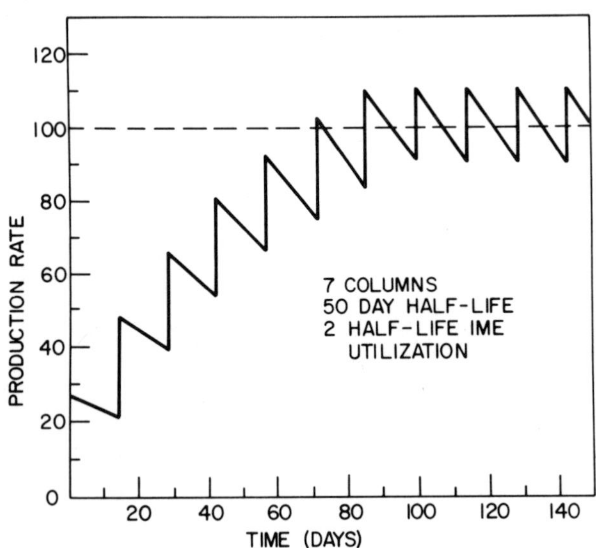

Fig. 1. Production Rate for Multiple Column System.

Realistic constraints for an immobilized lactase system might be 40° (because of microbial control problems) to 50°C with a fixed operating time such as 300 days. Assuming a half-life of 60 days at 50°C, initial activity of 450 units/g at 50°C, and a 22 hour operating day, calculations of total productivity were made for various operating policies. The productivity was defined as the pounds of lactose processed at 80% conversion per pound of immobilized enzyme. The reported values for activation and deactivation energies were used. Fig. 2 shows the calculated productivity as a function of temperature for isothermal operation for several operating cycle times. For operating cycle times much above 400 days, the optimal temperature was the minimum, 40°C. As the cycle time decreased, the optimum temperature rose (about 43.5°C for 200 days).

A somewhat more attractive operating strategy is to maintain a constant temperature (the lowest possible) for a certain time and then raise the temperature gradually to maintain constant conversion and flow rates. The time it takes to go from 40° to 50°C at constant conversion and flow rate was estimated at 168 days, using an iterative computer program. Thus, operating 132 days at 40°C followed by constant conversion operation for 168 days gave a 300 day operating cycle as shown in Fig. 3. Here

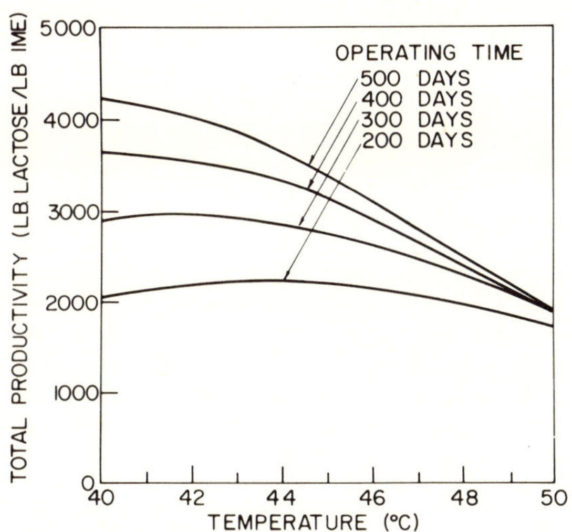

Fig. 2. Productivity as a Function of Temperature for Isothermal Operation.

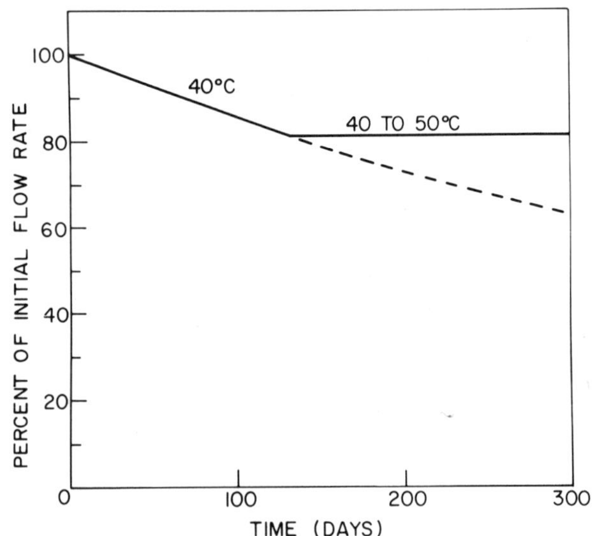

Fig. 3. Flow Rate as a Function of Time for 300 Day Operating Cycle.

the total productivity was 3095 versus 2947 lb/lb IME for the best isothermal case.

The optimal temperature policy involved three distinct phases. The reactor first was operated at the minimum temperature, in this case 40°C. Then constant conversion operation was carried out (2nd phase) until the maximum temperature was reached, 50°C (third phase). The total productivity was written as the sum of the productivity for each of these three phases, differentiated with respect to the time interval and set equal to zero. There was only one unknown, since the total time was given and the time at constant conversion was known. The equation was then solved for the time at 40°C (or 50°C) and the other interval calculated. The three successive operating intervals were 101, 168, and 31 days resulting in 3131 lb/lb IME productivity. The flow rate as a function of time is shown in Fig. 4. An approximation of this operating strategy is also shown in Fig. 4, with 2°C temperature intervals. The productivity is almost as high at 3128 lb/lb IME. As the operating time cycle was increased, the constant conversion and 50°C operating phases contributed less percentage wise to the total productivity.

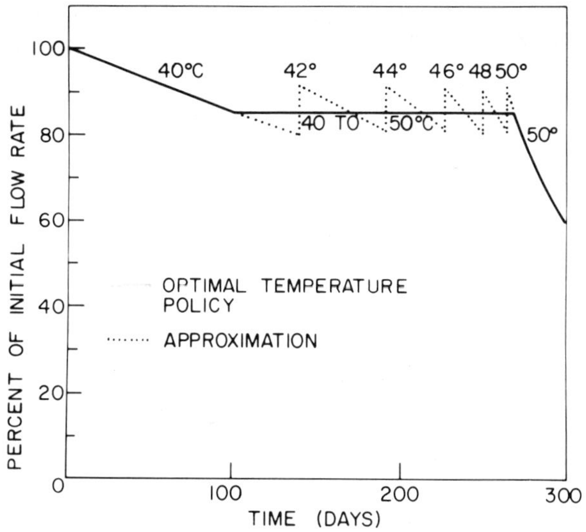

Fig. 4. Optimal Temperature Policy.

For actual system operation it also seems likely that the 40°C and constant conversion policy of Fig. 3 would be more attractive than the optimal policy. The additional one percent productivity for the optimal policy would probably not compensate for the wide variation in productivity during the 50°C operating phase.

ACKNOWLEDGMENTS

Most of the experimental data reported here was obtained by Mr. Raymond E. Lindner and Dr. James R. Ford. The porous silica was made by D.L. Eaton by a procedure developed with R.A. Messing.

REFERENCES

1. CHARLES, M., COUGHLIN, R.W., PARUCHURI, E.K., ALLEN, B.R. & HASSELBERGER, F.X. *Biotechnol. Bioeng.* *17*:203, 1975.
2. OLSON, A.C. & STANLEY, W.L. *J. Agr. Food Chem.* *21*: 440, 1973.
3. STANLEY W.L. & PALTER, R. *Biotechnol. Bioeng.* *15*: 597, 1973.

4. WEETALL, H.H., HAVEWALA, N.B., PITCHER, W.H., JR., DETAR, C.C., VANN, W.P. & YAVERBAUM, S. *Biotechnol. Bioeng. 16*:295, 1974.
5. WEETALL, H.H., HAVEWALA, N.B., PITCHER, W.H., JR., DETAR, C.C., VANN, W.P. & YAVERBAUM, S. *Biotechnol. Bioeng. 16*:689, 1974.
6. PITCHER, W.H., JR. *Dairy Scope 18*: March, 1974.
7. FORD, J.R. & PITCHER, W.J., JR. in "Proceedings--Whey Products Conference 1974," U.S.D.A. ERRC Publ. No. 3996:104, 1974.
8. WEETALL, H.H. & HAVEWALA, N.B. in "Enzyme Engineering," (L.B. Wingard, Jr., ed.), Wiley, New York, 1972, p. 249.
9. Standard Analytical Methods of the Member Companies of the Corn Refiners Association, First Revision, 5/27/68, E-26.
10. COUGHLIN, R.W., CHARLES, M., ALLEN, B.R., PARUCHURI, E.K. & HASSELBERGER, F.X., paper 17b, AIChE 66th Annual Meeting, Philadelphia, 1973.
11. HAVEWALA, N.B. & PITCHER, W.H., Jr. in "Enzyme Engineering," vol. 2 (E.K. Pye and L.B. Wingard, Jr., eds.), Plenum Press, New York, 1974, p. 315.

REACTION ENGINEERING ASPECTS OF CONTINUOUS OPERATION WITH BIOCATALYSTS

C. Wandrey* and E. Flaschel**

Technische Universitat Clausthal-Zellerfeld*
Federal Republic of Germany and
Ecole Polytechnique Federale de Lausanne**
Switzerland

This contribution is concerned with reaction engineering aspects using immobilized as well as soluble enzymes (1, 2). Continuous operation is possible with both types of enzyme, as shown in Fig. 1.

The production of L-methionine from a racemic mixture of N-acetyl-D,L-methionine was used as an example. The stereospecific hydrolysis of N-acetyl-L-methionine is catalysed by the enzyme acylase (Fig. 2). The product can be isolated easily due to the limited solubility. Enzyme

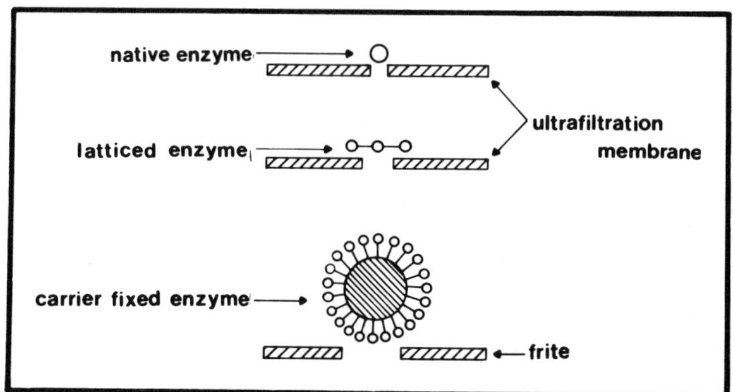

Fig. 1. Different types of enzyme catalysts for continuous operation.

engineering aspects were studied experimentally using a carrier-fixed type of acylase as well as native acylase in different type reactors.

We measured the equilibrium constant by following the forward and reverse reaction. The temperature dependency of the equilibrium was measured too. The reaction was exothermic, 7.9 kJoule/mol. No buffer was necessary due to the favorable pk values of the reactants. As catalyst we used native acylase in a membrane reactor and acylase fixed to an aluminium silicate carrier. The influence of the catalyst concentration was measured. The molecular weight of the catalyst was found to be 91,000 ± 6,000 using an ultracentrifuge. Quantitative data on the deactivation of native and carrier-fixed acylase were obtained. Thus, catalyst costs per unit weight of product could be calculated. The pH optimum of the reaction was found to be precisely 7.0. Numerous kinetic measurements were carried out to evaluate a kinetic model useful under technically relevant conditions. The amount of information could be increased a lot since continuous polarimetry was used for analysis of the reactor output.

Some of the kinetic results are summarized in Fig. 3. The reaction rate R is given as a function of conversion. The four solid lines represent four steps of a model development, valid for native acylase in a continuous stirred tank reactor (CSTR) using an ultrafiltration membrane for enzyme retention. The dashed straight line represents the convective term in the mass balance. The point of intersection represents the operating point of the reactor. The top line was calculated with kinetic data from the literature indicating a quantitative conversion. For the next line below the equilibrium was taken into account. In a third step the substrate inhibition measured from initial rate experiments was also considered.

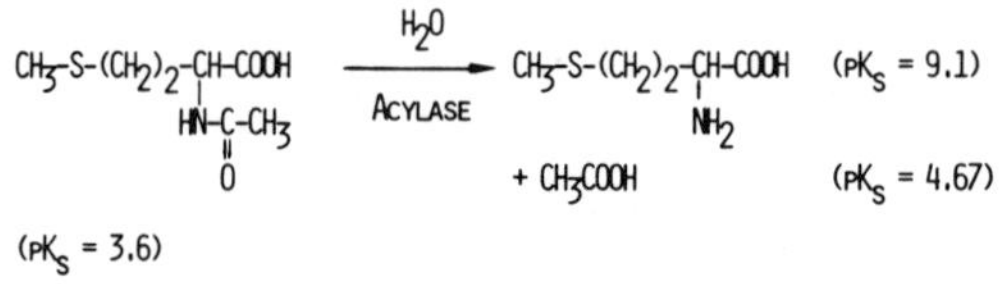

$$CH_3\text{-}S\text{-}(CH_2)_2\text{-}CH(NH\text{-}C(=O)\text{-}CH_3)\text{-}COOH \xrightarrow[\text{Acylase}]{H_2O} CH_3\text{-}S\text{-}(CH_2)_2\text{-}CH(NH_2)\text{-}COOH \quad (pK_S = 9.1)$$

$$+\ CH_3COOH \quad (pK_S = 4.67)$$

$(pK_S = 3.6)$

Fig. 2. Reaction example: stereospecific hydrolysis of N-acetyl-L-methionine by means of hog kidney acylase. Conditions: pH 7.0, 37°C.

Finally the product inhibition measured from integral runs was used to complete the model so that one obtained a sufficient fit of the experimental data (shown by circles). The intersections of the dashed vertical lines with the abscissa represent the achievable conversion predicted by the different models; this showed that insufficiently developed models led to erroneous predictions.

We also used a membrane reactor. This could be regarded as a CSTR, where the enzyme was retained in the reaction vessel by means of a suitable frit, while the small reactants could leave the system nearly without any obstruction. By means of a flow-through polarimetric cuvette and an automatic polarimeter, the signal was used to regulate an enzyme pump to deliver new enzyme to the reactor and maintain a constant productivity. A small amount of new enzyme was needed continuously due to the deactivation of the catalyst. We constructed a bench scale membrane reactor using hollow fibers as the frit. This apparatus had a productivity of 1 kg/day. The crystallization of L-methionine was observed. In our experiments about 350 mg acylase per square meter of membrane

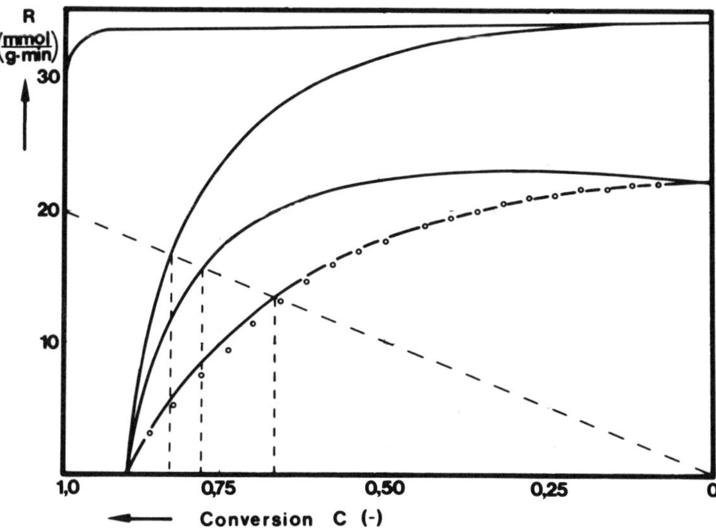

Fig. 3. Operating points for different kinetic models in a CSTR (substrate concentration 0.8 $mol_{D,L}/l$, 37°C, pH 7.0).

area were lost during the starting up procedure. After 25 residence times the retention was practically quantitative.

Another important aspect is the effectiveness of the native catalyst in the membrane reactor. This effectiveness is less than 100% due to some adsorption of the catalyst on the membrane. In Fig. 4 the reaction rate is shown as a function of the fraction S_O (feed concentration) to S_{OL} (reactor exit concentration) for "free" native enzyme and native enzyme in a membrane reactor. The ratio of these two rates for the same fraction is defined as the effectiveness. This value decreases slightly with decreasing fraction due to the increased mass transport limitation for the adsorbed catalyst.

Membrane reactors are useful not only for production purposes but also for kinetic measurements with native enzymes. This is illustrated by means of the problem of an optimal effector. For acylase cobalt normally is used as the effector. For environmental reasons we substituted calcium for cobalt. The cobalt was removed from the enzyme by means of ethylenediaminetetraacetate (EDTA) as the complex forming agent.

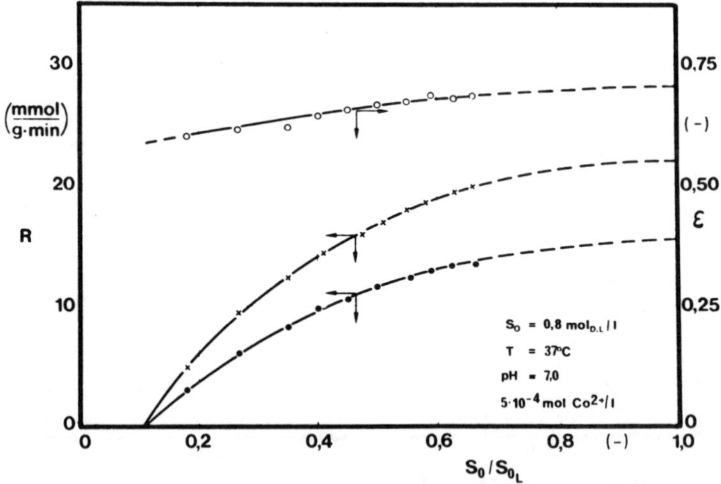

Fig. 4. Reaction rate of "free" enzyme (x-x), of native enzyme in a membrane reactor (●-●), and effectiveness (o-o) as a function of the fraction.

In our experiments with the membrane reactor an effectiveness of 70% was found due to some adsorption and deactivation of the enzyme at the membrane. To compare these results with the performance of an immobilized enzyme in a fixed bed reactor, a carrier-fixed acylase supplied by Boehringer Mannheim was used. The effectiveness with respect to the protein used was 5.8%. The catalyst retention was 100%. The loss of activity due to the reactor performance was 100% per residence time for the first alternative, less than 0.1% per residence time for the second alternative, and zero for the third. The loss of activity due to the kinetic deactivation was 2%/day, 3.44%/day, and 0.56%/day, respectively.

It should be emphasized that for reaction engineering purposes the effectiveness, the retention and the stability of an enzyme catalyst must be known under well defined conditions if the catalyst costs per unit weight of product are to be calculated.

ACKNOWLEDGMENTS

The authors thank D. Jaworek, Boehringer Mannheim Biochemica-Werk Tutzing, Federal Republic of Germany, who supplied the carrier-fixed acylase. The financial assistance of the Bundesministerium fur Forschung und Technologie of the F.R. Germany is gratefully acknowledged.

REFERENCES

1. ZABORSKY, D. "Immobilized Enzymes," Chemical Rubber Company Press, Cleveland, 1973.
2. WANDREY, C. "Reaktionstechnische Untersuchungen an Enzymkatalysatoren zur Entwicklung kontinuierlicher Verfahren," Habilitationsschrift, Tech. Univ. Hannover, F.R. Germany, 1977.

ECONOMICAL ASPECTS OF CONTINUOUS OPERATION WITH BIO-CATALYSIS

E. Flaschel* and C. Wandrey**

Ecole Polytechnique Federale de Lausanne*
Switzerland

Technischen Universitat Hannover**
Federal Republic of Germany

Since synthetic membranes for ultrafiltration are available in high-area-modules, continuous operation with native enzymes in membrane reactors will become an interesting alternative to continuous operation with covalently bound enzymes even on an industrial scale. Continuous operation has been studied with native and carrier-fixed acylase used for optical resolution of methionine by stereospecific hydrolysis of N-acetyl-D,L-methionine (1,2). In this report the data obtained are used to evaluate the economical aspects of different operational methods.

MODEL DESCRIPTION

For comparing the economy of reaction systems two models were combined: a reactor model and an economic model. The reactor model used a 4-parameter kinetic expression (2):

$$S_o \frac{dU}{dt} = R = \left[\frac{V_{max}(E)(ETA)}{1+S_o(2-U)/k_{in}}\right]\left[\frac{S_o(1-U) - S_o{}^2U^2/K}{k_m(1+S_oU/k_{ic}) + S_o(1-U)}\right] \qquad \text{(Eq. 1)}$$

See Table 1 for the definition of the symbols and Table 2 for values of the kinetic parameters.

TABLE 1

SYMBOLES AND ABREVIATIONS

Symbol	Meaning	Units
E	enzyme concentration	g/l
ETA	efficiency factor	-
FDE	linear deactivation factor	min^{-1}
F_v	flux	l/min
K	equilibrium constant	mmol/l
k_m	Michaelis-Menten constant	mmol/l
k_{ic}	product inhibition constant	mmol/l
k_{in}	substrate inhibition constant	mmol/l
KE	cost of enzyme	DM
KR	cost of raw material	DM
KT	cost of separation	DM
KG	total cost of product	DM/kg
KFE	price of enzyme	DM/g
KFR	price of substrate	DM/kg
KFT	price of thermal energy	DM/l
MP	molecular weight of product	kg/mol
MS	molecular weight of substrate	kg/mol
P_c	solubility of product at 100°C	mol/l
P_v	solubility of product at 0°C	mol/l
PR	productivity	kg/min
R	reaction velocity	mmol/l/min
S_o	L-substrate concentration level	mmole/l
U	degree of conversion	-
V_{max}	maximal reaction velocity	mmole/g/min
τ	mean residence time	min
t	time	min

TABLE 2

Parameters of the Kinetic Model

Enzyme	ETA	V_{max}	K	k_m	k_{in}	k_{ic}
	-	$\frac{mmol}{g \cdot min}$	$\frac{mmol}{\ell}$	$\frac{mmol}{\ell}$	$\frac{mmol}{\ell}$	$\frac{mmol}{\ell}$
Native acylase*	0.7	34.0	2750	0.77	1450	2.8
Immobilized acylase**	1.0	0.182	2750	1.00	3000	2.2

*Serva 10720
**Boehringer Mannheim, special preparation on porous clay, 250-315 μ

ETA was an efficiency-factor, which involved the decrease of initial enzyme activity in the membrane reactor due to concentration polarization. Calculations were carried out for CSTR, two- and three-staged cascades, and plug flow reactors.

The economic model contained information about the proportional cost of product formation. The process discussed here consisted of a reactor, an evaporator wherein the product-stream was concentrated to saturation with L-methionine at 100°C (P_C = 0.72 mol/l), and a crystallizer. The leaving concentration of product P_V was 0.18 mol/l. Re-racemisation and further product recovery was neglected in all cases, leading to a productivity of

$$PR = F_V(S_O)U(MP)\frac{P_C - P_V}{P_C} \qquad \text{(Eq. 2)}$$

For calculating the total cost per unit weight of product, three main parts were considered: raw materials, catalyst, and separation. The time dependent substrate cost was given by

$$\frac{d(KR)}{dt} = F_V(2S_O)MS(KFR) \qquad \text{(Eq. 3)}$$

S_o refers only to concentrations of the L-enantiomer. It has to be doubled in the case of racemic substrate. The cost per unit weight of product resulted from dividing Eq. (3) by Eq. (2)

$$ROH = \frac{2MS\,(KFR)\,P_c}{MP\,(P_c-P_v)\,U} \qquad \text{(Eq. 4)}$$

To evaluate the cost of catalyst, the measured deactivation constants (FDE) (2) were taken into account. The cost per unit time is given by Eq. (5)

$$\frac{d(KE)}{dt} = F_v(E)\,\tau\,(FDE)\ KFE \qquad \text{(Eq. 5)}$$

Therefore the cost per unit weight of product was

$$ENZ = \frac{FDE\,(KFE)\,P_c}{MP\,(P_c-P_v)} \quad \frac{(E)\,\tau}{S_o(U)} \qquad \text{(Eq. 6)}$$

The separation cost involved the evaporation of solvent at 100°C.

$$\frac{d(KT)}{dt} = F_v\left(1 - \frac{S_oU}{P_c}\right) \quad KFT \qquad \text{(Eq. 7)}$$

The cost per unit weight of product was

$$TRE = \frac{(I-S_o)\,(U/P_c)\ \ KFT\,(P_c)}{S_o(U)\ \ MP\,(P_c-P_v)} \qquad \text{(Eq. 8)}$$

The total proportional cost per unit weight of product was the sum of the mentioned individual costs:

$$KG = ROH + ENZ + TRE \qquad \text{(Eq. 9)}$$

For further calculations the following cost-factors were used, based on 1977 prices in the F.R. Germany: KFR = 10 DM/kg; KFT = 0.075 DM/l; KFE = 15 DM/g (for hog-kidney acylase, activity 27 mmol/g/min at S_o = 200 mmol/l, 37°C, pH 7, Co^{+2} = 0.5 mmol/l).

The loss of activity was measured during continuous operation. It was found to be FDE = 0.0344 l/day for native acylase in membrane reactors and 0.0056 l/day for carrier-fixed acylase in tube reactors.

RESULTS AND DISCUSSION

The cost of raw material depended only on the degree of conversion. The separation cost naturally depended on conversion as well as on the concentration level of substrate. The enzyme cost was inversely proportional to productivity, therefore depending on the type of reactor and thus on the type of catalyst (See Eq. (6)). The cost optimum was near equilibrium conversion due to the comparatively small influence of the catalyst cost in membrane reactors compared to the substrate cost.

To compare the enzyme cost of native and carrier fixed acylase the price for the carrier fixed type was at first calculated on the basis of protein content only. The enzyme protein in the carrier fixed type of catalyst was used with an efficiency of 5.77% under standard conditions (2). The membrane reactor (CSTR) was superior to a plug flow reactor with carrier fixed enzyme until 85% conversion and to a cascade of two membrane reactors even up to 91%.

Taking into account the different effectivity (ETA) and stability (FDE), the cost of catalyst per unit activity was 31.66 times higher when a carrier fixed type of catalyst was used, referring to initial rates under standard conditions. It should be stated that the value would be 1.97 even if no costs for the carrier and the preparation and no profits were taken into account. Thus, although the stability of the carrier fixed type was about 6 times higher, the efficiency was about 12 times smaller. Presently the factor 1.97 is multiplied by 16 to get 31.66, showing the amount paid for immobilization know-how.

To get the optimal reaction conditions, the reaction determining variables must be optimized to find the lowest achievable product cost. In general, the conversion (U) and the initial substrate concentration (S_o) are taken as independent variables. These variables were optimized according to the method of Hooke-Jeeves (3). The mean residence time (τ) or ($E\tau$) was the dependent variable. With a given set of parameters, the total cost (KG) was minimized according to:

$$KG = f(S_o, U)_{KFR,\ KFE,\ KFT,\ FDE,\ ETA} \qquad \text{(Eq. 10)}$$

It appeared that the optimum lay near equilibrium conversion, and at relatively high initial substrate concentration. The resulting residence time had to be optimal to minimize catalyst cost with respect to substrate and separation costs. Beside KFR, the parameters ETA and FDE had a strong influence on cost. The last two also influenced operating conditions. When ETA was low, the productivity was naturally low. When FDE was low, the enzyme cost was low; and the optimum was determined by the costs for substrate and separation; this gave high residence times and decreased productivity.

To compare the performance of membrane reactors (native acylase) with a plug flow reactor (immobilized acylase) we at first assumed no increase in enzyme price during fixation. The results show that the total proportional cost for product was lower in the case of a three stage cascade than in the case of a fixed bed using immobilized enzyme. With the earlier mentioned factor of 16 for price increase during immobilization, the cost for carrier fixed enzyme is higher than for native enzyme, even in a single membrane reactor behaving like a CSTR.

For technical scale application of new enzymes or enzyme systems, it might be a better strategy to test continuous operation with the native enzyme in membrane reactors, before trying to get it immobilized in a way sufficient for industrial purposes.

ACKNOWLEDGMENTS

The authors wish to thank the Bundesministerium fur Forschung und Technoligie of GFR for financial support and Boehringer Mannheim, Biochemica Werke Tutzing for carrier fixed acylase.

REFERENCES

1. WANDREY, C., FLASCHEL, E., SCHUGERL, K., *Chem. Ing. Tech 49(3)*:257, 1977.
2. WANDREY, C., Inaugural dissertation, Tech. Univ. Hannover, F.R. Germany, 1977.
3. HOOKE, R. & JEEVES, T.Z. *J. Assoc. Comp. Mach 8(2)*: 212, 1961.

PRESSURE DROP ACROSS COMPRESSIBLE BEDS

K. Buchholz and B. Godelmann

Dechema-Institut
Frankfurt (Main)
Federal Republic Germany

Columns with fixed beds of immobilized enzymes or other proteins are commonly used in affinity chromatography and as reactors for continuous operation. Most carrier particles for protein immobilization are deformable or compressible by application of pressure. For non-compressible beds at low Reynolds-numbers, <10, the pressure drop p is a linear function of flow rate; and the flow resistance is constant. With deformable particles in fixed beds p is a non-linear function of bed height h and volumetric flow rate v (1). This is shown in Fig. 1 where the flow resistance increased with p, finally leading to instability and occlusion. It is evident that at $p \approx 100$ mbar the pressure drop increased very rapidly with the flow rate, and occlusion occured beyond about 500 mbar. The maximal flow rate depended on h as expected. Reproducibility was difficult, depending on pretreatment and hysteresis. The flow through a compressible bed must take into account varying hydraulic radii r_h in the flow path, caused by particle deformation. The pressure drop then becomes a function of h,v, and r_h. From the law of Hagen-Poiseuille p= f(h) = (constant)(f(v)). Thus, it is possible to determine the application range of bed height, flow rate, and residence time, for a given carrier by one type of experiment, either from p=f(h) or from p= f(v). The results may be summarized through the specific flow resistance α as a function of p; by analogy to filtration theory (2)

$$\alpha = dp/d(hw)\mu \qquad \text{(Eq. 1)}$$

$$\bar{\alpha} = p/(hw\mu) \qquad \text{(Eq. 2)}$$

where $\bar{\alpha}$ is a mean flow resistance for an integral bed, w is the linear flow rate, and μ is the viscosity. For a given p the residence time or h and w can be calculated if $\bar{\alpha}$ is known as a function of p.

A qualitative understanding of the observed phenomena is possible using a simple schematic model based on elastic deformable spheres as carrier particles with a modulus A defined by $\Delta r/r_p = A\ p$ (3). p = f(h) was calculated by approximation and compared with f(v), which was experimentally determined. The approximation was done by calculation of the pressure drop in the first layer of spheres (in a dense column packing) according to Hagen-Poiseuille such that

$$p_1 = 8\mu v h_1/(n_q \pi\ r_{h1}^4) \qquad \text{(Eq. 3)}$$

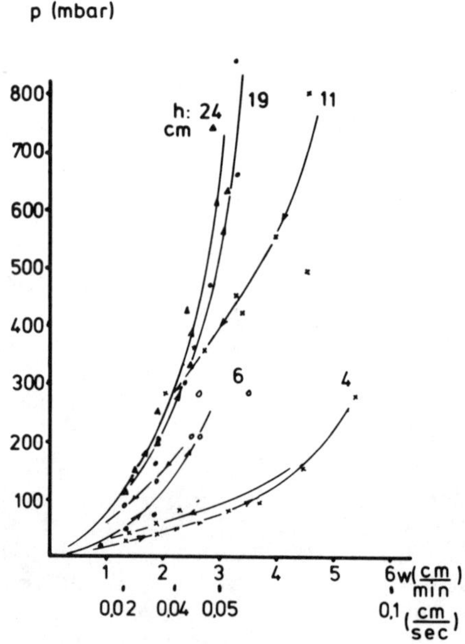

Fig. 1. Total pressure drop in columns with Sepharose CL 6B for different bed heights h (d_p 100-200 µm) (experimental).

Here h_1 is the height of the first layer and n_q the number of hydraulic radii r_h per cross section of column F. Note that

$$r_h = a2r_p \, E/(1-E) \qquad \text{(Eq. 4)}.$$

where E is the void volume.

The deformation of the particles Δr_{p2} in the second layer of speres by p_1 became

$$\Delta r_{p2}/r_p = A \, p_1 \, F/n_q \, q6 \qquad \text{(Eq. 5)}$$

where r_p is the particle radius. The force pF was applied on definite sections q at the interface of two particles with n_q particles in the cross section F. The reduction of the height h_2 and void volume E_2 became

$$h_2^2 = 4(r_p - \Delta r_{p2})^2 - \frac{4}{3} r_p^2 \qquad \text{(Eq. 6)}$$

$$h_o^2 = 4r_p^2 - \frac{4}{3} r_p^2 \qquad \text{(Eq. 7)}$$

$$E_2 = 1 + (h_o/h_2)(E_0 - 1) \qquad \text{(Eq. 8)}$$

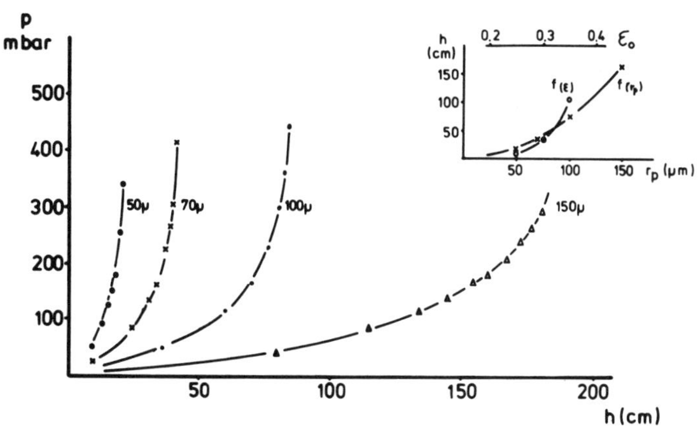

Fig. 2. Pressure drop as a function of h for different particle radii (calculated). Inset: Maximal bed length h_m, for p_{max}=200 mbar, function of particle radius r_p and void volume E_o.

It was assumed that the particle volume remained constant during deformation. A model with compressible particles gave similar results for p=f(h) with different values for A.

The fluid-dynamic radius in the second layer was

$$r_{h2} = a2\ r_p\ \varepsilon_2/(1-\varepsilon_2) \qquad \text{(Eq. 9)}$$

The pressure drop in the second layer p_2 was then calculated, and so on. Fig. 2 shows the result of a calculation with data for Sepharose CL 6B. The main pressure drop and finally occlusion occured at the end of the fixed bed; also shown is the influence of r_p and ε_o (which depended on the particle size distribution).

REFERENCES

1. CHEN, L.F. & TSAO, G.T. *Biotechnol. Bioeng.* *18*:1507, 1976.
2. TILLER, F.M. *Filtration Separation, 1975*:No 4, 386.
3. BORCHARD, W., personal communication.

HETEROGENEOUS KINETICS OF TWO-SUBSTRATE ENZYMIC REACTIONS UNDER DIFFUSIONAL LIMITATIONS

J.M. Engasser,* P. Hisland,* P.R. Coulet, and F. Paul

Laboratoires des Sciences du Genie chimique*, CNRS
Nancy, France and
Laboratoire de Biologie et Technologie des Membranes, CNRS
Villeurbanne, France

When a two-substrate enzymatic reaction is catalyzed by a bound enzyme, the diffusion of both substrates can affect the heterogeneous kinetics of the enzyme. The effect of external diffusion has been theoretically analyzed using the steady-state rate equation

$$v = h_A\ ([A]-[A]_o)=h_B([B]-[B]_o)=$$

$$\frac{V\ [A]_o\ [B]_o}{K_{AB}+K_B\ [A]_o+K_A[B]_o + [A]_o\ [B]_o}$$

where h_A and h_B are the substrate transport coefficients; V, K_A, K_B and K_{AB} the kinetic parameters; [A] and [B] the bulk substrate concentrations; and $[A]_o$ and $[B]_o$ their surface concentrations. The magnitude of diffusional limitations for A and B are expressed by the two substrate moduli:

$$u_A = \frac{V}{h_A\ K_A} \qquad u_B = \frac{V}{h_B\ k_B}$$

Their relative contributions on the enzyme kinetics depends on the value of the affinity ratio:

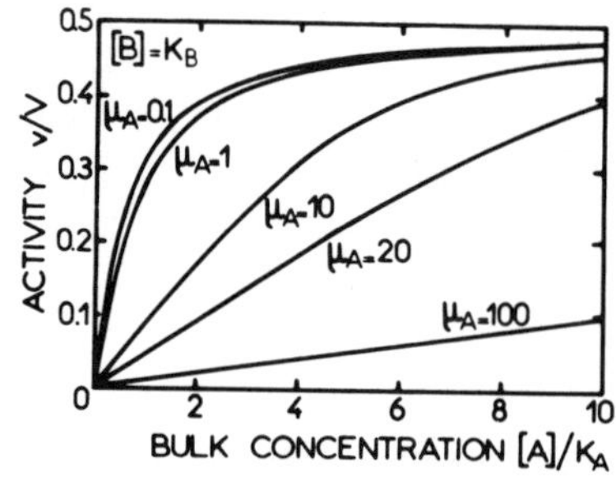

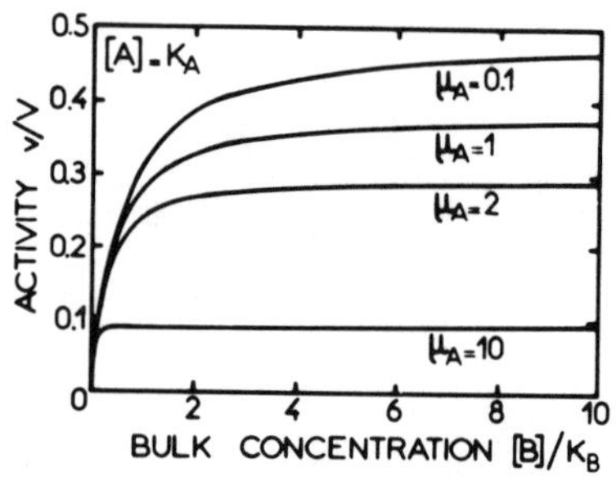

Fig. 1. Bound enzyme activity dependences on substrate concentrations for $\xi_{B/A} = 0.001$ and at different extent of diffusional limitations.

$$\xi B/A = \frac{u_B}{u_A} = \frac{h_A \; K_A}{h_B \; K_B} \simeq \frac{K_A}{K_B}$$

which indicates that diffusional limitations are predominant for the high affinity (low concentration) substrate.

In the case of two substrates with different affinities, diffusional limitations affect very differently the activity dependences on the two substrate concentrations, as illustrated in Fig. 1. Whereas for the high affinity substrate, A, diffusional effects are the same as for a single substrate reaction (1), for the low affinity substrate, B, the observed maximal activity is decreased by the diffusional limitations of the other substrate. Consequently, diffusional limitations yield opposite effects on the half maximal activity substrate concentrations, namely an increase in $[A]_{0.5}$ but a decrease in $[B]_{0.5}$.

These theoretical results have been experimentally verified by kinetic studies of collagen bound aspartate amino transferase (2) and sorbitol dehydrogenase (unpublished).

REFERENCES

1. ENGASSER, J.M. & Horvath, C. in "Applied Biochemistry and Bioengineering," vol. 1 (L.B. Wingard, Jr., E. Katchalski-Katzir, and L. Goldstein, eds.), Academic Press, New York, 1976, p. 127.
2. ENGASSER, J.M., COULET, P.R. & GAUTHERON, D.C. J. *BIOL. CHEM.* in press.

MEMBRANE REACTOR-SEPARATOR FOR CONTINUOUS ENZYMATIC REGENERATION OF ATP

Colin R. Gardner

Centre de Recherche Merrell International
Strasbourg, France

A potentially useful application of cell-free enzyme systems is the production of biologically important compounds (1,2). Biosynthetic reactions require the input of chemical energy usually in the form of ATP; but a continuous supply of ATP at current prices would make such reactions prohibitively expensive. During the reaction ATP is degraded to ADP or AMP. We have shown previously (2-4) that it is economically feasible to use cell-free enzymatic systems to regenerate ATP from ADP or AMP using adenylate kinase to convert AMP plus ATP to ADP and acetate kinase to react ADP and acetylphosphate to form ATP and acetate. We have already demonstrated several methods by which the acetate might be removed (3).

Here I report preliminary results on the preparation of a composite membrane to act as a support for immobilized enzymes and as a selective transporter of acetate ions under the influence of an electrical potential gradient. The incorporation of such a membrane into a thin channel electrodialysis unit is shown in Fig. 1. The composite membrane consists of three layers: 1. a microporous, asymmetric nylon membrane on which enzymes can be immobilized after activation of the nylon by triethyl-oxonium-tetrafluoroborate (TOTFB) (5); 2. an ultrathin film of a cation-exchange polymer; and 3. a thin anion-exchange membrane.

Acetate kinase was immobilized on microporous asymmetric nylon films and its activity measured (3) both in

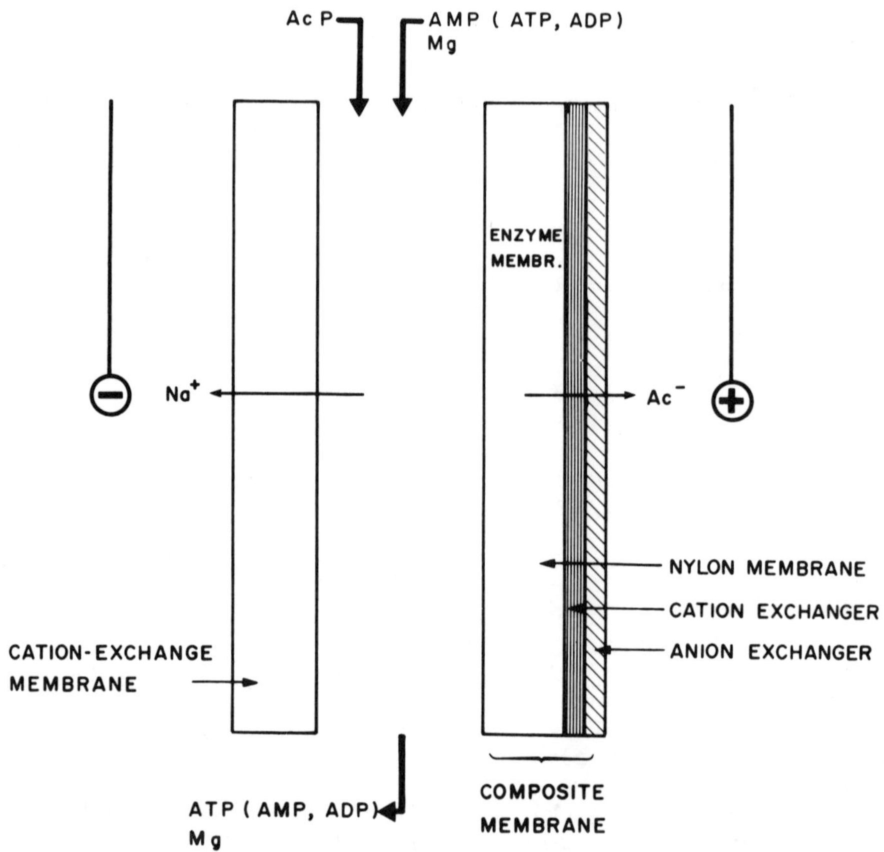

Fig. 1. Thin channel electrodialysis unit for continuous enzymatic regeneration of ATP with simultaneous removal of the by-product, acetate.

a stirred vessel and in a filtration cell. The rate of acetate production was found to be 10^{-8} mole/cm^2/sec. Bipolar membranes were constructed from cation-and anion-exchange polymers and their transport characteristics measured. In an electrodialysis cell such membranes were shown to be highly selective transporters of acetate relative to ATP, ADP and AMP. The results of one such study comparing the transport rates of acetate and ATP are shown in Fig. 2. At a current density of 2 mamps/cm^2 and under the conditions defined in Fig. 2, the flux of acetate was found to be 10^{-8} mole/cm^2/sec. This flux is probably due

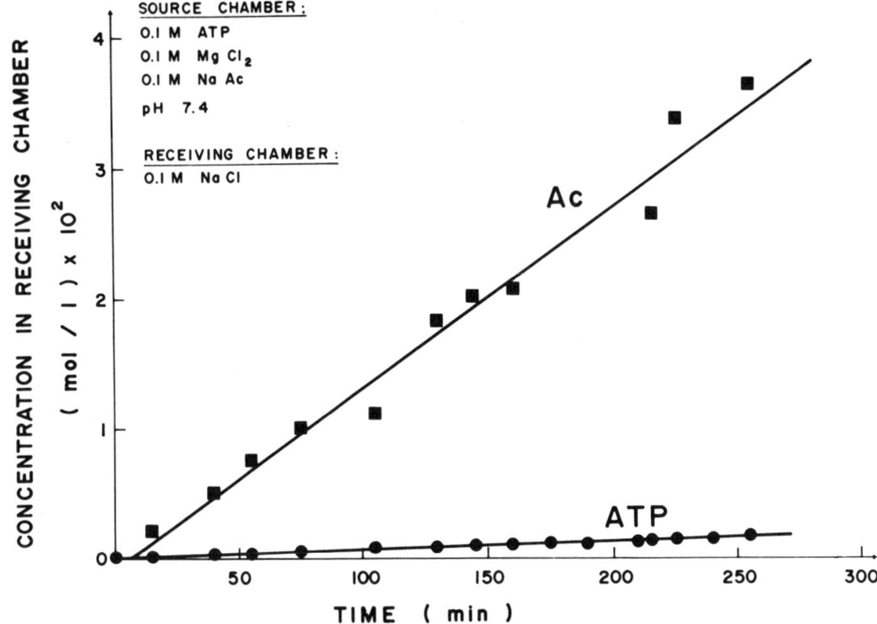

Fig. 2. Electrodiffusion of acetate and ATP across a typical bipolar membrane prepared as described in the text.

to the mechanisms: 1. electrodiffusive transport of acetate, and 2. exchange diffusion of acetate and chloride.

The results obtained so far have shown that it is possible to prepare microporous, asymmetric nylon membranes and to attach active acetate kinase to them. Bipolar membranes of the nature described exhibit highly selective transport of acetate relative to ATP, ADP and AMP. Combination of these two membranes into a composite film and incorporation into a thin channel electrodialysis unit should provide a membrane reactor-separator device for regeneration of ATP from ADP with simultaneous acetate removel. Co-immobilization of adenylate kinase and acetate kinase should extend this capability to regeneration from AMP.

ACKNOWLEDGMENTS

This work was supported in part by a Science Council Research Fellowship to the author at the Chemistry Department, University of Aberdeen, Scotland.

REFERENCES

1. HAMILTON, B.K., MONTGOMERY, J.P., & WANG, D.I.C. in "Enzyme Engineering" vol. 2, (E.K. Pye and L.B. Wingard, Jr., eds.), Plenum Press, New York, 1974, p. 153.
2. WHITESIDES, G.M. in "Applications of Biochemical Systems in Organic Chemistry," (J.B. Jones, C.J. Sih and D. Perlman, eds.), Wiley, New York, 1976, p. 901 and 917.
3. GARDNER, C.R., COLTON, C.K., LANGER, R.S., HAMILTON, B.K., ARCHER, M.C. & WHITESIDES, G.M. in "Enzyme Engineering" vol. 2 (E.K. Pye and L.B. Wingard, Jr., eds.), Plenum Press, New York, 1974, p. 209.
4. LANGER, R.S., HAMILTON, B.K., GARDNER, C.R., ARCHER, M.C. & COLTON, C.K. *A.I.Ch.E. J.* *22*:1079, 1976.
5. MORRIS, D.L., CAMPBELL, J., & HORNBY, W.E. *Biochem. J.* *147*:593, 1975.

IMMOBILIZED GLUCOSE-OXIDASE/CATALASE: DEACTIVATION IN A DIFFERENTIAL REACTOR

J. E. Prenosil

Technical Chemical Laboratorium
Federal Institute of Technology
Zurich, Switzerland

The immobilized enzyme system glucose-oxidase/catalase (GOD/CAT) is a three-phase system exhibiting many characteristics similar to conventional heterogeneous catalysis. The low solubility of oxygen in water at normal conditions indicates that for all practical concentrations of glucose, oxygen will be the rate limiting substrate.

Reaction kinetics and catalyst deactivation were studied in a batch differential loop reactor (DBR). The 10% glucose solution was continuously saturated by oxygen in a 1 liter thermostated vessel and circulated through the DBR, 30 mm diameter. The reaction rate was measured by automatic titration of the gluconic acid produced by the reaction. A high pressure reactor also was built to enable the study of the reaction at high dissolved oxygen concentrations.

The enzyme GOD (Merck, 12 U/mg), which contains CAT as a natural impurity, was immobilized on pumice or TiO_2 (d = 0.4 - 1.0 mm) with phenylenediamine glutaraldehyde polymer (PAG). A microscopic investigation showed that the pumice carrier was impregnated with the PAG well inside the pores, whereas the TiO_2 particles had only a thin surface coating.

The initial reaction rates were measured in a preliminary characterization of the catalyst. It was found that the overall reaction rate increased with the circulation flowrate, whereas the specific reaction rate (yield per pass through the catalyst bed) decreased.

The effect of gas composition was investigated using air or pure oxygen for aeration. The same catalyst was used in a BSTR for comparison. The ratio of the initial rates for oxygen and air was found to be $R_0/R_{Air} = 1.92$. This was in agreement with the kinetic expressions for the soluble enzyme (1). In the DBR this ratio depended on the flowrate and was always higher suggesting the presence of external diffusional resistance. The experimental data obtained in the pressure reactor (up to 2 atm) further confirmed this idea. The reaction rate was found to be pseudo-first order with respect to oxygen: $R = k_c a\, O_2^B$ The value of 0.42 sec^{-1} for the external mass transfer coefficient k_c agreed very well with the estimated value of 0.45 sec^{-1} from the Chilton-Colburn analogy.

A series of experiments was done to study the performance of the catalyst as a function of time. The catalyst with pumice as carrier was found to be less stable than the one with the TiO_2 carrier. The latter did not show any significant drop of activity for the first 30 hours of operation at normal conditions. However, the increasing oxygen concentration at higher pressure had an adverse effect on the enzyme stability. Although the initial reaction rates increased with pressure according to the pseudo-first order expression, the enzyme deactivation rate rose even faster making the net performance worse than at the normal pressure. On the other hand the pressure reactor might be used for fast testing of different catalyst preparations. It was observed that the activity of the partially deactivated catalyst could be restored to some extent by addition of soluble catalase. This led to the hypothesis that only the catalase was irreversibly denaturated; and a mathematical model with such an assumption was developed. The results of the mathematical simulation corresponded well with the experimental results.

REFERENCES

1. ATKINSON, B. & LESTER, D.E. *Biotechnol. Bioeng.* *16*: 10, 1974.

EFFECT OF DESIGN PARAMETER ON THE PERFORMANCE OF IMMOBILIZED GLUCOSE ISOMERASE REACTOR SYSTEM

S.M. Kim, S.B. Lee, and D.Y. Ryu

The Korea Advanced Institute of Science
Seoul, Korea

We have recently reported that the reaction kinetics of glucose isomerase showed unique characteristics due to the reversible nature of the reaction.(1) The rate expression may be written as:

$$R = [(V_{mf}/K_{mf})\ S - (V_{mb}/K_{mb})\ P] \ / \ [1+(S/K_{mf})+(P/K_{mb})]$$

where, R is the reaction rate, V_{mf} and V_{mb} the maximum reaction rates for forward and reverse reaction respectively, K_{mf} and K_{mb} the Michaelis Menten constants for forward and reverse reaction respectively, S the substrate concentration, and P the product concentration. We have also determined the optimal operating conditions in terms of the space time, τ, and substrate concentration of the feed, S_o, under a given set of optimal reaction conditions and constraints.

At an optimal space time, one may have a wide range of superficial velocities with the varying height to diameter (H/D) ratios of the reactor, while maintaining a constant value of the space time. The change in superficial velocity, V_s, will influence the performance of the reactor system due to its effect on the mass transfer rate, the reaction rate, and Michaelis Menten constant. In this work, we have studied the effect of parameters related to the reactor design H/D ratio on the performance of the reactor system.

The experimental methods and materials used in this study are basically the same as that used in our earlier

work (1), except for the enzyme (Sweetzyme Type S, NOVO Industri). In order to determine the effect of V_s on the apparent kinetic constants at a constant τ, the H of the immobilized enzyme and the flow rate, Q, were varied by the same factor while maintaining the D constant. To determine the apparent values of V_{mf}, K_{mf}, V_{mb}, and K_{mb}, glucose and fructose were used for the forward and reverse reactions respectively; and the following equations were used: (a) forward reaction

$$S_o\ (X) = K_{mf}(\ln(1-X)) + V_{mf}\ (H/V_s)$$

and (b) reverse reaction

$$S_o\ (X) = K_{mb}\ (\ln\ (1-X)) + V_{mb}\ (H/V_s)$$

The S_o was varied in the range of 0.2-2.0 M. From the plots of (S_o X) versus ln (1-X), the apparent values of K_{mf}, V_{mf}, K_{mb}, and V_{mb} were determined (Fig. 1). The values of K_{mf} and V_{mf} go through a maximum at a low V_s, and they decrease exponentially with increasing V_s. The values of K_{mb} and V_{mb}, however, vary only slightly as the V_s changes. The influence of the H/D ratio on the per-

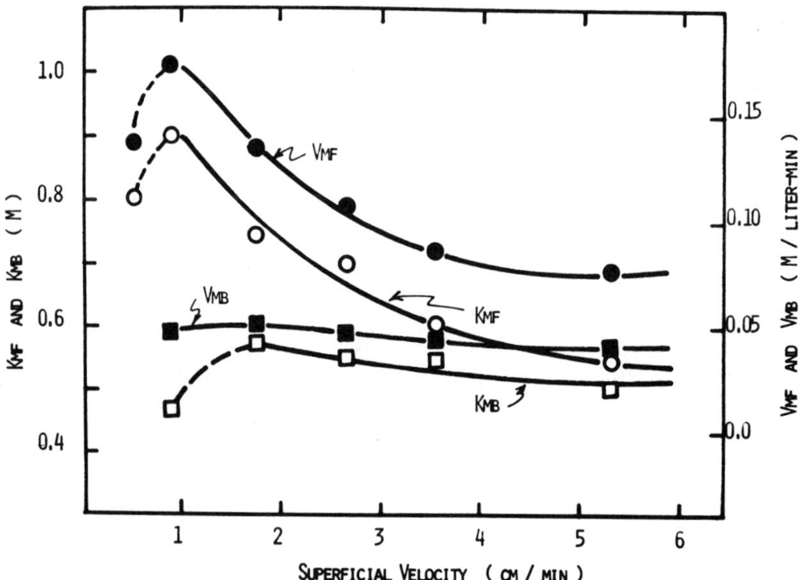

Fig. 1. The effect of superficial velocity on kinetic constants.

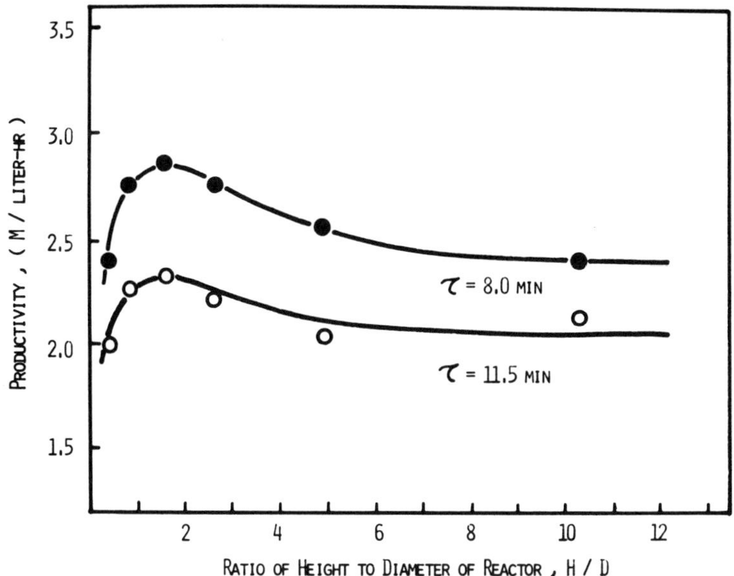

Fig. 2. The effect of reactor design parameter, H/D ratio, on the productivity.

formance of the reactor was examined (Fig. 2). It was found that the productivity in moles of product/l-hr went through a maximum at about 1.5 H/D ratio and decreased significantly as the H/D ratio was increased. The fact that there existed a maximum productivity was, in large part, due to the trend of K_{mf} and V_{mf} as shown in Fig. 1.

Based on these results one may conclude that the reactor geometry, namely the H/D ratio, is also a very important factor for the optimal design and best performance of the immobilized glucose isomerase reactor system under a given set of operating conditions.

REFERENCE

1. RYU, D.Y., CHUNG, S.H. & KATOH, K. *Biotechnol. Bioeng.* *19*:159, 1977.

IMMOBILIZATION OF ACETATE KINASE ON FUNCTIONALIZED SOLID-CORE POLYMERIC BEADS

B.A. Solomon, C.C. Chen, and C.K. Colton

Department of Chemical Engineering
Massachusetts Institute of Technology
Cambridge, Massachusetts, USA

We present here a method for immobilizing enzymes on a new support material useful in various reactor configurations: very small (1 to 3 μm diameter), high specific surface area (6 m^2/g), solid acrylate-based polymeric beads having carboxylic acid moieties (0.4 meq/g) on the surface (Matrex 101, Amicon Corp.). Acetate kinase (AcK) was selected because of its potential use in regeneration of adenosine triphosphate (ATP) (1).

In a typical preparation, 0.5 g dried beads were stirred at 80°C for one hr in 6.0 ml distilled H_2O containing 1.0 g polyethyleneimine of MW 1800 (PEI-18, Dow Chemical Co.) so as to provide a covalently bound, highly active primary amino functionalized surface (0.5 mmole NH_2/g). The beads were washed in distilled H_2O and 0.1 M borate buffer, pH 9.0, and dried *in vacuo* for 24 hours at 25°C. Next, a highly reactive aldehyde surface was formed by reaction of the PEI-coated beads with 10 ml of 12% (w/v) glutaraldehyde in 0.1 M borate buffer, pH 9.0 at 25°C for varying periods of time. The reaction was stopped by repeated washing with cold 0.1 M phosphate buffer, pH 7.5, after which the beads were centrifuged, the supernatant decanted, and the beads washed with distilled H_2O and then dried *in vacuo* for one hour at 4°C. The takeup of ornithine from solution in contact with the aldehyde-functionalized beads was used as a measure of the surface density of reactive aldehyde groups. The washed and dried beads were incubated with ornithine in 0.1 M borate buffer, pH 9.0. After one hr, the reaction was stopped by addition of $NaBH_4$ to reduce unreacted aldehyde groups and unstable Schiff's bases formed with the ornithine. The solution concentration of ornithine was determined by reacting it with 2,4,6 - trinitrobenzene-sulfonic acid, followed by spectrophotometric assay of the product. The amount taken up by the beads was calculated by mass balance.

AcK was coupled to the aldehyde-functionalized beads by incubating 18 units of enzyme (220 units/mg protein) with various amounts of beads in 300 μl of 0.1 M phosphate buffer, pH 7.5, containing 10 mM substrates (acetyl phosphate and adenosine diphosphate) for varying periods of time. The reaction was quenched by (a) rapid addition of an equal volume of 1.0 M glycine in the phosphate buffer, (b) centrifugation and separation of the beads from solution, and (c) addition of $NaBH_4$ to reduce any unreacted aldehydes or Schiff's bases. The reaction was carried out at pH 7.5 because of inherent instability of AcK at higher pH values. When desired, the number of available aldehyde groups on the surfaces of the beads was reduced prior to coupling with the enzyme by a preliminary reaction with glycine. This preattenuation with glycine was carried out for 5 min, after which the beads were washed prior to enzyme exposure. Soluble AcK activity was determined by measuring the rate of production of product ATP with a coupled enzymatic assay involving hexokinase and glucose-6-phosphate dehydrogenase (2). Immobilized AcK activity was similarly measured in a continuous spectrophotometric assay in which the beads were suspended in a cuvette at a concentration of 5 $\mu g/cm^3$ (3,4).

The uptake of ornithine by the aldehyde-functionalized beads is shown in Fig. 1. For reaction times less than 60 min, the binding capacity increased with reaction time because of the increased surface density of reactive aldehyde groups. The decreased binding capacity observed at longer times results from inter-bead crosslinking through glutaraldehyde and oxidation of the aldehyde. Also shown in Fig. 1 is the pH-dependence of the ornithine reaction during incubation with aldehyde-functionalized beads prepared by a one hr activation of the PEI-coated beads. The rate increased with pH because ornithine (pK_2=8.69) becomes a stronger nucleophile; the data suggest that immobilization might be most effective if carried out at a pH of 9.0 or higher.

Fig. 2 shows the effect of reaction conditions on AcK immobilization. With a 5 min coupling time and in the

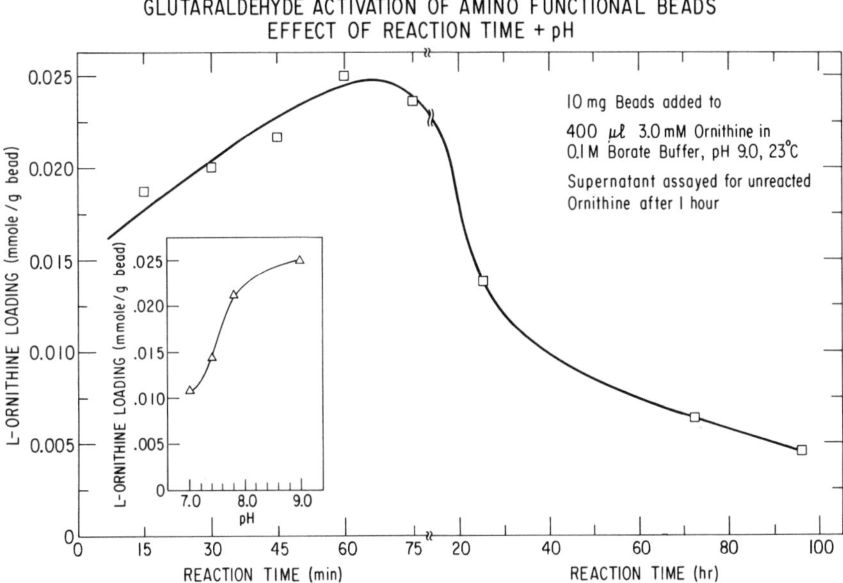

Fig. 1. Effect of reaction time and pH on the glutaraldehyde activation of PEI-coated beads as determined by L-ornithine loading.

absence of glycine preattenuation, virtually all of the enzyme activity was removed from solution, but only 10% was recovered as active enzyme on the beads (Fig. 2a). This percentage was even lower with longer coupling reaction times. The low yield may have resulted from an excessive number of bonds between surface aldehyde groups and the enzyme, as suggested by Datta *et al.* (5). By assuming 41,000 MW and 3.2 nm molecular diameter, one calculates that a monolayer of active enzyme corresponds to about 14 units on 5 mg beads. The maximum bound activity measured was 11.4 units, or about 63% of the enzyme initially added to the solution. Fig. 2b shows that a 5 min enzyme coupling time at the optimum glycine preattenuation concentration gave the maximum recovery of bound activity. The effect of the enzyme to bead ratio is shown in Fig. 2c for experiments in which the enzyme added to solution was held constant and the weight of the beads was varied between 1 and 10 mg. The open circles represent the total activity of bound enzyme, and the solid circles represent the normalized activity of bound enzyme (units/mg bead). Although the normalized activity was higher when only 2 mg of beads were used, more than 40% of the

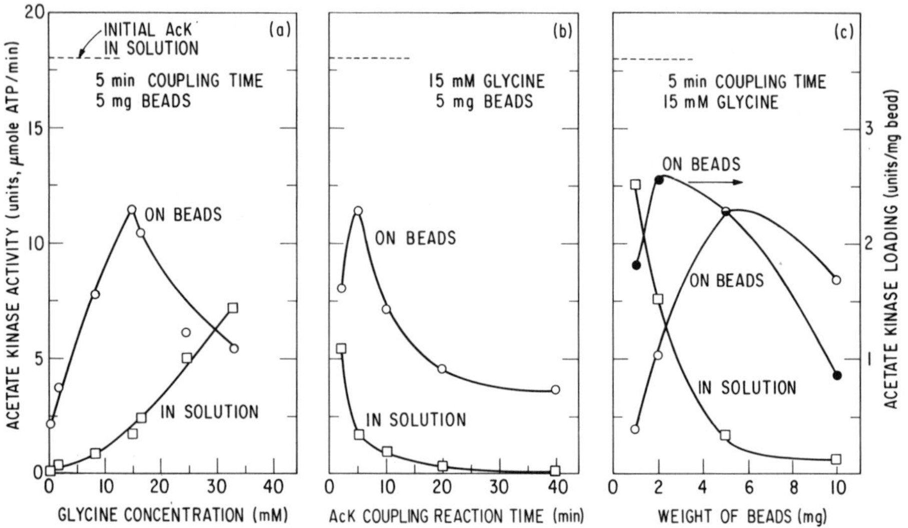

Fig. 2. Effect of reaction conditions on AcK immobilization when incubated with the aldehyde-functionalized PEI-coated beads at pH 7.5 and 37°C.

activity remained in solution, whereas only about 10% was left when 5 mg of beads were used.

The immobilization technique provides a pellicular support which is not subject to diffusive limitations under reaction conditions. Under optimum conditions, over 2000 units of enzyme, corresponding to about 10 mg active enzyme, were bound per gram of bead.

ACKNOWLEDGMENTS

This work was supported in part by NSF grant GI-34284 and by a grant from Amicon Corporation.

REFERENCES

1. LANGER, R.S., HAMILTION, B.K., GARDNER, C.R., ARCHER, M.C. & COLTON, C.K. *AIChE J.* *22*:1079, 1976.
2. LANGER, R.S., GARDNER, C.R., HAMILTON, B.K. & COLTON, C.K. *AIChE J.* *23*:1, 1977.
3. NEMET, M., Ph.D. Thesis, MIT, Cambridge, 1976.
4. CHEN, C.-C., M.S. Thesis, MIT, Cambridge, 1976.
5. DATTA, R., ARMIGER, W. & OLLIS, D.F. *Biotechnol. Bioeng.* *15*:993, 1973.

Session III
STUDY OF MODIFIED, STABILIZED, AND IMMOBILIZED ENZYMES

Chairmen: H. Fasold and G. Royer

SURFACE MODIFICATION OF PROTEINS

Roland R. Reiner and Helga Doring

Battelle-Institut e.V.
Frankfurt am Main, Federal Republic of Germany

The covalent bonding of proteins to soluble polymers is closely related chemically to immobilization. Reactions similar to the activation of polymers are used. Polymers with a chemical structure similar to those used for solid phases and surfaces (gels, membranes, coatings) may be used. The disadvantage of using soluble polymers instead of insoluble compounds is the more laborious purification needed after polymer activation and reaction with the protein. The excessive reagent and unreacted protein have to be separated by precipitation (often with organic solvents), gel filtration, ultrafiltration or dialysis.

For bonding of proteins to soluble polymers it is possible to choose one of the following three ways: reaction of the protein with an activated soluble polymer, reaction with an insoluble polymer and subsequent solubilization, or *in situ* formation of the polymer by covalent bonding of copolymerizable monomers to the protein and copolymerizing the product with soluble monomers. Some of the important properties of the soluble polymer are the number of active groups, location of active groups in the polymer chain, molecular weight and chain length, and adsorption to protein surface. Key reaction conditions include the concentration of polymer and protein, the molar ratio of polymer to protein, and solvent and salt concentration especially the presence of chaotropic solutes. Depending on these properties and conditions the reaction may lead to three types of soluble protein conjugates (products): (a) single-point binding, especially end-on binding as with polyethylene glycol (no cross-links); (b) multi-point binding of polymer molecules on a single protein molecule (intramolecular cross-linking); and (c) intermolecular cross-linking.

The covalent bonding of soluble polymers to proteins may change the following protein properties: molecular weight; water solubility; thermal stability; resistance to denaturants; pH effects owing to changes in the microenvironment (non-covalent influences) or to cross-linking or the change of the free energy of different structures of the protein molecule; enzymatic activity owing to changes in the chain mobility or the probability of the most active structure; and biospecific surface properties such as recognition, binding to other proteins, reactions with other proteins, and incorporation in membranes.

If hydrophobic instead of hydrophilic soluble polymers are used, products with properties advantageous for use in organic solvents, at interfaces, or in hydrophobic matrices may be obtained.

Some proteins modified with soluble polymers are listed in Table 1.

GLYCOSIDATION OF ENZYMES WITH DEXTRANES

We investigated a relative simple and cheap method which consisted of bonding HIO_4-oxidized soluble polysaccharides to proteins under mild conditions. Dextran T10 (MW 10,000) was oxidized with HIO_4 according to the method for cellulose activation. The moles of HIO_4 per moles of Dextran TIO were 10/1.5, 10/4, and 10/12. The number of formyl groups introduced was analyzed with 4-amino-3-hydrazino-1,2,4-triazolthiol-5 (11). The polymer, purified by gel chromatography, was stable; and one batch was used for different couplings.

To prevent intermolecular cross-linking the maximum protein concentration during the coupling reaction was in all cases below 1 mg/ml. The reaction was carried out in 0.1 M phosphate buffer pH 6 except in the case of carboxypeptidase where 0.025 M borax buffer of pH 8 was used. The reaction temperature was 4°C; and the reaction time ranged between 2 and 6 days. No attempts were made to establish the shortest possible reaction time.

The product was purified by gel chromatography using Biogel P10. The protein losses during the reaction and

TABLE 1

SOLUBLE POLYMER-MODIFIED PROTEINS

Enzyme	Soluble Polymer	Ref.
α-Amylase (E.C. 3.2.1.2)	Polysaccharides (Dextran 2000, DEAE Dextran 2000, CM-cellulose)	(1)
Carbonic anhydrase (E.C. 4.2.1.1)	Polyacrylamide-co-polymer	(2)
Catalase (E.C. 1.11.1.6)	Polyethylene glycol	(3-5)
Chymotrypsin (E.C. 3.4.4.5)	Polysaccharides (Dextran T10, DEAE-cellulose, dextranase hydrolyzed insoluble dextran)	(6-8)
β-Glucosidase (E.C. 3.2.1.2.1)	Polysaccharides (Dextran T10, FITC-Dextran)	(6)
Kallikrein (E.C.3.4.21.8)	Polyvinylpyrrolidon	(3)
Lysozyme (E.C. 3.2.1.17)	Polysaccharides (Dextran T10)	(6)
Peroxidase (E.C. 1.11.1.7)	Polyacrylamide copolymer	(2)
Trypsin (E.C. 3.4.4.4)	Polyvinylpyrrolidone, Polysaccharide	(9,10)

TABLE 2

GLYCOSYLATED ENZYMES

Enzyme	DT10	Mol DT10 added	Bound DT10 Enzyme	Relative Activity (%)
RNase	1.5	10(10)(a)	6.5(8)(a)	225 (215)(a)(b)
RNase	4	3	1.5	200
α-Ch	1.5	15(15)	11(11)	510(490)(c)
α-Ch	4	5	3	540
α-Ch	4	15	11	430
α-Ch	12	15(15)	12(12)	440(460)
Try	1.5	15	11	130(d)
Try	4	5	3	100
LDH	1.5	25	7.5	75
LDH	4	15(45)	3.5(3.5)	70(120)
Carb.p.	1.5	10(20)	8(11)	80(75)
Carb.p.	4	3(6)	2(5)(e)	80(90)(f)
GOD	1.5	90(90)	6	185(180)
GOD	4	30(30)	3(3)	135(140)

(a) Figures in brackets give number of repeated experiments.

(b) Unbound DT10 in solution had no effect on activity.

(c) Shelf life in aquous solution at pH 3 was markedly increased compared with unsubstituted α-Ch.

(d) Shelf life at pH 8 equal to that of unsubstituted trypsin.

(e) Low glycosylated products precipitated during purification resulting in low yields; higher glycosylated products were obtained in good yields.

(f) Shelf life: unsubstituted enzyme was strongly precipitated after 6 days; substituted enzyme showed only marginal precipitation after 26 days under identical conditions.

the chromatographic analysis were normally below 15%. The amount of Dextran T10 bonded per enzyme was analyzed using anthron (12).

The enzymatic activities shown in Table 2 correspond to the activities of the native enzymes measured under identical conditions, as described in the following literature references: ribonuclease (bovine) (13), substrate cyclic-2'-3'- cytidine monophosphate; α-chymotrypsin (14), substrate N-benzoyl-L-tyrosine-ethyl ester; trypsin (15) substrate N-benzoyl-L-arginine-ethyl ester; lactate dehydrogenase (16), substrate pyruvate and NADH; carboxypeptidase (17), substrate hippuryl-L-phenylalanine; glucosidase (18), substrate D-glucose.

The results of the investigation are summarized in Table 2. Since only relative activities were measured, no definite answer on the nature of the enzyme activation can be given. The comparison with BrCN binding of DT10 to the same enzymes confirmed the superiority of the described method.

REFERENCES

1. WYKES, I.R., DUNNILL, P. & LILLY, M.D. *Biochem. Biophys. Acta 250*:522, 1971.
2. EPTON, R., MARR, G. & MORGAN, G.J., British Patent Application (Koch-Light), 1976.
3. VON SPECHT, B.U., WAHL, M., KOLB, H.J. & BRENDEL, W. *Arch. Int. Pharmacodyn. Ther. 213*:242,1975.
4. ABUCHOWSKI, A., MCCOY, I.R., PALCZUK, N.C., VAN ES, T. & DAVIS, F.F., *J. Biol. Chem. 252*:3582, 1977.
5. ABUCHOWSKI, A., VAN ES, T., PALCZUK, N.C. & DAVIS, F.F., *J. Biol. Chem. 252*:3578, 1977.
6. VEGARND, G. & CHRISTENSEN, T.B., *Biotechnol. Bioeng. 17*:1391, 1975.
7. AXEN, R., MYRIN, P.A. & JANSON, J.C., *Biopolymers 9*: 401, 1970.
8. O'NEILL, S.P., WYKES, I.R., DUNNILL, P. & LILLY, M.D. *Biotechnol. Bioeng. 13*:319, 1971.
9. VON SPECHT, B.-U., SEINFELD, H. & BRENDEL, W. *Hoppe Seylers's Z. Physiol. Chem. 354*:1659, 1973.
10. MARSHALL, J.J. & RABINOWITZ, M.L., *J. Biol. Chem. 251*:1081, 1976.
11. DICKINSON, R.G. & JACOBSEN, N.W., *Chem. Comm. 1799*: 1719, 1970.

12. SCOTT, J.R., TROY, A. & MELVIN, E.H., *Anal. Chem. 25*: 1655, 1953.
13. CROOK, E.M., MATHIAS, A.P. & RABIN, B.R., *Biochem. J. 74*:234, 1960.
14. RICK, W., in "Methoden der enzymatischen Analyse I," (H.U. Bergmeyer, ed.) Verlag Chemie, Weinheim, 1970, p. 989.
15. BERGMEYER, H.U., GAWEHN, K. & BRASSL, M. in "Methoden der enzymatischen Analyse I," (H.U. Bergmeyer, ed.) Verlag Chemie, Weinheim, 1970, p. 476.
16. Boehringer Mannheim GmbH, *Biochemica Information I*: 121, 1973.
17. FOLK, J.E. & SCHIRMER, E.W., *J. Biol. Chem. 238*: 3884, 1963.
18. BERGMEYER, H.U., GAWEHN, K. & GRASSL, M. in "Methoden der enzymatischen Analyse I," (H.U. Bergmeyer, ed.) Verlag Chemie, Weinheim, 1970, p. 416.

STAPHYLOCOCCAL NUCLEASE IMMOBILIZED ON AGAROSE GEL

Jose M. Guisan and Antonio Ballesteros

Instituto de Catalisis y Petroleoquimica
Consejo Superior de Investigaciones Cientificas
Madrid, Spain

The enzyme micrococcal nuclease (ribonucleate (deoxyribonucleate) 3'-nucleotidohydrolase, EC.3.1.4.7) from *Staphylococcus aureus* has been extensively studied by Anfinsen and collaborators (1). They have elucidated the primary structure, the catalytic mechanism and details of the active centre. The X-ray structure of this enzyme at 2Å resolution is also known (2). We report here the immobilization on different supports and the analysis of the properties of the immobilized enzyme.

EXPERIMENTAL

Materials included CNBr-activated Spharose 4B from Pharmacia, nuclease from Worthington, *E. coli* K12 transfer RNA from General Biochemicals, calf thymus DNA (10^6MW) from Sigma, and thymidine 3'-phosphate-5'-(p-nitrophenyl phosphate) (nitrophenyl-pdTp) from Ash Stevens, Inc. (Detroit). All other reagents were from commercial sources.

CNBr-activated Sepharose 4B and nuclease, in 0.1 M Na_2CO_3-$NaHCO_3$ buffer, pH 9.0, were mixed and left at 4°C for 24 h with gentle stirring. After that time more than 95% of the enzyme was immobilized, as judged from the specific activity remaining in the supernatant. In control experiments without Sepharose, the activity of the enzyme remained at 100%. The Sepharose-nuclease was filtered and washed several times with the same buffer.

The remaining activated groups in the support that had not reacted with the enzyme were blocked by 1 h immersion of the derivative in 1 M ethanolamine pH 8.0. The Sepharose-nuclease was then washed several times alternately with 0.1 M acetate buffer, pH 4.0, containing 1 M NaCl, and 0.1 M borate buffer, pH 8.8, containing 1 M NaCl, and finally with distilled water.

Assay of the soluble and insoluble enzyme was carried out at 30°C by following the increase in absorbance at 260 nm of substrates (heat-denatured DNA, tRNA) (3) or at 330 nm of nitrophenyl-pdTp (4), using a Zeiss PMQII spectrophotometer equipped with a 2 cm pathlength cuvette with magnetic stirrer.

RESULTS AND DISCUSSION

All the measurements of activity were made at the maximal stirring speed of the Zeiss device in order to minimize the external diffusional limitations. The amount of immobilized derivative in the assay was always main-

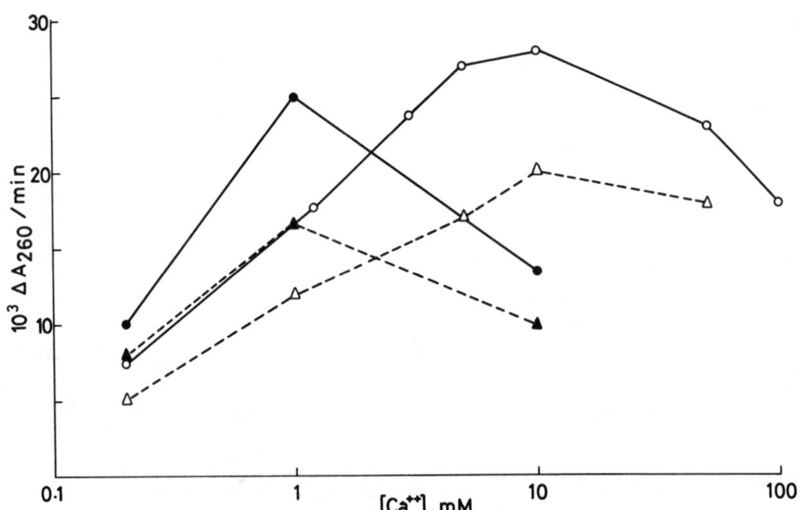

Fig. 1. Effect of calcium concentration on soluble (---) and insoluble (——) nuclease activity, at pH 8.8 (○ and △) and at pH 10.0 (● and ▲). The insoluble derivative contained 780 μg/enzyme/ ml gel. DNA concentration: 0.1 mg/ml.

tained in the region of linearity between activity and amount of derivative.

Fig. 1 shows the dependence of enzymic activity on calcium concentration. Both the soluble and the insoluble enzyme behaved similarly, with less Ca^{++} required for maximal activity at higher pH. As with the soluble enzyme (3), the optimum pH for enzymatic activity was dependent on the Ca^{++} concentration. Increasing the calcium concentration from 1 to 100 mM shifted the maximum to a lower pH. Fig. 2 shows the increase in specific activity when the amount of immobilized nuclease was reduced several thousand-fold. In Fig. 2 the catalytic efficiency is defined as the ratio of the specific activity of the immobilized enzyme to that of the soluble one. Some of the derivatives prepared have also been assayed at 6°C, the catalytic efficiency being higher at this temperature (Fig. 2). (The specific activity of the soluble nuclease decreased 32-fold from 30 to 6°C). It is apparent from the figure that when the derivative contained more than 400 μg of enzyme/ml of gel, the efficiency of the immobilized species remained practically constant, possibly

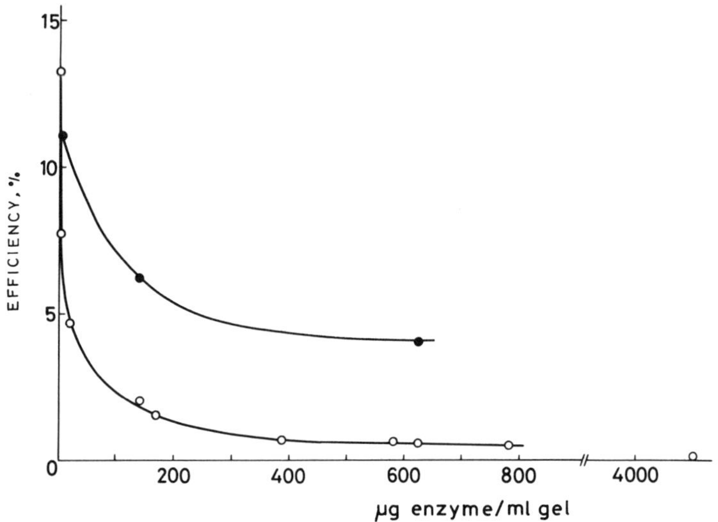

Fig. 2. Variation of the catalytic efficiency of the insolubilization as a function of the amount of nuclease immobilized per unit volume of the gel. ● 6°C; O 30°C. Assay conditions: 10 mM Ca^{++}, Tris buffer pH 8.8, 100 μg DNA/ml.

because the diffusion of the substrate became the rate determinant step.

It was not possible to measure the activity of the immobilized derivatives with very small nuclease content. Even working with as little as 0.5 µg enzyme/ml gel, we were not able to get a catalytic efficiency higher than 13.2%. The present data favored the idea that the enzyme was losing a significant proportion of its activity in the covalent binding to the Sepharose. However, immobilization of the enzyme on Sepharose greatly enhanced the stability; the derivative maintained 80% of its activity after two months at 4°C.

In Fig. 3 the initial activities of soluble and insoluble enzyme are compared as a function of DNA concentration. The amounts of soluble and insoluble enzyme were chosen by trials so that the higher ordinate values of both curves were the same (i.e., approx. 72 in this case). It is well known that one can shift from diffusional to reaction rate control by working well outside the conditions of optimal enzyme activity such that the rate of reaction becomes the slowest step. If the experiments of

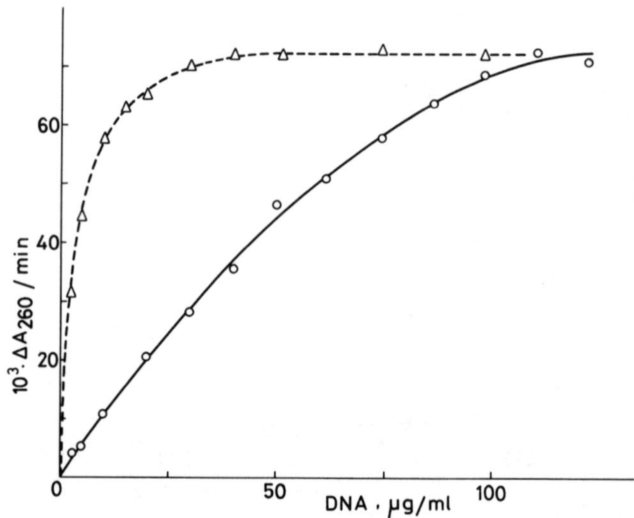

Fig. 3. Effect of DNA concentration on native (- -Δ- -) and insoluble (- O -) nuclease activity. Assay conditions: 10 mM Ca^{++}; Tris buffer pH 8.8; 140 µg enzyme/ml gel.

Fig. 3 are repeated at pH 9.5 and a Ca^{++} concentration of 100 mM, both sets of data follow a single curve. With a substrate intermediate in size between DNA and tRNA, polyadenylic acid of MW 10^5, then two curves again are observed. (J.M. Guisan, unpublished).

Cuatrecasas *et al.* (4) reported that nitrophenyl-pdTp (MW 613) was an excellent substrate for routine studies with the nuclease. The spontaneous rate of nitrophenyl-pdTp hydrolysis was negligible if the temperature was not higher than 24°C, the pH below 10, the Ca^{++} concentration below 0.1 M, and the substrate concentration not too high (4). All experiments were run at 24°C with these conditions. A plot of the effective reaction rate against the reaction rate divided by the bulk concentration of the substrate is shown in Fig. 4 (Eadie-Hofstee plot). Following the reasoning of Regan *et al.* (5), when diffusional effects are insignificant a straight line is obtained; this is the case of the soluble enzyme (J.M. Guisan, unpublished). Fig. 4 clearly shows that at low nucleotide concentrations diffusion is controlling the overall rate.

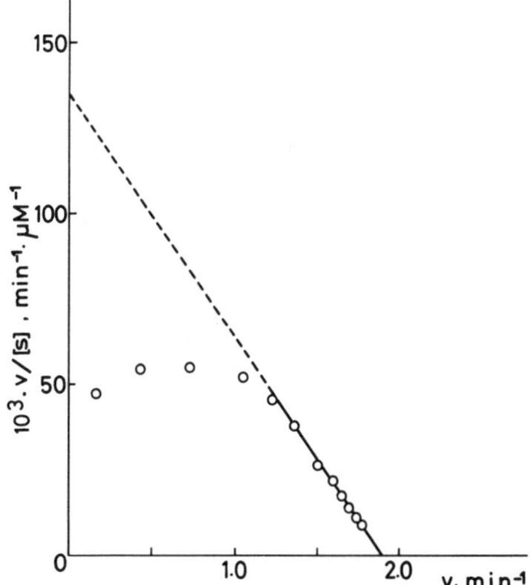

Fig. 4. Eadie-Hofstee plot of the specific activity of insoluble nuclease with nitrophenyl-pdTp concentration. Assay conditions: as in Fig. 3.

ACKNOWLEDGMENTS

We thank M.C. Ceinos for skillful technical assistance, V.M. Fernandez for helpful discussions, and C. Montuenga for critical reading of the manuscript. This work has been supported by the Spanish Comision Asesora de Investigacion Cientifica y Tecnica.

REFERENCES

1. ANFINSEN, C.B., CUATRECASAS, P. & TANIUCHI, H. in "The Enzymes," 3rd Edit., vol. 4 (P.D. Boyer, ed.), Academic Press, New York, 1971, p. 177.
2. COTTON, F.A. & HAZEN, E.E. JR. in "The Enzymes," 3rd Edit., vol. 4, (P.D. Boyer, ed.), Academic Press, New York, 1971, p. 153.
3. CUATRECASAS, P., FUCHS, S. & ANFINSEN, C.B. *J. Biol. Chem.* *242*:1541, 1967.
4. CUATRECASAS, P., WILCHEK, M. & ANFINSEN, C.B. *Biochemistry* *8*:2277, 1969.
5. REGAN, D.L., LILLY, M.D. & DUNNILL, P. *Biotechnol. Bioeng.* *16*:1081, 1974.

ENZYME COLLAGEN MEMBRANES IN REACTORS AND FOR ANALYTICAL PURPOSES

Daniele C. Gautheron and Pierre R. Coulet

LBTM - CNRS
Universite Claude Bernard
Villeurbanne, France

In 1971 in order to simulate the influence of the biological membrane microenvironment on embedded enzymes, we grafted glutamate dehydrogenase (EC 1.4.1.3) to handmade ultra-thin collagen films (1). The enzyme kept its oligomeric structure and regulations. The Centre Technique du Cuir (Lyon) supplied us with films of highly polymerized, undegraded, insoluble collagen, prepared in industrial conditions as a by-product of the leather industries (2,3).

Covalent attachment of enzymes was performed after activation of carboxyls either by acyl-azide formation (4) or by the use of a water-soluble carbodiimide (EDC) or the Woodward's reagent "K" (5). The yield in retained activity being similar, the acyl-azide method was selected for all further work because of its low cost and general use for all proteins. The activated films can be washed thoroughly before the enzymes spontaneously bind; thus, the enzyme never comes into contact with damaging chemicals (6).

BIOREACTORS

A comparative study of the properties of free and bound amyloglucosidase was conducted to estimate the diffusional restrictions of the bound system and to optimize the performances of enzymatic films in reactors (7). Our acyl-azide method was adapted to obtain the best enzyme surface activity; and the homogeneity of distribution of the coupled enzyme on the membrane surface was checked. The grafted enzyme exhibited an exceptional stability.

Two different types of reactors were built and tested. The helicoidal film reactor is shown in Fig. 1. The reactor was made of a hollow ALTUGLAS cylinder with a 108 ml reaction chamber, containing two strips of enzymatic films rolled in helicoidal coiling. The magnetic stirring speed was adjusted to ensure good mixing of the reaction medium. Tracer studies proved that the reactor behaved as a well-stirred CSTR.

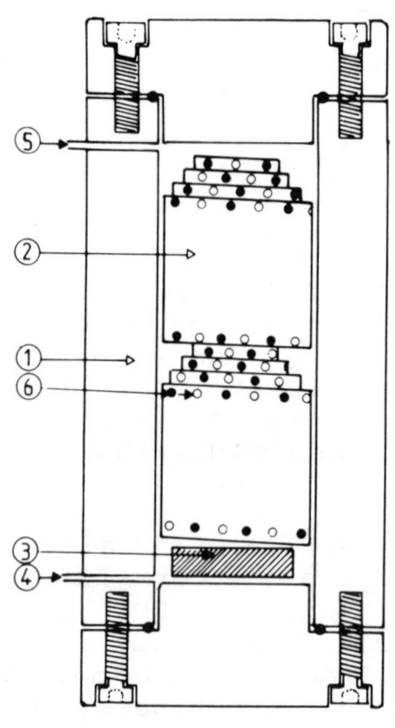

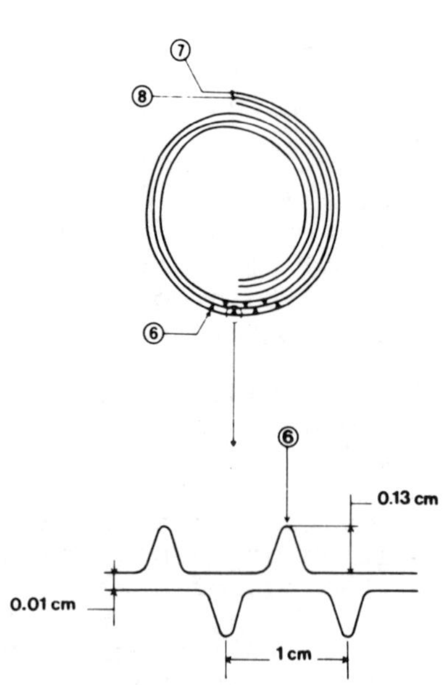

REACTOR SCHEME

1 REACTOR BODY
2 HELICOIDAL COILING
3 MAGNETIC BAR
4 INLET
5 OUTLET

COILING SECTION VIEW AND PLASTIC DETAILS

6 SPIKE
7 PLASTIC TAPE
8 ENZYMATIC COLLAGEN MEMBRANE

Fig. 1. Helicoidal film reactor.

In view of extending our system to the semi-pilot scale, it appeared interesting to use piles of flat enzymatic sheets, since reactors of the plug-flow type can often behave more efficiently than a CSTR. A thin layer flow reactor was built with 0.4 mm thick amyloglucosidase membranes (8), which consisted of two blocks of ALTUGLAS fitted with feeding and drain-off pipes (Fig. 2). A 67.5 ml rectangular reaction volume lay between the blocks. The substrate solutions passed through the reactor in thin layers of 350 μm thickness (horizontal unidirectional flow). In a typical experiment two enzymatic sheets were used with a total activity of 362 units. The plug-flow material balance equation (9) was applied to the reactor. The relationship between XS_{in} and ln (1-X) in Fig. 2 was linear only within a limited range of concentration and flow rate. An apparent K_m was calculated from the slope of the linear part of the graph (see below).

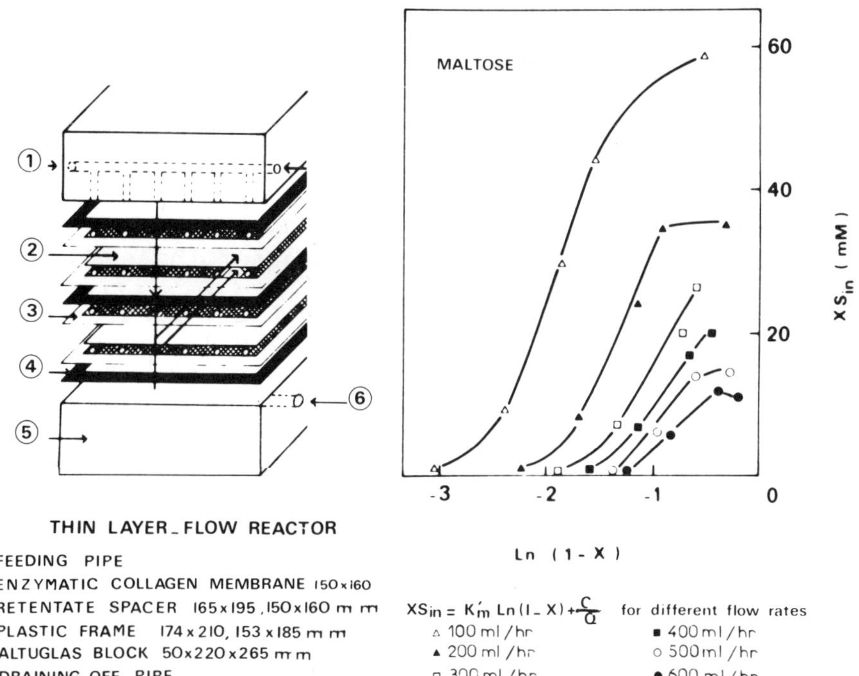

Fig. 2. Thin layer flow reactor.

The reactors were fed with maltose and then with soluble starch (average MW = 22,500). For a 37 ml/hr flow rate the conversion was practically complete in the range 1-125 mM maltose; with increased flow rates and substrate concentrations the conversion decreased (CSTR). In the thin layer flow reactor the conversion was complete in the range 1-56 mM maltose at 40 ml/hr. With increased flow rate (up to 680 ml/hr) and concentration (up to 125 mM maltose), the conversion decreased to 0.20.

The CSTR loaded with amyloglucosidase membranes was fed with concentrated liquors of maltose or soluble starch (30% w/v). Recycling allowed a complete conversion of maltose into glucose in 48 cycles; under the same conditions, soluble starch was converted 66%. Recycling in the thin layer flow reactor at 260 ml/hr (space time 6.6 min) achieved full conversion of 30% maltose (w/v); 90% of soluble starch was converted, probably due to the presence of small branched units, as observed with the free enzyme.

With a freely stirred two substrate enzyme-membrane, an opposite variation of affinity was exhibited for each substrate (11,12). Since our coupling mainly secured a surface attachment of the enzymes onto the membrane (13), only external diffusional limitations were taken into consideration. For a freely stirred amyloglucosidase membrane, the K_m of maltose was not affected; although it increased five-fold with soluble starch. In the CSTR the calculated K_m for maltose was very close to that of the free enzyme or stirred membrane. With the thin layer flow reactor the diffusion was dependent on the flow rate; and at 600 ml/hr the maltose K_m increased three-fold.

Comparison of the intrinsic stability of free and bound aspartate aminotransferase (EC 2.6.1.1) demonstrated that the binding itself greatly increased the stability (10). Collagen-bound amyloglucosidase (EC 3.2.1.3) kept its full activity much longer than the soluble enzyme upon storage. In the CSTR running in continuous operation at 40°C, the conversion remained constant for at least 18 days even at low surface activity of the amyloglucosidase membranes (5 enzyme units).

ANALYTICAL APPLICATIONS

Enzymatic stamps and tapes can be used routinely for series of micro-assays. The necessary enzymatic activity is obtained by cutting the adequate area or separating the stamp from a dotted sheet. The enzymatic films can be lyophilized. With several enzymes of analytical interest (aspartate aminotransferase, lactate and malate dehydrogenase, creatinekinase, hexokinase, glucose oxidase, etc...) 200 to 300 micro-assays can be run with a single enzymatic strip 2.5 by 1.5 cm.

We also tested the abilities of our glucose oxidase collagen membranes combined with a platinum anode to estimate glucose. Amperometry at 650 mV was used to detect the hydrogen peroxide produced, which was directly proportional to the oxidized glucose. The performances of our glucose electrode is described elsewhere (14).

REFERENCES

1. JULLIARD, J.H., GODINOT, C. & GAUTHERON, D.C. *FEBS Lett.* *47*:295, 1971.
2. COMTE, P.P. *Brev. Invention Francais 1* (*568*):829, 1967.
3. PICHON, G.J. & PIAT, B.A. *Brev. Invention Francais 1* (*596*):789, 1968.
4. COULET, P.R., JULLIARD, J.H. & GAUTHERON, D.C. *Brev. Invention Francais* (*A.N.V.A.R.*) *2* (*235*):133, 1973.
5. LEE, K.H.K., COULET, P.R. & GAUTHERON, D.C. *Biochimie* *58*:489, 1976.
6. COULET, P.R., JULLIARD, J.H. & GAUTHERON, D.C. *Biotechnol. Bioeng.* *16*:1055, 1974.
7. BRILLOUET, J.M., COULET, P.R. & GAUTHERON, D.C. *Biotechnol. Bioeng.* *19*:125, 1977.
8. BRILLOUET, J.M., COULET, P.R. & GAUTHERON, D.C. *Biotechnol. Bioeng.* *18*:1821, 1976.
9. LILLY, M.D., HORNBY, W.E. & CROOK, E.M. *Biochem. J.* *100*:718, 1966.
10. ENGASSER, J.M. & COULET, P.R. *Biochim. Biophys. Acta* *485*:29, 1977.
11. COULET, P.R., GODINOT, C. & GAUTHERON, D.C. *Biochim. Biophys. Acta* *391*:272, 1975.
12. ENGASSER, J.M., COULET, P.R. & GAUTHERON, D.C. *J. Biol. Chem.*, *252*:7919, 1977.
13. COULET, P.R. & GAUTHERON, D.C. in "Analysis and Control of Immobilized Enzyme Systems," (D. Thomas and J.P. Kernevez, ed.) North Holland, Amsterdam, 1976.
14. THEVENOT, D.R., COULET, P.R. & GAUTHERON, D.C., this volume.

ANALYSIS OF THE REACTION KINETICS OF SINGLE ESTERASE-SEPHAROSE BEADS BY MICROFLUOROMETRY

O. Hannibal-Friedrich and M. Sernetz

Institut fur Biochemie
Giessen, Federal Republic of Germany

The evaluation of immobilized enzyme kinetics by conventional techniques often relies on batch reactor experiments where equations are derived from hypothetical properties of single beads (1). Batch reactor experiments however can provide only indirect evidence for these hypothetical properties. For direct evidence methods have to be applied which allow for examination of the reaction kinetics of single immobilized enzyme beads. In the folll-lowing it is shown that microfluorometry of the turnover of fluorogenic substrates may be a suitable direct method.

As a model system to study diffusional effects under steady state conditions, the turnover of fluoresceine diacetate (2) by an esterase from pig liver (EC 3.1.1.1, Boehringer) immobilized on CNBr-Sepharose 4B (Pharmacia) was chosen. Esterase-Sepharose-beads were trapped in a wedge-shaped chamber in focus on a microfluorometer (3). The substrate was pumped continuously through the chamber at high speed (60-90 ml/h or > 3 cm/sec), to minimize external diffusional effects. The high sensitivity of the microfluorometer (4) allowed us to determine the total product mass as well as the distribution within a single bead. The fluorescence intensity of the product tail behind the bead served as a measure for the overall reaction rate.

Assuming spherical particle, uniform enzyme distribution, steady state conditions, Michaelis-Menten kinetics, and noncompetitive product inhibition, the kinetic equation describing the influence of diffusion on the system is given by

$$\frac{d^2 s}{dr^2} + \frac{2}{r}\frac{ds}{dr} = \frac{\sigma \quad s}{(1+\alpha s)(1+\alpha\beta(1-s))}$$

where s is the ratio of internal to surface substrate concentration S_o; r is the radial position; α is the relative surface concentration S_o/K_m; β is the effective inhibition constant $\beta = (D_s/D_p)(K_m/K_I)$; and σ is the Damkohler number. The boundary conditions are given elsewhere (5). From the equation the product distributions were calculated for various conditions. The resulting total mass and the overall reaction rate were highly correlated and yielded similar hyperbolic plots except for high concentrations. These results were confirmed by measurements on single beads (Fig. 1) (6). Determinations of apparent K_m values were made for different radii; extrapolation to zero radius gave a true K_m for the immobilized enzyme.

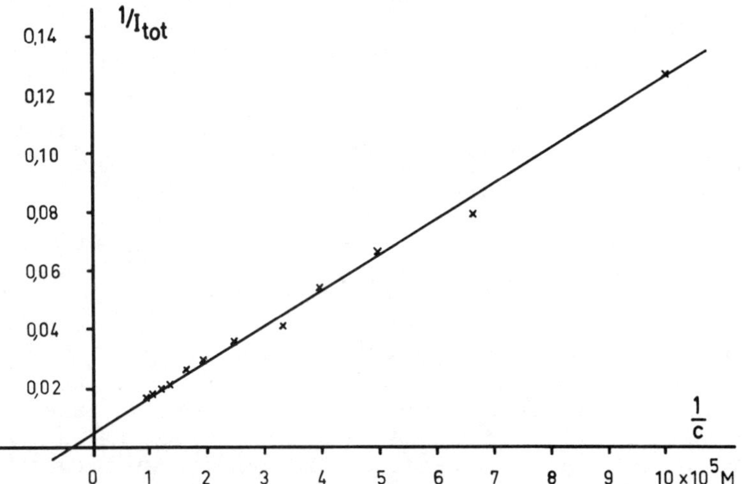

Fig. 1. Measurement of the total fluorescence intensity of a single Esterase-Sepharose-bead as a function of the fluoresceine diacetate concentration in the surrounding solution; double reciprocal plot.

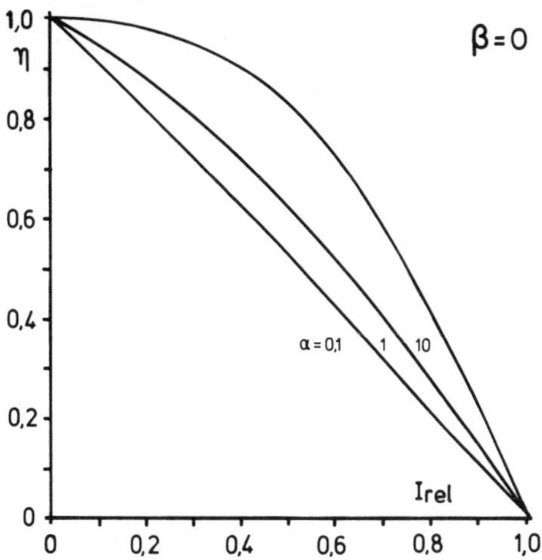

Fig. 2. Effectiveness factor η versus relative intensity for the case of no inhibition (β=0) with α as parameter.

Effectiveness factors could be determined only indirectly, because the case of no diffusional limitation could not be tested experimentally. Instead a relative intensity was defined as the ratio of product mass within the bead under the given condition to that of maximum diffusional limitation. This was simulated by measuring the fluorescence of the beads in a product solution of equal concentration. An effectiveness factors could then be determined directly from a correlation plot (Fig. 2), if the values of α and β were known.

The development of new fluorogenic substrates and their application to systems of other geometries and chemical behaviour should allow more general use to be made of this method.

ACKNOWLEDGMENT

This work was supported by a grant from the Deutsche Forschungsgemeinschaft.

REFERENCES

1. LASCH, J. *Mol. Cell. Biochem.* 2:79, 1973.
2. ROTMAN, B. & PAPERMASTER, B.W. *Biochemistry* 55:134, 1966.
3. SERNETZ, M. & PUCHINGER, H., *Meth. in Enzymology* 44: 373, 1976.
4. SERNETZ, M. & THAER, A. in "Fluorescence Techniques in Cell Biology," (A Thaer & M. Sernetz, eds.), Springer, Berlin, 1973, p. 41.
5. ENGASSER, J.M. & HORVATH, C. *J. Theoret. Biol.* 42: 137, 1973.
6. HANNIBAL-FRIEDRICH, O.H., thesis, University of Giessen, 1977.

A NEW KINETIC METHOD FOR ANALYSING THE PARAMETERS OF IMMOBILIZED ALDOLASE

Veronika Jancsik, Ferenc Bartha, and Jurgen Lasch*

Institute of Biochemistry
Hungarian Academy Science
Budapest, Hungary, and
Institute of Physiological Chemistry,*
Martin Luther University, Halle
Saale, German Democratic Republic

Rabbit muscle aldolase was attached to Sepharose 6B by CNBr activation. Under optimal conditions the yield was 70-90% for protein, but only 20-50% for enzyme activity, as measured immediately after coupling. This activity decreased to about 5% during a few days storage at +5°C, and then remained stable for some weeks. From initial velocity measurements we calculated the K_{Mapp} value, which did not significantly differ from the K_M of the soluble enzyme.

The experiments were carried out in a CSTR reactor. The product concentration was determined at the outlet at a constant feed rate and under stationary conditions.

For the theoretical considerations, the kinetic mechanism of the reaction catalyzed by immobilized aldolase was taken as follows:

$$E + S \underset{k_{-1}}{\overset{k_1}{\rightleftharpoons}} EPQ \underset{k_{-2}}{\overset{k_2}{\rightleftharpoons}} EQ + P; \quad EQ \underset{k_{-3}}{\overset{k_3}{\rightleftharpoons}} E + Q$$

where: S = fructose-1,6-diphosphate (FDP), P =

glyceraldehyde-3-phosphate, Q = dihydroxy-acetone-phosphate. EPQ and EQ are the corresponding enzyme-complexes.

Let us feed an FDP solution of molarity $[S_o]$ into the reactor of volume V, which contains active enzyme of concentration E_o at a feeding rate u. If the reactor performance is stationary, then the following relationship holds between the measurable product concentration, [P], and the reactor parameters:

$$\frac{[S_o]}{[P]} - 1 = \frac{K_{Mapp} + \frac{[P]}{K_{eq}} \left[\frac{k}{k_{-1}} (K_2 + [P]) + k \left(\frac{1}{k_{-2}} + \frac{1}{k_{-3}} \right) + \frac{V}{u} V_{max} \right]}{\frac{V}{u} V_{max} - [P] \left[1 + \frac{k}{k_3} \frac{[P]}{k_2} \right]}$$

where $K_{Mapp} = \frac{k_3}{k_2 + k_3} \left[\frac{k_{-1} + k_2}{k_1} \right]$, $\frac{1}{k} = \frac{1}{k_2} + \frac{1}{k_3}$

$K_1 = \frac{k_{-1}}{k_1}$; $K_2 = \frac{k_2}{k_{-2}}$; $K_3 = \frac{k_3}{k_{-3}}$, $K_{eq} = \frac{K_2 K_3}{K_1}$,

$V_{max} = k\ E_o$

Let us suppose that the reactor parameters satisfy the following criteria:

$$\frac{V}{u} V_{max} >> [P]; \quad \frac{K_{Mapp}}{[P]} \quad Keq; \quad \frac{K_{Mapp}}{K_1} \quad [P]$$

Then a linear relationship exists between $1/[P]^2$ and $1/([S_o] - [P])$:

$$\frac{1}{[P]^2} = \frac{u}{V} \frac{1}{k_3 E_o K_2} + \frac{1}{K_{eq}} \left[1 + \frac{u}{VE_o} \left(\frac{1}{k_{-3}} + \frac{K_{eq}}{K_3} \frac{K_{Mapp}}{k} \right) \right] \frac{1}{[S_o] - [P]}$$

Our experimental results are in good agreement with this prediction.

The k and K_{Mapp} values can be determined from initial velocity measurements. Knowing the values of K_{eq} and K_3, and determining the slope and intercept of the straight line $1/P^2$ vs $1/(S_o-P)$ obtained in a relatively simple reactor experiment, one has six independent quantities for the calculation of the six unknown rate constants k_1, k_{-1}, k_2, k_{-2}, k_3 and k_{-3}.

We suggest that a similar approach can also be applied to other immobilized enzymes catalysing reversible reactions.

ACKNOWLEDGMENT

We are grateful to Dr. K. Kretschmer from the Halle Institute for constructing the reactor and for his continuous interest and advices.

INTERACTION OF ENZYMES WITH SYNTHETIC POLYMERS

T. Keleti, V. Jancsik, M. Nagy[+] and E. Wolfram[+]

Enzymology Department, Institute of Biochemistry
Hungarian Academy of Sciences and
Department of Colloid Science[+]
Lorand Eotvos University
Budapest, Hungary

We have studied the kinetic consequences of the interactions of non-charged soluble polymers with enzymes. As polymers, poly-(vinyl-alcohol) (PVA) and poly-(N-vinyl-pyrrolidone) (PVP) appeared to be most suitable because their molecular weight and structure could be varied over a wide range. The enzymes studied were D-glyceraldehyde-3-phosphate-dehydrogenase (GAPD) and aldolase.

The effects of polymers on aldolase are very complex. We tested various PVA and PVP preparations differing in molecular weight (1). Some PVA preparations were treated with glutaraldehyde to introduce 2, 11 or 22 intramolecular cross-linkages per polymer chain. Several aldolase preparations differing from each other in age, specific activity, and stability were examined. With some polymer-aldolase combinations the maximal velocity of the aldolase reaction decreased above a given polymer concentration. This was most pronounced with polymers of low molecular weight and with those having many cross-links. At certain polymer-aldolase combinations a marked increase in maximal velocity occurred at low polymer concentrations.

The Michaelis constant of fructose-1,6-diphosphate for aldolase increased, decreased, or showed no change as a function of polymer concentration.

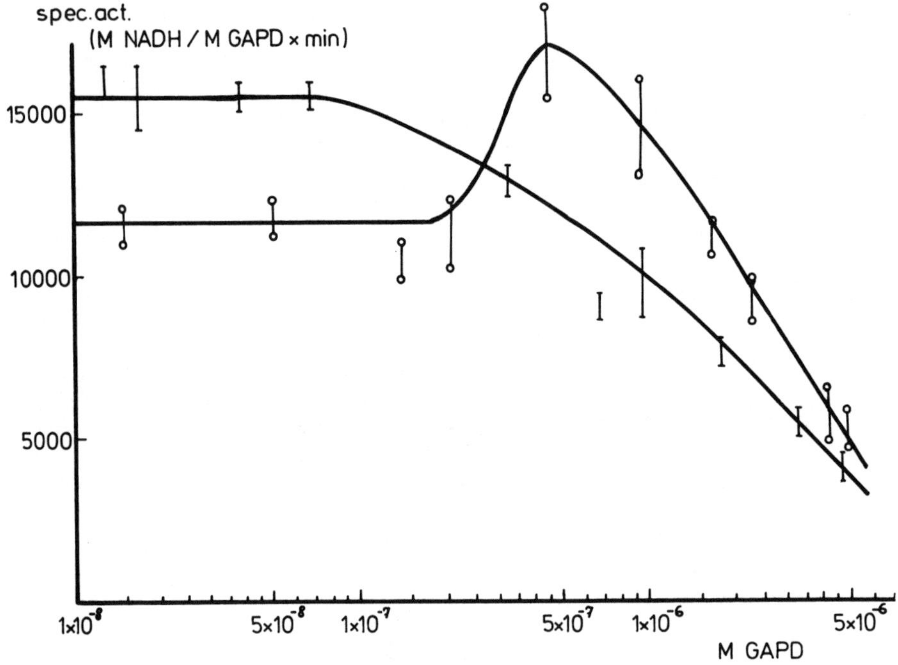

Fig. 1. Specific Activity of GAPD. Specific activities were calculated from initial velocity measurements carried out by stop-flow techniques. Reaction mixtures were as follows: 2.5×10^{-2} M tris buffer, pH 7.5, containing 5×10^{-4} M mercaptoethanol (I) or 1.5% poly-(N-vinylpyrrolidone) solution, pH 7.5, containing 5×10^{-4} M mercaptoethanol (0), 2 mM glyceraldehyde-3-phosphate, 2 mM NAD, 10 mM inorganic phosphate.

Above a certain level the specific activity of GAPD in buffer decreased with increasing enzyme concentration. This could be attributed to association-dissociation phenomena (2). One could find polymer concentrations which altered the pattern shown in Fig. 1. It seemed that the interaction of the same polymer with different oligomeric forms of GAPD was markedly different.

These results suggest that the unspecific enzyme-macromolecule interactions were influenced considerably by the colloidal state of both interacting partners.

REFERENCES

1. JANCSIK, V., KELETI, T., BICZOK, GY., NAGY, M., SZABO, Z., & WOLFRAM, E. *J. Mol. Catal.* *1*:137, 1975-76.
2. KELETI, T., BATKE, J., OVADI, J., JANCSIK, V., & BARTHA, F. in "Advances in Enzyme Regulation," vol. 15 (G. Weber, ed.), Pergamon Press, New York, 1977, p. 233.

PROTEOLYSIS OF IMMOBILIZED LACTATE DEHYDROGENASE IN THE PRESENCE OF PYRUVATE AND NADH

G.P. Royer, S. Ikeda and T. Lee

Department of Biochemistry
The Ohio State University
Columbus, Ohio, USA

We have described the use of enzyme immobilization in a flow system as a means to detect substrate induced conformational changes of enzymes (1,2). The enzyme and substrate were mixed just prior to contacting a column of activated solid support. The enzyme forms bound in the presence and absence of substrate were then studied in chemical modification and kinetic experiments. In our new approach we study the susceptibility of a bound enzyme to proteolysis in the presence and absence of substrates. Susceptibility to proteolysis is a useful property in the study of conformational changes of proteins (3). However, the conformation of the ES complex has not been studied because of the difficulty in maintaining saturating levels of substrate.

Rabbit muscle lactate dehydrogenase (LDH) was attached to CL-Sepharose via a spacer arm ($-NH(CH_2)_6NHCO(CH_2)_2SS(CH_2)_2CO-$) (4). The fixed enzyme was digested in the presence of substrates by subtilisin BPN. The use of a flow system permitted the maintenance of saturating levels of substrates. Proteolysis was followed by the loss of activity of the enzyme column. The time course of proteolysis in the presence of pyruvate or NADH alone did not differ from that of the control (Fig. 1). However, when both substrates were present, the enzyme became more susceptible to proteolysis: the rate of proteolysis increased by at least 40% with both substrates present. The rate in the presence of acetyl pyridine NADH(AP-NADH in Fig. 1) was retarded as expected

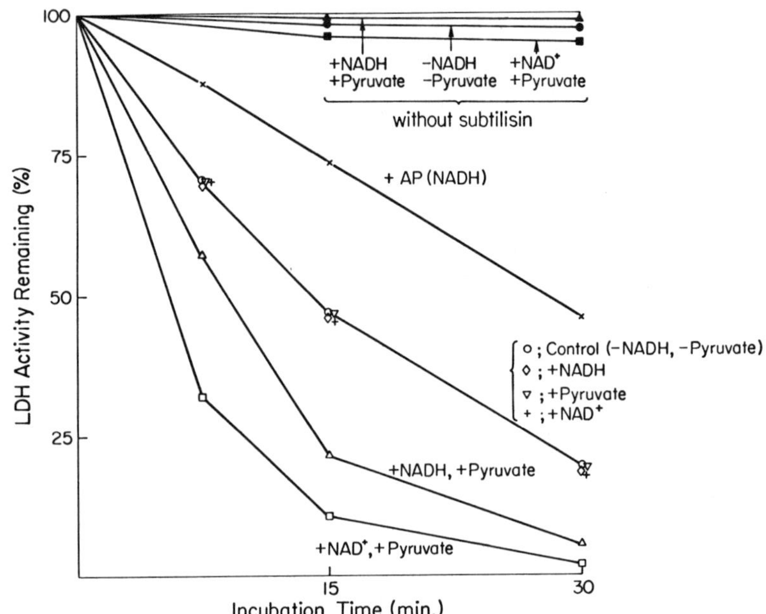

Fig. 1: Effect of Subtilisin Digestion on Immobilized Lactate Dehydrogenase Activity. (-) without, (+) with.

from solution studies (5). The abortive ternary complex was also more susceptible to proteolysis. These findings suggest that the productive ternary complex (pyruvate-NADH-LDH) is conformationally different from the free enzyme and binary complexes under optimal catalytic conditions. (Supported by grants from the National Institutes of Health (GM 19507) and the Pierce Chemical Co. G. P. R. is a recipient of a Public Health Service Research Career Development Award (1-K4-GM-00051).)

REFERENCES

1. ROYER, G.P. & UY, R. *J. Biol. Chem.* *248*:2627, 1973.
2. UY, R., LIU, V.S.H., & ROYER, G.P. *J. Solid Phase Biochem.* *1*:51, 1976.
3. CITRI, H. *Advances in Enzymology* *37*:397, 1973.
4. ROYER, G.P., IKEDA, S., & ASO, S. *FEBS Letters* *80*:89, 1977.
5. MCKAY, R.H. & KAPLAN, N.O. *Biochim. Biophys. Acta* *52*:156, 1961.

SYNTHESIS AND KINETIC PROPERTIES OF A NEW NAD^+ DERIVATIVE CARRYING A VINYL GROUP

M. Muramatsu, I. Urabe, Y. Yamada, and H. Okada

Department of Fermentation Technology
Osaka University
Osaka, Japan

A new NAD^+ derivative carrying a vinyl group, NAD^+-N^6-[N-(N-acryloyl-1-methoxycarbonyl-5-aminopentyl)-propioamide (M-NAD^+) was synthesized as is summarized in Fig. 1. The NAD^+ derivative was converted to a water soluble macromolecular NAD^+ derivative (P-NAD^+) by co-polymerization with acrylamide.

High coenzyme activities compared to free NAD^+ and NADH were obtained for M-NAD^+ (71-86%) and M-NADH (66-87%) with yeast alcohol dehydrogenase, horse liver alcohol dehydrogenase, rabbit muscle lactate dehydrogenase and pig heart malate dehydrogenase. Lower but substantial relative coenzyme activities were obtained for P-NAD^+ (18-33%) and P-NADH (28-60%) with yeast alcohol dehydrogenase, horse liver alcohol dehydrogenase and rabbit muscle malate dehydrogenase. But lactate dehydrogenase for pig heart failed to use P-NAD^+ and P-NADH as a cofactor.

Kinetic studies by using yeast alcohol dehydrogenase and lactate dehydrogenase were carried out, and the kinetic constants obtained are illustrated in Table 1. From these results the following conclusions were obtained: (a) the modification of NAD^+ at N^6 position of the adenine ring resulted in a decrease in the V_m value possibly due to configurational alterations of the binary or ternary coenzyme complexes of the enzymes resulting in reduced coenzyme activity, but the decrease in the activity was partially compensated by the decrease in K_a, K_b and K_{ia} values which were also caused by the modification; (b)

TABLE 1

KINETIC CONSTANTS FOR THE REACTION OF NAD^+, M-NAD^+ AND P-NAD^+ WITH YEAST ALCOHOL DEHYDROGENASE AND RABBIT MUSCLE LACTATE DEHYDROGENASE.

Enzyme	Coenzyme	V_m (μM/min)	K_a (mM)	K_b (mM)	K_{ia} (mM)
Alcohol dehydrogenase	NAD^+	14.3	0.5	29	0.6
	M-NAD^+	5.5	0.15	6	0.7
	P-NAD^+	2.6	0.6	19	2.0
Lactate dehydrogenase	NAD^+	8.3	0.25	11	0.7
	M-NAD^+	3.2	0.04	4	0.4

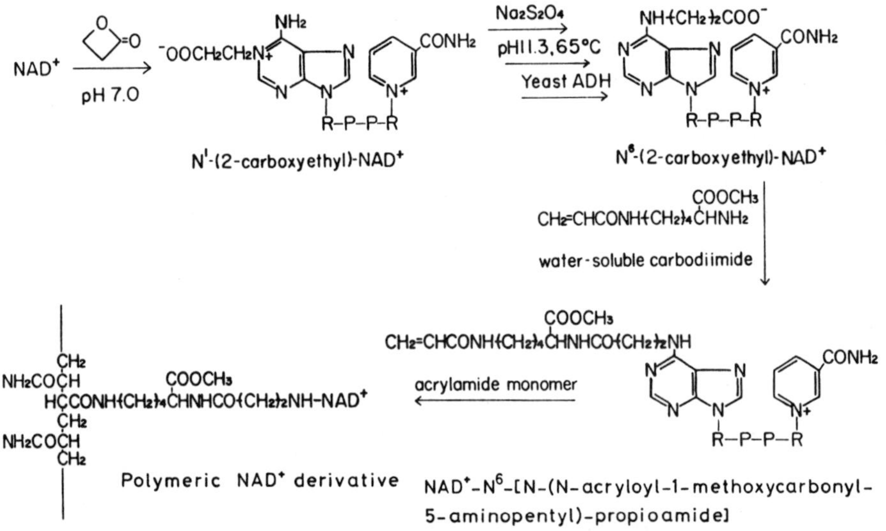

Fig. 1. Synthesis and polymerization of an NAD^+ derivative carrying a vinyl group. ADH=alcohol dehydrogenase R=ribose; P=phosphate

the polymerization of M-NAD^+ with acrylamide caused increases in K_a, K_b and K_{ia} values and a further decrease in the V_m value which decreased the coenzyme activity to below that of M-NAD^+.

ACKNOWLEDGMENT

This work was partly supported by Japanese Ministry of Education grant No. 247094.

ISOCYANIDE DERIVATIVES OF POLYSACCHARIDES AS SUPPORTS FOR ENZYME IMMOBILIZATION

A. Freeman, M. Sokolovsky and L. Goldstein

Department of Biochemistry
Tel Aviv University,
Tel Aviv, Israel

Polymers containing isocyanide functional groups offer versatile supports for the covalent fixation of proteins and other biologically active molecules *via* four-component condensation reactions involving amine, carboxyl, aldehyde and isocyanide, (1-6) as follows:

$$R^1\text{-COOH} + R^2\text{-NH}_2 + R^3\text{CHO} + \text{CH-}R^4 \longrightarrow R^1\text{CON}(R^2)\text{CH}(R^3)\text{CONHR}^4$$

In this reaction, carried out at neutral pH, the polymeric support supplies the isocyanide component, and the protein supplies either the amine or the carboxyl components. A water-soluble aldehyde, e.g. acetaldehyde, and the missing fourth component are added to the aqueous medium. The isocyanide functional groups can also be transformed into many other types of chemically reactive species which could be employed for covalent binding of proteins (2-4).

This communication describes the preparation of derivatives of polysaccharides and other hydroxylic polymers containing pendant isocyanide functional groups, attached to the polymer through stable ether bonds. The method is based on a two-step procedure involving (a) ionization of some of the hydroxyl groups on the polymer by treatment with a strong base (t-butoxide) in a polar aprotic solvent (dimethylsulfoxide) and (b) introduction of isocyanide side chains by nucleophilic attack of polymeric alkoxide ions on a low-molecular-weight isocyanide containing a good leaving group in the ω-position, e.g. 3-tosyl-1-isocyanopropane. The overall procedure and preparation

TABLE 1

PREPARATION AND CHARACTERIZATION OF ISOCYANIDE DERIVATIVES OF VARIOUS POLYSACCHARIDES

Polymer	Reaction Mixture[a] Tos$(CH_2)_3$NC (M)	Isocyanide Content[b,c] (equiv. x 10^6)/g polymer	Bound Trypsin Total (mg/g)	Bound Trypsin Active (mg/g)	Bound Trypsin % of Total
Cellulose powder Mesh 230-270 (56-63 μ)	0.05	31	11.6	6.4	55
Sephadex G-75	0.05	19	8.8	8.8	100
	0.10	30	15.1	10.5	70
Sephadex G-150	0.05	12	12.7	8.8	70
	0.10	33	16.5	11.0	66
Sephadex G-200	0.05	12	15.7	12.3	78
	0.10	15	15.4	11.6	75
CL-Sepharose 4B	0.05	68	61.6	24.3	40
	0.10	91	91.3	28.5	30

a All contained 500 x 10^{-6} mole t-butoxide/g polymer.
b Determined from amount of bound Gly-Leu-NH_2
c Cellulose powder polymer had 42 x 10^{-6} equiv./g polymer determined titrimetrically.

of the isocyanide reagent are summarized as:

$$\text{}\!\!\rangle\!\text{-OH} \xrightarrow[\text{DMSO}]{\text{t-ButO}^-} \text{}\!\!\rangle\!\text{-O}^- \xrightarrow[\text{DMSO}]{\text{TosO(Ch}_2)_3\text{NC}} \text{}\!\!\rangle\!\text{-O-(CH}_2)_3\text{NC}$$

$$HO(CH_2)_3NH_2 \xrightarrow{HCOOEt} HO(CH_2)_3NCHO \xrightarrow{TosCl} TosO(CH_2)_3NC$$

Table 1 summarizes data on the preparation and characterization of isocyanide derivatives of various polysaccharides and some of their protein conjugates. The procedure presented here can serve as an alternative to the cyanogen bromide method of activation.

REFERENCES

1. GOLDSTEIN, L., FREEMAN, A. & SOKOLOVSKY, M. in "Enzyme Engineering," vol. 2 (E.K. Pye. and L.B. Wingard eds.), Plenum Press, New York, 1974, p. 97.
2. GOLDSTEIN, L., FREEMAN, A. & SOKOLOVSKY, M. *Biochem. J.* *143*:497, 1974.
3. FREEMAN, A., SOKOLOVSKY, M. & GOLDSTEIN, L. *J. Solid-Phase Biochem.* *1*:261, 1977.
4. FREEMAN, A., GRANOT, R., SOKOLOVSKY, M. & GOLDSTEIN, L. *J. Solid-Phase Biochem.* *1*:275, 1977.
5. GOLDSTEIN, L., FREEMAN, A., BLASSBERGER, D., GRANOT, R. & SOKOLOVSKY, M. in "Proceedings Symposium on Biotechnological Application of Proteins and Enzymes," (Z. Bohak and N. Sharon, eds.), Academic Press, New York, in press.
6. BLASSBERGER, D., FREEMAN, A. & GOLDSTEIN, L. *Biotechnol. Bioeng.*, in press.

ENZYMES ON MAGNETIC SUPPORTS

P.J. Halling and P. Dunnill

Department of Chemical and Biochemical Engineering
University College London
London, U.K.

The potential advantages and practical problems in the use of non-porous magnetic supports for immobilized enzymes have been described (1). We now report progress in dealing with three problems.

Enzymes on the support surface lost activity fairly rapidly, a process associated with leaching of metal ions. Controlled roasting in air was found to oxidise a thin surface layer of the Ni support powder to give Ni-NiO, which showed greatly enhanced resistance to corrosion. Chymotrypsin immobilized to Ni-NiO showed a half-life of activity of 190 days on storage at 25°C, comparable to that with good conventional supports.

Chymotrypsin immobilized to Ni-NiO by adsorption and cross-linking was found to be rapidly desorbed in the presence of substrates. Covalent immobilization-via 3-aminopropyltriethoxy-silane (APS), using glutaraldehyde, $CSCl_2$, or CNBr to react with the amino group, gave only a limited improvement in the stability of the linkage to the surface. This could be accounted for by the rapid elution of APS from the surface in water. Heat treatment of Ni-NiO-APS stabilized the APS linkage somewhat, but not the resulting immobilized enzyme. However, after many APS links were coupled together by depositing a layer of poly-glutaraldehyde, enzyme could be stably attached by reaction with surface aldehyde groups. The product had a half life of 350 hr under operational conditions. The mechanism of action of APS in this system is discussed in terms of

Plueddemann's reversible hydrolysable model (2), developed in relation to composite materials.

Smaller support particles are desirable to give larger specific surface areas and hence higher specific enzyme activities. They should also be magnetically soft to discourage agglomeration in the reactor. A Mn-Zn soft ferrite powder with a particle size of 0.2-0.3 μ was found to be resistant to metal leaching and to support chymotrypsin with high storage stability. Using the APS and polyglutaraldehyde coupling method, immobilized enzyme with an operational half life of 230 hr was produced.

REFERENCES

1. MUNRO, P.A., DUNNILL, P. & LILLY, M.D. *Biotechnol. Bioeng.* *19*:101, 1977.
2. PLUEDDEMANN, E.P. *Adhesives Age* *18* (*6*):36, 1975.

OXIRANE-ACRYLIC BEADS, PREPARATION 2878-C

D.M. Kraemer, K. Lehmann, H. Pennewiss and H. Plainer

Research Laboratory
Roehm GmbH
Darmstadt, Federal Republic Germany

By a special Rohm process (1) oxirane-acrylic beads were prepared by copolymerization of glycidylmethacrylate and/or allylglycidylether, methacrylic amide and methyhylenene-bismethacrylamide. Due to the neutral nature of these components, the matrix of the beads is electroneutral, and so is the binding mechanism via the oxirane groups.

The beads are macroporous, as can be shown by electron microscopy; and they show a water regain of 2.5 ml/g of dry beads. This water regain (or swelling) is independent of pH (0.5 to 12.5) and ionic strength; the beads are morphologically and chemically stable at these conditions, even if exposed to them for several weeks. The mechanical stability upon stirring is excellent; and high flow rates are observed upon filtration or column-chromatographic processes.

The oxirane-group content is 1000 μmol/g dry beads, as determined by Axen's thiosulphate method (2).

The binding of proteins can be achieved at a pH-range from 7.0 to 9.0, which is optimal for enzyme stability. Approximately 100 mg of protein can be bound to 1000 mg dry beads, equaling about 30 mg of protein/ml of gel (wet beads). Recovery of enzyme activity upon binding is 60% with penicillin-amidase (*E. coli*); the product exhibits 260 U/g wet gel. Trypsin-Oxirane-Acrylic beads

(21 mg Trypsin/ml gel) show 212 I.U./ml with BAEE as a substrate (pH 8.0; 25°C) and 20 U/g with Casein (high molecular weight substrate, 37°C, pH 8.0). Similar good results are obtained with RNase, using RNA from yeast as substrate. The binding capacity for serum albumin, IgG and IgM is 100 mg/g dry gel.

The stability of the bound enzymes upon reuse proves to be better than with the respective Agarose-derivatives (CNBr-activated Agarose and epoxy-activated Agarose).

REFERENCES

1. British Patent No. 1,329,062 Rohm GmbH; German Offenlegungsschrift 2,237,316, Rohm GmbH; German Offenlegungsschrift 2,263,289, Rohm GmbH.
2. SUNDBERG, L. & PORATH, J., *J. Chromatog.* *90*:97, 1974.

IMMOBILIZATION OF 3β-HYDROXYSTEROID DEHYDROGENASE ISOLATED FROM *Streptomyces griseocarneus*

A. Szentirmai, G. Kerenyi, and M. Natonek

Department of Microbiology
Research Institute for Pharmaceutical Chemistry
Budapest, Hungary

A *Streptomyces griseocarneus* strain was selected from 600 actinomyces tested to transform 3β,17α,21-trihydroxy-5-pregnen-20-one-3,21-diacetate into 17α,21-dihydroxy-4-pregnen-3,20-dione (compound Reichstein-S). This strain gave not only the highest yield of compound Reichstein-S, but the steroid transforming activity was not reduced in the presence of water immiscible organic solvents. This biosynthetic method, designed to obtain compound Reichstein-S by nonproliferating bacterium cells, has been patented (1).

For the propagation of *S. griseocarneus* different kinds of nitrogen sources were suitable. They could be natural organics (soybean meal, corn steep liquor, meat extract), or inorganics (NH_4NO_3, $(NH_4)_2SO_4$); but the culture had no steroid transforming activity in the presence of solely inorganic nitrogen sources. Some transforming activity was found in the presence of natural organic nitrogen sources, but the highest 3β-hydroxysteroid dehydrogenase activity and steroid acylase activity were detected when the medium contained soya extract. *S. griseocarneus* living cells were able to metabolize steroid compounds, but the total amount of steroid compound was never decreased in the presence of water immiscible organic solvents. In the latter case only 3β-hydroxysteroid dehydrogenase, isomerase, and steroid acylase activity were detected; all of the other steroid transforming activities (NAD-linked dehydrogenase, oxygenase, etc.) were completely inhibited.

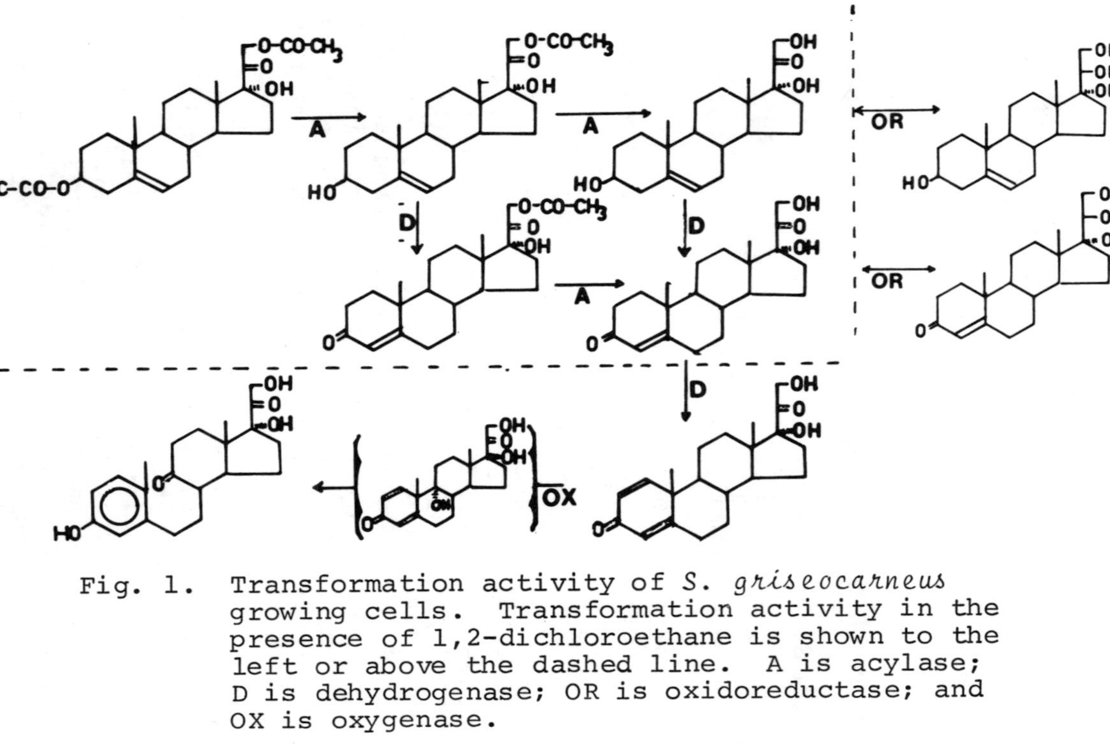

Fig. 1. Transformation activity of *S. griseocarneus* growing cells. Transformation activity in the presence of 1,2-dichloroethane is shown to the left or above the dashed line. A is acylase; D is dehydrogenase; OR is oxidoreductase; and OX is oxygenase.

3β-Hydroxysteroid dehydrogenase was purified from the crude extract of sonicated cells by a procedure involving ammonium sulfate fractionation, QAE-Sephadex A-50 chromatography and Sephadex G-100 gel filtration (2). The 3β-hydroxysteroid dehydrogenase activity and isomerase activity were inseparable. Probably only one bifunctional enzyme catalyzes the overall reaction, the dehydrogenization of the 3β-hydroxy group and the isomerization of the Δ^5-3-keto steroid. However the turnover of isomerase was 30 times greater than that of the dehydrogenase. The purified enzyme had no characteristic visible absorption spectrum. NAD,NADP,FAD,FMN never influenced the activity of this enzyme, which was able to oxidize 3β-hydroxysteroids, such as cholesterol, sitosterol, dehydroepiandrosterone, any 3β-hydroxy-5αH-androstane, or pregnane. However, the transforming ac tivity was completely inhibited by any kind of large substituent on the α-side of the steroid nucleus, e.g. 17α-methyl or 17α-acetoxy group. The reduced enzyme was reoxidized by molecular oxygen. The molecular weight was estimated at 36,000 by Sephadex gel filtration.

The purified enzyme was immobilized through the free aldehyde groups of freshly prepared m-phenylenediamine-glutardialdehyde polymer (3) in the presence of substrate dehydroepiandrosterone and Mg^{++}ions. The coupling efficiency was 52%. The transforming activity of the fixed enzyme decreased by about 15% after 8 experiments; but the cell bound enzyme activity decreased by 75% under the same conditions.

The chemistry is summarized in Fig. 1.

REFERENCES

1. HUNGARIAN PAT. No. 151580; 1962.
2. KERENYI, G. *Acta Microbiol. Acad. Sci. Hung.* 22:487, 1975.
3. GERMAN PAT. No. 2407840; 1973.

A METHOD TO PREPARE BEAD-SHAPED IMMOBILIZED ENZYME

Koji Kawashima

National Food Research Institute
Shiohama Kotoku
Tokyo, Japan

A new method to entrap enzymes in a polymer matrix is presented. In this method ionizing radiation was applied to polymerize a water soluble synthetic monomer and a polymer in the presence of an enzyme. More than twenty compounds were selected as the polymerizing reagent, including acrylic acid, several acrylates (Na-, K-, Ca-, Mg-), vinyl pyrrolidone, polyvinyl pyrrolidone, polyvinyl alcohol, acrylamide, and divinyl sulfone. The enzyme and polymerizing reagent were mixed and quickly frozen at -78°C and then irradiated at 50 to 500 Krad by gamma rays. Polymerization was conducted in air. By this method the immobilized enzyme was prepared in bead, membrane, bag, and tube form having high enzymic activities. Some were colored black or gray.

When the bead shaped immobilized enzyme is to be prepared, the mixture of enzyme and polymerizing substance is injected into a precooled solvent like hexane, toluene, xylene, ethyl ether, or butanol but not into methanol, ethanol, or acetone. The size of the bead was controlled by changing the injection speed (20μ to 10mm in diameter). The surface of the bead was not smooth but had numerous small hollows, which were formed when small ice crystals in the frozen polymer melted. The hollows existed from the surface to the inside of the bead, which might help explain the quick flow of substrate within the polymer.

Bead shaped immobilized glucose oxidase (GOD) and other enxymes (30-40 mesh size) were prepared and packed in a small column (1.3 mm diameter, 40-120 mm long) and placed in a Technicon Autoanalyzer II. The immobilized GOD worked more than twenty days, during which time over 10,000 glucose samples were analyzed.

Invertase was immobilized; and the retained activity was 92-18%, depending on the bead size (smaller beads had higher activity) and on the enzyme concentration. Immobilized invertase was more stable against heat treatment than was the native enzyme. No big difference in K_m between the immobilized and native invertase was noted. 30% of the sucrose solution was decomposed completely by the bead shaped immobilized invertase column for up to 47 days (space velocity 1 hr^{-1}, 50°C).

Some acrylates were suitable for the immobilization of enzymes requiring metal ions. For example, magnesium acrylate was favorable for the immobilization of glucose isomerase, NAD kinase, and pyruvate kinase. Microorganisms and multienzymes also were easily immobilized by this method.

The method has the advantages of (a) no enzyme inactivation due to chemical catalyzers; (b) little heat inactivation; (c) spongy immobilized enzyme preparations having large surface areas can be made; (d) being cheap, quick, and easy; (e) physically stable; and (f) killing immobilized microorganisms by the additional irradiation without seriously inactivating the enzyme.

So far invertase, glucose oxidase, glucose isomerase, catalase, d-aminoacid oxidase, acylase, fumarase, lactate dehydrogenase, pyruvate kinase, urease, lipase, NADH oxidase, peroxidase, NAD kinase, glucoamylase, and protease have been immobilized with high retention of enzyme activities.

IMMOBILIZATION OF HYDROGENASE FROM *ALCALIGENES EUTROPHUS* H 16 ON GLASS BEADS

P. Egerer, E. Schleicher and H. Simon

Technical University Munich
Munich, Federal Republic Germany

This hydrogenase, a flavo-enzyme, has interesting properties. It transfers electrons directly to NAD^+ and other artificial acceptors such as methylviologen or benzylviologen. Furthermore it is rather insensitive to oxygen (1). Therefore we immobilized the hydrogenase on a silanized CPG-10 carrier (48.6 nm, 120-200 mesh) by different coupling methods (2).

Experiments were carried out with a ten-fold enriched enzyme. Enzymatic activities were determined by spectrophotometric assays in anaerobic cuvettes measuring the rate of NADH production in H_2-saturated 0.05 M Tris-HCl buffer pH 7.8, 0.8 mM NAD^+. The results of a series of experiments are shown in the Table.

Miscellaneous experiments showed better results with glutaraldehyde coupling as well as much lower efficiency with CPG-10 (24.1 nm, 200-400 mech) in spite of the double surface area per unit weight. Using an enzyme preparation of maximal specific activity (54 U/mg protein) it should be possible to obtain an activity of about 800 U/g of CPG carrier. Hydrogenase from *Clostridium pasteurianum* was immobilized on glass by the same procedures with a yield not higher than 6% (3). Storage conditions under air over a two month period at 0°C showed nearly no loss of activity for the immobilized hydrogenase from *Alcaligenes* compared with a 50% loss for the free enzyme. At -16°C only 10 to 20% of the original activity of the immobilized as well as the free enzyme could be detected. Furthermore, the thermostability of the hydrogenase could be improved by a factor of 5-7 by means of the immobilization.

Schneider and Schlegel (1) found that this hydrogenase was rapidly and irreversibly deactivated by incubating the enzyme solution with NADH. Comparative deactivation studies with the immobilized as well as the free enzyme under different ratios of concentrations NAD^+/NADH revealed an increased stability of the glutaraldehyde-immobilized hydrogenase by a factor of about 7.

TABLE

RESULTS OF DIFFERENT COUPLING METHODS AND CONDITIONS FOR THE IMMOBILIZATION OF HYDROGENASE

Coupling method	Units of hydrogenase added/g carrier*	Activity of immobilized hydrogenase (%)**	Units of hydrogenase detected in the supernantant and wash water
Aminoalkyl carrier and water soluble carbodiimide	205	42.2	1.2
	410	30.0	0.5
	513	41.7	0.3
	410	43.4***	0.3
Succinyl carrier and water soluble carbodiimide	200	32.5	3.0
	410	9.0	68.0
	500	9.3	120.0
Glutaraldehyde	213	89.0	0.3
	411	50.0	9.0
	521	41.0	50.0
Diazotization	202	26.0	--

*1 unit = 1 µmole NADH/minute; 1 mg protein ≅ 8 units.
**All the immobilized hydrogenase samples were stored in 0.05 M Tris-HCl, pH 7.8.
***Double amount of water soluble carbodiimide.

The apparent K_M for NAD^+ was 1.3 mM for the glutaraldehyde-immobilized hydrognase, i.e about 2 times higher than that for the free enzyme. The immobilized enzyme was successfully used for stereospecific hydrogenations with catalytic amounts of NAD^+.

REFERENCES

1. SCHNEIDER, K. & SCHLEGEL, H.G. *Biochim. Biophys. Acta* *452*:66, 1976.
2. WEETALL, H.H. & FILBERT, A.M., *Meth. in Enzymology* *34*:59, 1972.
3. LAPPI, D.A., STOLZENBACH, F.E., KAPLAN, N.O. & KAMEN, M.D. *Biochem. Biophys. Res. Comm.* *69*:878, 1976.

HEMISYNTHESIS OF AMINOGLYCOSIDES

F. Le Goffic, J.F. Le Bigot, S. Sicsic, and C. Vincent

Centre d'Etudes et de Recherches de Chimie
Organique Appliquee, CNRS
Thiais, France

It is well known (1) that one of the processes of inactivation of aminoglycosides by some bacteria is a structural modification of these antibiotics, e.g. N-acetylation, o-adenylylation, o-phosphorylation. The study of the structure activity relationship requires the preparation of specifically modified aminoglycosides at one of the amino or hydroxyl functions. The only way to obtain such specificity is by enzymatic reactions. We therefore used enzymes isolated from resisting bacterial strains, performing specific modification of aminoglycosides. We thus modified Gentamicin, Sisomicin and Tobramycin.

A cofactor is needed for reactions catalyzed by N-acetyl transferase (AAC 3, AAC 2') or 0-adenylyl transferase (0-4'-ANT). These sections as exemplified as follows:

$$RNH_2 + CoA\text{-}S\text{-}COCH_3 \xrightarrow{AAC} RNH\text{-}COCH_3 + CoA$$

$$ROH + ATP \xrightarrow[0\text{-}4'\text{-}ANT]{} R\text{-}OAMP + PPi$$

In order to allow the production of large amounts of modified aminoglycosides, we immobilized the enzymes, which are otherwise very unstable. Our attempts to immobilize the purified enzymes failed. Hence we immobilized crude extracts including very low amounts of our enzymes.

Among all our assays, only reticulation with glutaraldehyde (2) gave good results. The AAC was immobilized using 1 ml of crude enzymatic extract, 300 mg of DEAE cellulose, and 7 μl of glutaraldehyde (25% in water). The immobilization of 0-4'-ANT was done using 1 ml of crude enzymatic extract, 100 mg of T500 Dextran, and 1.4 μl of glutaraldehyde (25% in water). In both cases we obtained resins with good activities and increased enzyme stability

TABLE 1

ENZYME STABILITIES AT 4°C

Enzyme	Soluble	Immobilized
AAC 3	1 month	4 months
AAC 2'	15 days	3 months
0-4'-ANT	5 days	50 days

The N-acetylaminoglycosides were produced in packed bed reactors 1.6 cm x 17 cm. From 0.256 mmole of AcCoA and 0.5 mmole of Gentamicin C, 220 mmole of N_3-acetyl gentamicin was produced at 90% yield. The AcCoA was regenerated chemically (3), using polyvinyl-4 pyridine/ divinyl benzene 2% as catalyst. 0-4'-adenylylaminoglycoside was produced in an ultrafiltration cell; to 75 ml of buffered solution of enzymatic dextran were added 0.035 mmole of Tobramycin and 0.215 mmole of ATP. After 30 min, the yield reached 100%.

REFERENCES

1. PRICE, K.E., GODFREY, J.C. & KAWAGUCHI, H. in "Advances in Applied Microbiology," vol. 18 (W.W. Umbreit, ed.), Academic Press, New York, 1974, p. 191.
2. BROUN, G., THOMAS, D., GELFF, G., DOMURADO, D., BERJONNEAU, A.M. & GUILLON, C. *Biotechnol. Bioeng.* *15*:359, 1973.
3. LE GOFFIC, F., SICSIC, S. & VINCENT, C. *Tetrahedron Ltrs.* *33*:2845, 1976.

Session IV
NEW MEDICAL APPLICATIONS FOR ENZYMES

Chairmen: G. Brunner and T. Chang

ENZYME-POLYETHYLENE GLYCOL ADDUCTS:

MODIFIED ENZYMES WITH UNIQUE PROPERTIES

F. F. Davis, A. Abuchowski, T. van Es, N.C. Palczuk, R. Chen, K. Savoca and K. Wieder

Departments of Biochemistry and Zoology and
Bureau of Biological Research
Rutgers University
New Brunswick, New Jersey, USA

There are many clinical disorders which, in theory might be treated by the use of appropriate enzymes. However, there are serious problems associated with enzyme therapy. First, inadequate amounts of human enzymes are available; and most foreign enzymes are immunogenic in humans. They can be administered only a few times. Second, the circulating life in the blood may be very short even on the first injection (often a matter of a few minutes). And third, the cost of highly purified enzymes may be too great to allow general use.

We concluded that the immunological problem would first have to be dealt with if the potential for large-scale therapeutic use of enzymes were to be realized. We directed our activities to the development of a general method for rendering enzymes from any source nonimmunogenic. Our approach has been to attach nonimmunogenic soluble polymers to enzymes in an effort to minimize or eliminate their immunogenicity.

The polymer selected was polyethylene glycol (PEG). This polymer is linear, hydrophilic by virtue of hydrogen-bonded water at the ether oxygens, uncharged, flexible, and available in a variety of molecular weights. We use the monomethoxylated form which has only one end free for activation and coupling. PEG with a molecular weight of 5,000 daltons is commonly used, though PEG of 1900 daltons also is effective. The PEG is activated for coupling by reaction with cyanuric chloride and is then covalently attached to enzymes (Fig. 1).

To test the concept, we first attached PEG of 1900 daltons to bovine serum albumin (BSA), a monomeric protein. The molar mass of BSA is 66,210 daltons. Attachment of PEG to 56 of the available amine groups yielded an adduct of about 174,000 daltons. The PEG-BSA was shown to be incapable of causing antibody formation in rabbits, and to have extended blood circulating life. The PEG-BSA was not affected by the presence in the blood of antibodies to BSA (1).

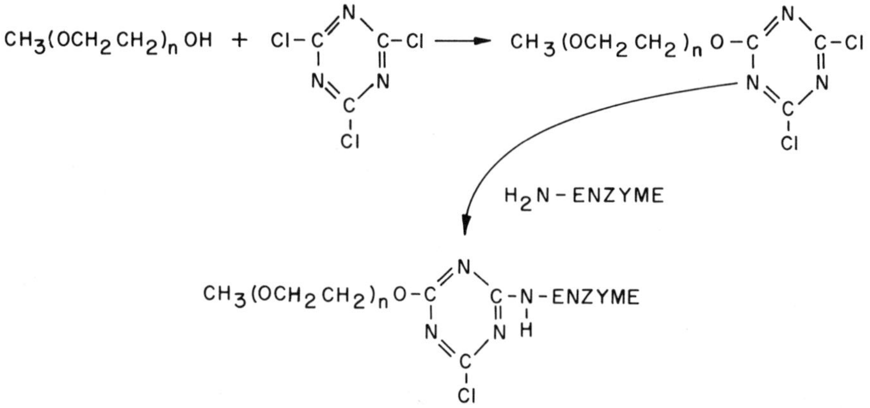

$CH_3(OCH_2CH_2)_nOH$ + cyanuric chloride (Cl–C₃N₃–Cl, Cl) ⟶ $CH_3(OCH_2CH_2)_nO$–C₃N₃(Cl)₂

H_2N-ENZYME

⟶ $CH_3(OCH_2CH_2)_nO$–C₃N₃(Cl)–N(H)-ENZYME

Fig. 1. Activation of monomethoxypolyethylene glycol and coupling to an enzyme.

PEG adducts were prepared for four tetrameric enzymes: bovine liver catalase (2), bovine liver arginase, and uricase from hog liver and *Candida utilis*. Table 1 shows the extent of modification and activity of each of the enzymes.

These PEG-enzymes were nonimmunogenic in the rabbit by intramuscular injection with complete Freund's adjuvant, followed by Ouchterlony (3) and complement fixation (4) tests. Their blood circulating lives in mice are shown in Figs. 2 and 3. Each enzyme was injected intravenously three times a week on a 2-, 2-, and 3-day schedule into 20 mice. Immediately after the first injection, the 13th injection (4 weeks), the 26th injection (8 weeks), and the 39th injection (12 weeks), the level of enzyme in the blood was monitored over a period of 48 or 72 hr.

TABLE 1

PROPERTIES OF PEG-ENZYMES

Enzyme	M.W.	PEG attached (daltons)	Enzymic Activity (% of unmodified enzyme)
Catalase	242,000	220,000	95
Arginase	120,000	245,000	65
Hog liver uricase	128,000	141,000	29
C utilis uricase	128,000	216,000	23

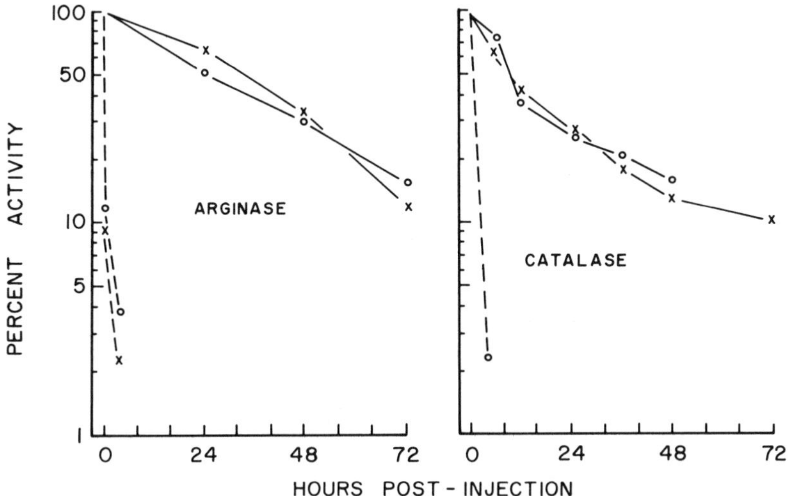

Fig. 2. Blood circulating life in the mouse of arginase and catalase. (0) first injection, (X) 39th injection; solid lines PEG-enzymes, dashed lines unmodified enzymes.

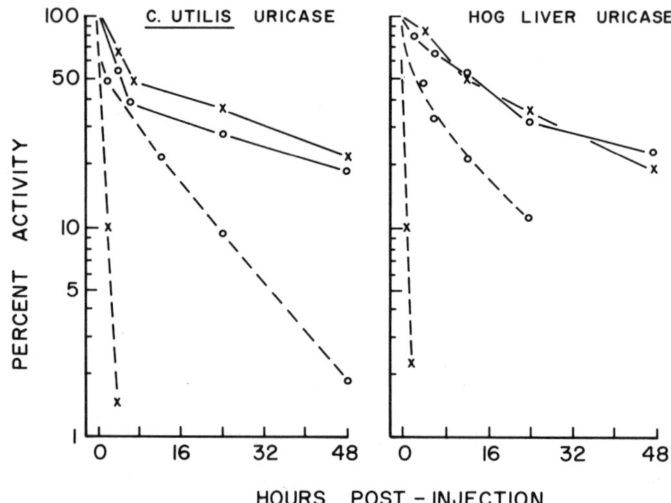

Fig. 3. Blood circulating life in the mouse of uricase from *C. utilis* and hog liver. (0) first injection, (X) 39th injection; solid lines PEG-uricases, dashed lines unmodified uricases.

As shown in Fig. 2, the circulating lives of both PEG-arginase and PEG-catalase were essentially identical following the first and 39th injections (the circulating lives after the 13th and 26th injections were also similar to the first-day injection). The unmodified enzymes, in contrast, were cleared rapidly from the blood even on the first injection, and were cleared even more rapidly when, after a week or so, antibodies formed against them.

Successful modification of an enzyme from different sources is shown in Fig. 3. PEG-uricase preparations from hog liver and *C. utilis* showed similar patterns of blood circulating life after repetitive injections. The slightly higher level of the *C. utilis* enzyme after the 39th injection may have been due to long-lived residual activity from previous injections.

The ability to carry out successful modification of a given enzyme from diverse sources, if a general phenomenon, will be of importance in effective enzyme therapy. Enzymes can be selected for therapeutic use on the basis of factors such as cost, and desirable catalytic properties such as low K_m, best pH optimum, and stability. It should be pointed out also that contaminating proteins as well as the enzyme are rendered nonimmunogenic by PEG attachment. Thus, reasonable amounts of contaminating protein can be tolerated in the enzyme preparations. The successful repetitive injection of oligomeric enzymes indicates that in general, exposure of previously hidden antigenic determinants by dissociation of subunits in the blood will not be a problem.

OTHER PROPERTIES OF PEG-ENZYMES

PEG-enzymes show interesting physical properties. They are soluble over an extended range of pH, are not held on ion-exchange resins, and move poorly in gel electrophoresis. In the ultracentrifuge, PEG-enzymes sediment more slowly than unmodified enzymes; although the adducts are much larger. These observations are consistent with the picture of an enzyme surrounded by a shell of PEG and its bound water, which causes profound changes in the hydrodynamic and other properties of the molecule.

K_m values were increased slightly in the PEG-enzymes we examined. For example, the K_m of uricase from hog liver was increased from 2.0×10^{-5} to 7.5×10^{-5} and *C. utilis* uricase from 4.8×10^{-5} to 5.6×10^{-5} by PEG attachment. V_{max} was decreased for PEG-enzymes with the possible exception of catalase. Large substrates were acted upon more slowly than small substrates. PEG-trypsin, for example, was essentially inactive against proteins, but fully active against small ester substrates. PEG-enzymes resisted protease digestion (2).

REFERENCES

1. ABUCHOWSKI, A., VAN ES, T., PALCZUK, N.C., & DAVIS, F.F. *J. Biol. Chem.* *252*:3578, 1977.
2. ABUCHOWSKI, A., MCCOY, J.R., PALCZUK, N.C., VAN ES, T., & DAVIS, F.F. *J. Biol. Chem.* *252*:3582, 1977.
3. OUCHTERLONY, O. in "Handbook for Experimental Immunology," (D.M. Weir, ed.), F.A. Davis Co., Philadelphia, 1967, p. 655.
4. DRESSER, D.W. & WORTIS, H.H. in "Handbook of Experimental Immunology," (D.M. Weir, ed.), F.A. Davis Co., Philadelphia, 1967, p. 1054.

CARRIER-BOUND DETOXIFYING ENZYMES FOR AN EXTRACORPOREAL DETOXIFICATION OF ENDOGENOUS AND EXOGENOUS TOXINS

G. Brunner

Department of Gastroenterology and Metabolic Diseases
University of Gottingen
Gottingen, Federation Republic Germany

Enzymes for many years have had a firm place in medicine for diagnostic purposes. Enzymes have been routinely measured in blood, urine, stool, and tissue samples. For therapeutic use only one group of enzymes has been developed for safe and efficient application: enzymes of digestion. Lipases, peptidases, and amylases applied orally could correct digestive insufficiency and have been given without side effects for many years.

One aim of scientists is to replace enzymatic function in cases of partial or total inborn or acquired lack of enzymes in blood and tissues, on either a temporary or a permanent basis. The great difficulty here lies in the immune reactions of the body to foreign proteins brought into contact with blood or tissues. Therefore, enzymes have to be applied in a state which makes immunoreactions impossible. Different approaches have been started by a number of investigators. Chang and Gregoriades (1,2) have encapsulated enzymes into lipid or polymer materials. Marconi (3) has entrapped enzymes in fibers; and Davis has attached enzymes to polyethylene glycol (4).

Our group has started yet another approach; enzymes of detoxification were covalently bound to the surface of solid carriers allowing direct contact of the enzyme with the substrates in the blood (5). One enzyme preparation has been developed for safe and efficient application in an extracorporeal device. UDP-Glucuronyltransferase was prepared in sufficient quantity as a carrier bound enzyme to eliminate completely phenolic substances occurring in severe liver disease within 4 hours.

The enzyme activity, and thus the capacity of such a device in serum or blood, can be predicted and calculated by the following equation:

$$V_{serum} = V_o \cdot f(T, pH, \mu([M_i^+], [X_j^-]), [Mg^{2+}], [UDPGA], [S])$$

In this equation the activity of the enzyme is shown as a function of temperature, pH, ionic strength, Mg concentration, cofactor UDPGA concentration, and the concentration of the toxins. In humans even in severe illness the temperature, pH, ionic strength, and Mg concentration are fairly constant. Thus, the enzyme activity is dependent only on the concentration of toxin and cofactor.

However glucuronidation alone is not sufficient for detoxification of endogenous intoxication, like liver failure. Initial steps of enzymatic hydroxylation using carrier bound cytochrome P-450 and NADPH-cytochrome P-450 reductase have already been reported (6). Stability and coenzyme combination have been improved but are still quite far from efficient application.

A new enzyme has been investigated for fixation and detoxification: glutathione transferase (ligandin). This is a soluble enzyme which has a double function. It is a carrier protein for many toxic substances, having a higher binding constant in most cases than albumin. It also catalyzes the binding of glutathione to toxins. The advantages of this enzyme are threefold: (a) the enzyme is not bound to cell particles and thus does not have to be detached from organelles; (b) the enzyme in accordance with its importance is present in large amounts in the cytosole of liver and kidney, consisting in up to 8% of liver protein; and (c) the necessary cofactor glutathione is very cheap and easy to prepare.

The combination of enzymes for hydroxylation, glucuronidation, and gluthathion binding will cover the detoxification of a large range of materials. As soon as all three of these enzymatic systems have been immobilized on the same large scale as with glucuronyltransferase, then applications in clinical medicine should be possible. The large scale preparation, combination, and detoxification with the following enzymes is being persued: UDP-glucuronyltransferase, cytochrome P-450 reductase, and gluthathion transferase.

REFERENCES

1. CHANG, T.M.S. & MALAVE, N. *Trans. Amer. Soc. Artif. Int. Organs* *14*:141, 1970.
2. GREGORIADES, G. *New Engl. J. Med.* *295*:704, 1976.
3. MARCONI, W., this volume.
4. DAVIS, F.F., ABUCHOWSKI, A. VAN ES, T., PALCZUK, C. CHEN, R., SAVOCA, K & WIEDER, K. this volume.
5. BRUNNER, G. & LOSGEN, H. in "Enzyme Engineering," vol. 3 (E.K. Pye & H.H. Weetall, eds.) Plenum Press, New York, 1978, in press.
6. BRUNNER, G. & LOSGEN, H. in "Artificial Organs," McMillan, London, 1977, p. 388.

BIOMEDICAL APPLICATIONS OF ENZYMATIC FIBRES

W. Marconi

Snamprogetti S.p.A.
Monterotondo (Roma), Italy

Mono or polyenzymatic systems immobilized in cellulose triacetate fibers were investigated in our laboratories for applications in the biomedical field. These systems were investigated for analytical uses, particularly for the correction of pathological situations that require the continuous monitoring of blood glucose. Another potentially important application is the elimination of toxic metabolites in some pathological situations, such as liver coma. To attain this it was necessary to develop biocompatible fibers; this was achieved through physically blending a suitable platelet antiaggregant to the polymeric matrix before spinning the fibers. These biocompatible fibers then were used with success for the elimination of ammonia and aromatic aminoacids from blood serum. A further improvement of this technique was the chemical binding of the employed platelet antiaggregant to various polymeric surfaces.

USE OF FIBRE-ENTRAPPED ENZYMES IN BIOMEDICAL ANALYSES

We have previously reported the use of fiber-entrapped enzymes in biomedical analyses, for both manual and automatic analyzers (1-2). We have experimented with an *in vitro* system for continuous monitoring of glucose under conditions similar to ones occurring *in vivo*. The apparatus consists of a reference thermistor inside a glass cell in order to compensate for possible temperature fluctuations of the flowing solution and a measuring thermistor with glucose oxidase and catalase fiber wrapped around it. The glucose solution flows continuously through the glass cell immersed in a thermostatic bath at 25°C. Two pumps, one for the buffer and the other for concentrated glucose solution, make it possible to vary the glucose concentration flowing through the cell. The thermistors are connected to a Wheatstone bridge. The amplified unbalance potential depends on the temperature difference between the thermistors, and therefore on the glucose concentration. The standard curve is linear up to 1 g/l of glucose concentration (Fig. 1).

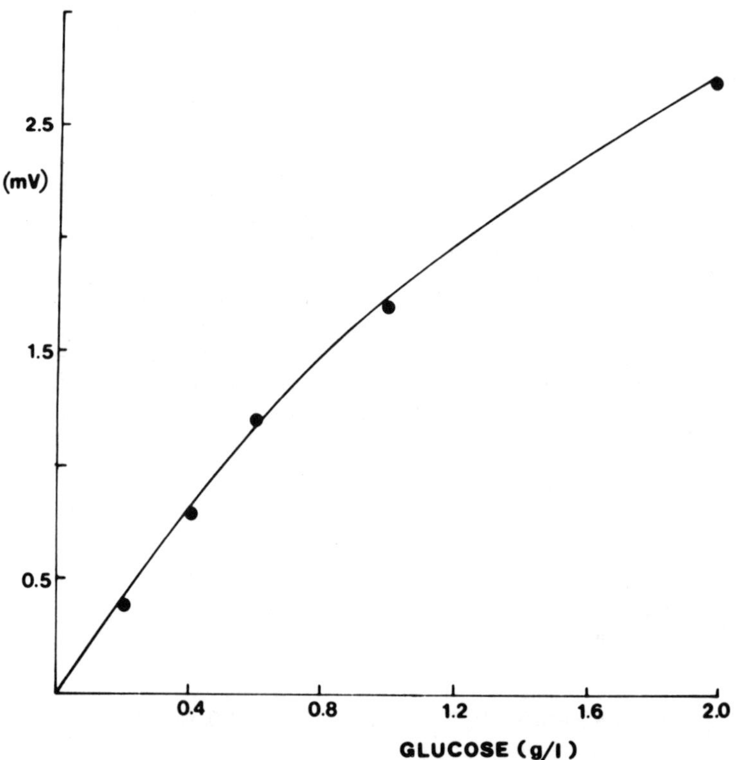

Fig. 1. Standard curve for the determination of glucose.

PHENYLALANINE AMMONIA LYASE

Phenylalanine ammonia lyase catalyses the transformation of phenylalanine into cinnamic acid. This enzyme can be a useful tool in the treatment of liver coma since phenylalanine is a very dangerous poison. The same enzyme also catalyzes the tyrosine transformation into coumaric acid.

This enzyme was extracted from *Rhodotorula glutinis* by sonification followed by protamine, ammonium sulfate, and sodium citrate fractionation to give a specific activity of 1.7 units/mg protein. The activity unit was defined as the amount of enzyme that produced 1 μmole of cinnamate per minute at 30°C, when L-phenylalanine was the substrate. The assay was done spectrophotometrically at 290 nm (cinnamate 10.6 x 10^3 and coumarate 7.1 x 10^3), as well as detected with an aminoacid analyzer. Fig. 2 shows the operational stability of phenylalanine ammonia lyase entrapped in fibers containing 4-5, diphenyl-2 bis (2-hydroxyethyl) amino oxazole (ditazol) as platelet antiaggregant agent. Fig. 3 shows the stability of the same

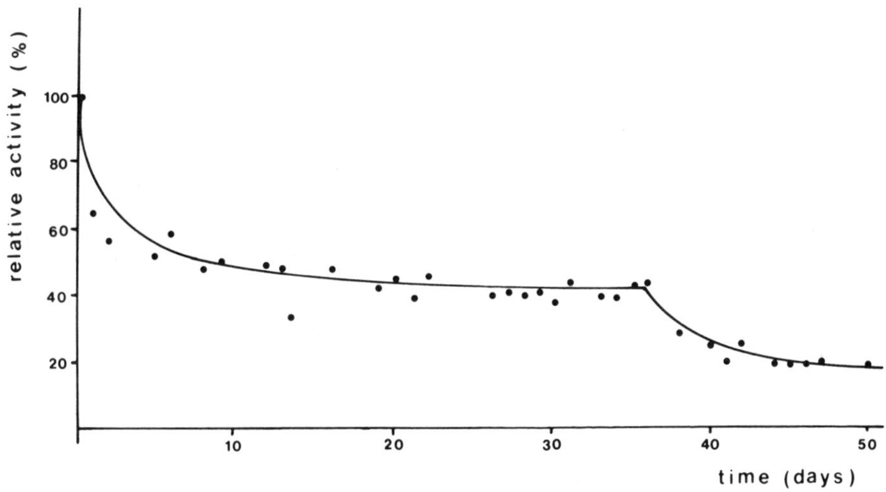

Fig. 2. Operational stability of phenylalanine ammonia lyase in 0.05 M phosphate buffer, pH 7.5, containing 10 μM phenylalanine.

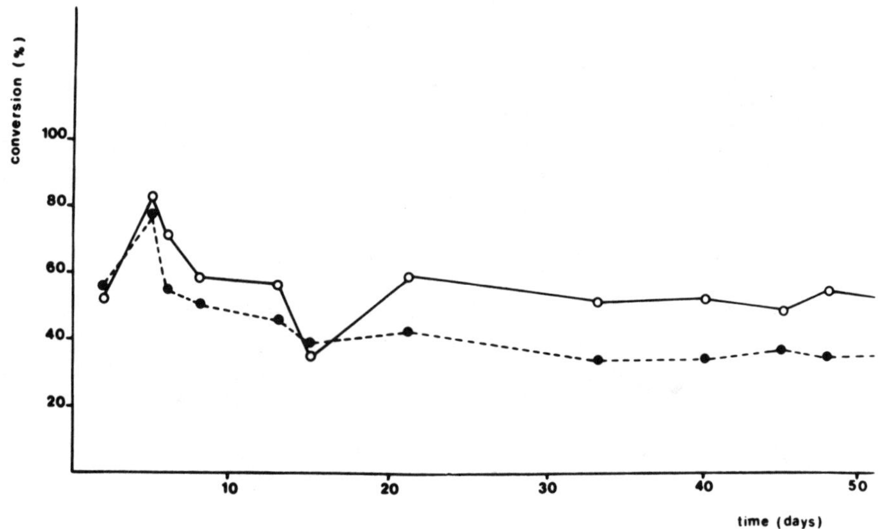

Fig. 3. Operating stability of fiber entrapped phenylalanine ammonia lyase on citrate treated blood. The conversion of phenylalanine (0——0) and thyrosine (●- - -●) after 60 min of reaction with fresh blood is plotted versus time. The blood was substituted when the substrates were exhausted.

fiber under operating conditions of 37°C, 25 ml of blood, and concentrations of substrates increased to 0.7 mM (10 times the physiological level. The results up to now seem to open promising perspectives for *in vivo* applications.

POLYENZYMATIC SYSTEM: ELIMINATION OF AMMONIA FROM BLOOD

Polyenzymatic thromboresistant fibers were prepared according to a previously published procedure (3). The enzymes entrapped in the fibers were alanine-dehydrogenase (10 mg/g CTA), lactate-dehydrogenase (30 mg/g CTA), and NAD^+ macromolecularized binding NAD^+ to a polyethyleneimine (PEI-NAD^+ 18 mg/g CTA). The overall cycle is shown in Fig. 4; L-alanine is produced, whereas the concentration of lactic acid and ammonia are decreased.

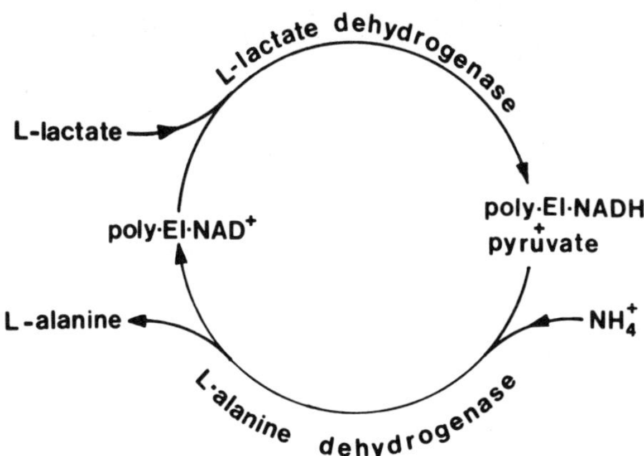

Fig. 4. Production of L-alanine from L-lactate and ammonia by L-alanine dehydrogenase, L-lactate dehydrogenase and poly-EI-NAD^+ entrapped in fiber.

Under experimental conditions (TRIS buffer 0.1 M pH 7; Li lactate 0.24 M; NH_4Cl 0.24 M; EDTA 1 mM) the enzyme fiber catalyzes the production of 16 μmoles/min/g dry fiber, corresponding to about 16 mg of removed NH_3/ hr/g dry fiber.

One gram of polyenzymatic fiber was inserted in an extracorporeal by-pass silicone tube (height 500 mm; internal diameter 7 mm). The fibers were fixed at the ends of the tube. The silicone tube was connected to nylon tubes, which were made biocompatible by chemical binding of ditazol on the inner surfaces. The silicone tubes had been previously impregnated with ditazol by evaporating an organic solution through their walls. A peristaltic pump was inserted in the circuit, and the system connected to the femoral artery and femoral vein of the medium size dog with the blood flow kept at about 4.6 ml/sec.

The ammonia level in the blood was increased by continuous perfusion of a physiological solution of ammonium chloride (20 g/l) into the femoral vein at a rate of 50 mg of NH_3/hr/kg body weight. Sampling of the blood was made at the inlet and outlet of the reactor; and the ammonia content was measured. The results for a typical experiment are shown in Table 1. The data demonstrate that 1 g of enzyme fibers, under conditions in which ammonia is not produced, should be able to reduce to normal values within 16 hours an ammonia content of 300 μg/100 ml in the human body (the total volume of fluid containing NH_3 has been considered 40 l).

TABLE 1

EXTRACORPOREAL BY PASS*

Time (min)	Artery (NH_4^+ μg/100 ml)	Vein (NH_4^+ μg/100 ml)	NH_4^+ Removal (mg/hr/g dry fiber)
30	165	145	3,0
60	145	104	6,2
120	230	182	7,2
150	152	104	7,2
180	165	145	3,0
240	152	98	10,6
270	258	**	-

*NH_4Cl perfusion 50 mg/h x Kg body weight; flow rate 4.2 ml/sec; dog weight 15 Kg.
**reactor has been removed. Blood content of ammonia in artery goes up to 258 μg/100 ml.

BIOCOMPATIBLE THROMBORESISTANT MATERIALS

We have investigated the possibility of containing biocompatible thromboresistant polymers by binding a platelet antiaggregant agent (for instance ditazol) onto polymeric surfaces of partially hydrolyzed nylon materials, thus containing free $-NH_2$ groups. Ditazol, which contains two hydroxyethyl groups, was reacted with phosgene to give the highly reactive bischloroformate derivative (Fig. 5). The reactive chloroformate groups can be used to form a covalent bond onto surfaces having, for instance, $-NH_2$ groups (Fig. 6). In order to obtain such

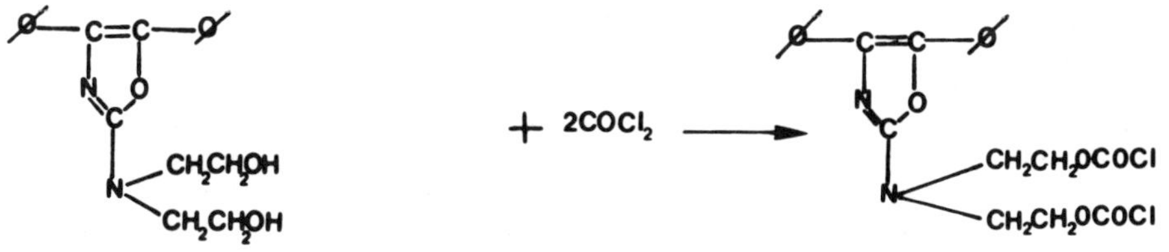

Fig. 5. Preparation of ditazol bischloroformate.

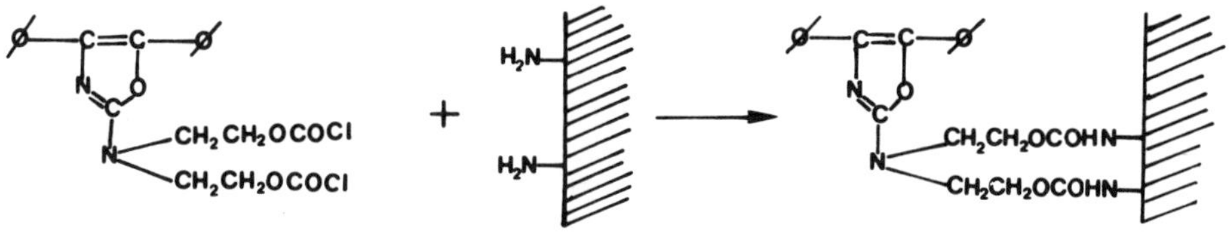

Fig. 6. Possible reaction scheme of ditazol bischloroformate with hydrolyzed nylon surfaces.

a surface samples of commercial nylon tubes and wires have been partially hydrolyzed with 3 N HCl and treated with ditazol bischloroformate (5% w/w) in dry acetone.

The ditazol content of these polymeric materials was determined by dissolving 350 mg of nylon in 6 N HCl and measuring the absorbance of the solution at 290 nm. As reference, a solution prepared by dissolving the same amount of treated nylon in 6 N HCl was used. A standard curve was prepared by dissolving known amounts of ditazol bischloroformate in 6 N HCl, containing nylon (60-70 mg/ 50 ml) and measuring the absorbance of the solution at 290 nm. It is remarkable that the UV spectra of ditazol and ditazol bischloroformate are practically coincident. The ditazol content was about 0.2%.

To test the biocompatibility intravenous catheters were prepared by winding a 0.25 mm nylon wire, treated with ditazol bischloroformate, around a fluorinated ethylene-propylene copolymer catheter, previously coated with a film of cellulose triacetate and ditazol. The nylon wire was uniformly wound around the catheter so that the probe was covered along its full length. A reference probe was made by winding a sample of untreated nylon wire around a similar catheter, covered with a film of cellulose triacetate and ditazol. The probes were inserted into the femoral veins of medium size dogs. During and after the operation the animal was heparinized. In an early experiment two probes were inserted for two weeks. When extracted the reference probe was covered with numerous thrombi, whereas no thrombi could be observed on the probe with the nylon wire covered with ditazol. These results seem to be important since they demonstrate that it is possible to obtain a biocompatible thromboresistant polymer by covalent binding of a platelet antiaggregant agent on its surface.

REFERENCES

1. MARCONI, W., BARTOLI, F., GULINELLI, S. & MORISI, F. *Process Biochem.* May, 1974.
2. CORDOVA, C., BARTOLI, F. & MORISI, F. *Quaderni SCLAVO*, in press.
3. BARTOLI, F., GIOVENCO, S., LOSTIA, O., MARCONI, W., MORISI, F., PITTALIS, F., PROSPERI, G. & SPOTORNO, G. *Pharmac. Res. Comm.* 9:521, 1977.

LIPOSOMES AS CARRIERS OF PROTEINS: POSSIBLE MEDICAL APPLICATIONS

Gregory Gregoriadis

Division of Clinical Investigation
Clinical Research Center
Harrow, UK

The use of proteins in therapeutic and preventive medicine is often associated with a number of difficulties. In inherited metabolic disorders, for instance, replacement therapy with enzymes foreign to the body can be immunogenic and promote hypersensitivity reactions. In addition, access of such enzymes to the target is often poor either because of inactivation by extracellular or intracellular agents or because of target inaccessibility. Some enzymes, such as glycohydrolases potentially useful in the treatment of lysosomal storage diseases, have been shown to act on non-target substrates located on cell surfaces or circulating glycoproteins, thus upsetting normal metabolism (1). In other instances administration of a protein through a particular route which would have been beneficial to the patient, is impossible because of the protein's properties. Thus, oral treatment of diabetes and of other peptide hormone deficiencies is prevented by the hormone's vulnerability in the gut.

Research efforts from this laboratory in the course of several years have produced considerable evidence that oral or parenteral administration of proteins and other agents in a variety of situations in pharmacology and medicine could be possible via a universal biodegradable carrier, namely the liposome (2). The suitability of liposomes (phospholipid vesicles) as carriers of active agents derives from the fact that phospholipids rearrange themselves in the presence of water to form one or more closed concentric spherical bilayers so that there is unrestricted entry of water and solutes. In this way substances are entrapped in the aqueous spaces formed. It is possible to modify the liposomal structure to promote stability or increase the volume of aqueous spaces (3, 4).

FATE OF ADMINISTERED LIPOSOME-ENTRAPPED PROTEINS

Liposomes remain stable in the circulation; and are eventually taken up by the reticuloendothelial system, namely the liver and spleen, probably as a result of their association with plasma α_2-macroglobulin (5). In addition to early uptake by the Kupffer cells, there is participation of the hepatic parenchymal and endothelial cells at a later stage. A small but measurable proportion of injected liposomes and their contents also reach lung, kidney, and brain tissues. Blockade of the liver and spleen with a population of excess empty liposomes can delay the uptake of a concurrently injected active liposomal population, prolong its circulation in blood, and allow its subsequent transcapillary passage to other areas. Following intraperitoneal injection liposomes, regardless of size, enter the circulation and reach the liver and other tissues. Even after intragastric administration, liposomes are capable of introducing insulin and glucose oxidase into the circulation.

Liposomes and their contents also are endocytosed and eventually localized in the lysosomal apparatus of cells (6). An additional recognized mechanism of cell-liposome interaction is fusion (7) of the respective membranes. Liposomes also can be made with IgG on their surface. With IgG against normal and malignant cells or desialylated fetuin, which has a specific affinity for the hepatic parenchymal cells, the liposomes were capable of selective association with the corresponding target cells *in vitro* (IgG) and *in vivo* (desialylated fetuin) (8). Application of the IgG liposome system *in vivo*, however, was presented with difficulties (9); although there was a modest improvement in the uptake of the liposomal drug by the target tumor, there was also augmented participation of the liver and spleen. The fact that the IgG used was raised against whole mouse tumor cells, which share antigens with normal mouse tissues, could be an explanation. It may be necessary either to prepare IgG raised against target specific antigens or to purify the IgG raised against whole cells by adsorption with liver and spleen homogenates.

LIPOSOME-ENTRAPPED ENZYMES IN THE TREATMENT OF LYSOSOMAL STORAGE DISEASES

Following our development of a method (10) for the rapid partial purification of glucocerebroside β-glucosidase from human placenta, an adult Gaucher's disease patient was treated with repeated intravenous injections of the liposomal enzyme, up to about 10,800 units/dose, over an 18 month period (11). Clinical observations immediately after injection suggested that the preparation had an effect. From about 30 min to 48 hr there was sleepiness, headache, nausea and difficulty in concentration. These transitory, largely central nervous system effects were attributed to the size of unsonicated liposomes because empty liposomes had a similar effect. However, these symptoms were not seen in another patient with Gaucher's disease treated similarly; and the effect of size was thought to be related to the clinical characteristics of the patient. One to 20 days after treatment there was abdominal pain, which was proportional to the amount of the entrapped enzyme given, and which lasted from one to several days before gradually diminishing. After several months of treatment, palpation of the abdomen revealed an overall reduction in liver size. Interestingly, this was associated with considerable relief of severe pressure symptoms in the left lower abdomen. We are now treating a second adult Gaucher's disease patient as well; and both patients are receiving small amounts of the liposomal enzyme at monthly intervals. Clinical response, however, is barely noticeable, most probably bacause only 200 units are given per patient. Should the absence of improvement continue in the next few months we may return to a more rigorous treatment.

LIPOSOMES AS CARRIERS OF INSULIN IN ORAL TREATMENT

Treatment of diabetes would be more physiological if insulin were to reach the liver through the portal circulation as this would imitate the pathway of the hormone secreted by the pancreas. Attempts to this end by administering insulin orally have failed owing to the vulnerability of the hormone in the gut. A number of reports indicated that some semi-synthetic phospholipids, dispersed as liposomes, could be hydrolyzed by pig pancreatic phospholipase A_2 only near their transition temperatures (12). Above or below these temperatures, the rate of hydrolysis was negligible. It was thought that insulin-containing liposomes composed of phospholipids with transition temperatures well below or above 37°C would be likely to survive in the gut. A variety of phospholipids were tested in rats. Rats in groups of five were treated intragastrically with insulin entrapped in liposomes composed of phospholipid and cholesterol (molar ratio 7:2). For each group of rats a group of five rats was treated similarly with free insulin. The values in Table 1 are expressed as percentages of those measured in individual rats within each group before treatment. In the case of dimyristoyl lecithin cholesterol liposomes, as little as 0.3 units of insulin diminished blood glucose levels in diabetic rats within 4 hr to 37% of the values before treatment (13). The effect was retained for up to 24 hr. Radioimmunoassay of insulin in the serum of normal rats undergoing similar treatment showed that most of the insulin carried into the blood was in the entrapped form (14). It is still unknown to what extent and by which route, i.e. portal circulation or lymphatics, liposomes carry their contents in the periphery. It is possible that much of the administered liposomal insulin escapes absorption although the extent of loss may be dependent on the lipid composition of the preparation and the physiological state of the animal. Our preliminary efforts with two diabetic patients have failed; and it seems that other factors peculiar to man may be involved.

LIPOSOMES AS IMMUNOLOGICAL ADJUVANTS

Attempts to establish whether liposomal entrapment of proteins was associated with prevention of immune response led to the unexpected finding that entrapment could induce antibody response (3). This finding was explored and the adjuvant property of liposomes was established (15). It was found that antibody titres in mice injected with liposome-entrapped diphtheria toxoid were severalfold greater than when the antigen was administered in the free form. Further, there were no granulomas formed at the site of injection. Liposomes as immunological adjuvants have now attracted the interest of pharmaceutical industries; and it seems that immunopotentiation via this system may be an early application on a commercial basis.

FUTURE PROSPECTS

An important prerequisite for success in the controlled delivery of proteins or other agents to target areas in the body is specificity. At present, injected liposomes can transport their contents to a limited variety

TABLE 1

BLOOD GLUCOSE CONCENTRATIONS IN DIABETIC RATS TREATED WITH LIPOSOME-ENTRAPPED INSULIN*

Liposomal phospho-lipid	Administered insulin (units)	Change in blood glucose concentration (% ± S.E.)					
		1 hr	P	4 hr	P	24 hr	P
PC	5.0	81.8 ± 1.6	<0.05	65.7 ± 1.8	<0.001	92.5 ± 5.1	N.S.
DOPC	2.9	62.0 ± 8.3	N.S.	45.6 ±12.2	<0.05	101.4 ±10.2	N.S.
DSPC	2.9	50.6 ± 7.1	<0.02	58.6 ± 3.9	<0.02	80.1 ± 8.4	<0.02
DPPC	2.9	70.8 ± 4.4	<0.02	45.1 ± 2.9	<0.01	94.4 ± 5.6	N.S.
DMPC	0.3	68.1 ± 5.3	<0.02	47.1 ± 3.3	<0.02	76.3 ± 6.9	<0.05
DMPC	0.3	62.9 ± 5.5	<0.001	37.1 ± 4.1	<0.001	58.6 ± 5.8	<0.001
DLPC	0.3	78.6 ± 5.8	<0.05	44.7 ± 5.2	<0.02	98.8 ± 5.8	N.S.

*P refers to probability of significant difference between values obtained with liposomal insulin and values obtained with free insulin (not shown). N.S. not significant; PC egg phosphatidylcholine; DOPC dioleoyl phosphatidylcholine; DPPC dipalmitoyl phosphatidylcholine; DSPC distearoyl phosphatidylcholine; DLPC dilauroyl phosphatidylcholine; DMPC dimyristoyl phosphatidylcholine (13).

of cells, mostly phagocytic; and the uptake of liposomes by such cells is so rapid that other cells which are also capable of endocytosing liposomes fail to take up their share. It is thus apparent that redirection of liposomes to alternative sites is dependent on two developments, namely the prevention or diminution of liver and spleen participation and the construction of liposomes which can be selectively recognized by target cells.

REFERENCES

1. GREGORIADIS, G., PUTMAN, D., LOUIS, L. & NEERUNJUN, D. *Biochem. J. 140*:323, 1974.
2. GREGORIADIS, G. *New Engl. J. Med. 295*:704 & 765, 1976.
3. GREGORIADIS, G. *Meth. in Enzymology 44*:218, 1976.
4. GREGORIADIS, G. *Meth. in Enzymology 44*:698, 1976.
5. BLACK, C.D.V. & GREGORIADIS, G. *Biochem. Soc. Trans. 4*:253, 1976.
6. WISSE, E., GREGORIADIS, G. & DAEMS, W.Th. in "The Reticuloendothelial System in Health and Disease: Functions and characteristics," (S.M. Reichard, M.R. Escobar & H. Friedman, eds.), Plenum Publishing Corporation, New York, 1976, p. 237.
7. PAPAHADJOPOULOS, D., POSTE, G. & MAYHEW, E. *Biochim. Biophys. Acta 363*:404, 1974.
8. GREGORIADIS, G. & NEERUNJUN, E.D. *Biochem. Biophys. Res. Comm. 65*:537, 1975.
9. GREGORIADIS, G., NEERUNJUN, D.E. & HUNT, R. *Life Sciences 21*:357, 1977.
10. BRAIDMAN, I.P. & GREGORIADIS, G. *Biochem. J. 164*:439, 1977.
11. BELCHETZ, P.E., BRAIDMAN, I.P., CRAWLEY, J.C.W. & GREGORIADIS, G. *Lancet ii*:116, 1977.
12. OP DEN KAMP, J.A.F., KAUERZ, M.T. & VAN DEENEN, L.L. *Biochim. Biophys. Acta 406*:169, 1975.
13. DAPERGOLAS, G. & GREGORIADIS, G. *Biochem. Soc. Trans. 5*:1383, 1977.
14. DAPERGOLAS, G. & GREGORIADIS, G. *Lancet ii*:824, 1976.
15. ALLISON, A.C. & GREGORIADIS, G. in "Recent Results in Cancer Research," (G. Mathe, I. Florentin & M.-C. Simmler, eds.), Springer Verlag, Heildenberg, 1976, p. 58.

GALACTOSE CONVERSION USING A MICROCAPSULE IMMOBILIZED MULTIENZYME COFACTOR RECYCLING SYSTEM

T.M.S. Chang and N. Kuntarian

Department of Physiology
McGill University
Montreal, Quebec, Canada

Immobilized enzymes, especially microencapsulated enzymes, have been used to replace enzyme deficiency in some forms of inborn errors of metabolism in animals (1,2). Still, the clinical use of immobilized enzymes for inborn errors of metabolism is not yet widely feasible. This is due partly to the fact that in most of the inborn errors of metabolism, complex multienzyme system with cofactor requirements are necessary. The further development of multienzyme systems with cofactor regeneration is thus essential (1,3). Basic studies carried out using microencapsulated multienzyme systems have demonstrated the feasibility of recycling cofactors like ATP, NAD and NADP (1,4,5). Galactosemia is usually due to a deficiency in galactose-1-P-uridyl-transferase or, less frequently, to a deficiency in galactokinase. Since galactokinase is more readily available, the present report is a feasibility study of microencapsulating this enzyme together with a cofactor recycling enzyme system for the conversion of galactose.

MATERIALS AND METHODS

The enzymes galactokinase (grade IV, from Yeast, 50 Units/mg), pyruvate kinase (type III, from rabbit muscle, 485 Units/mg and lactic dehydrogenase (type VI, from porcine heart, 390 Units/mg) were obtained from Sigma. Substrate and coenzymes included galactose, ATP, β-NADH, and phosphoenol pyruvate (trisodium salt) also from Sigma.

Collodion microcapsules were prepared by the updated interfacial precipitation method (2). In this procedure the pH of the acidic Tween 20 was adjusted to 7.5 by the addition of NaOH. Galactokinase (1 mg containing 50 Units) and pyruvate kinase (1 mg containing 480 Units) were added to 2.5 ml of a 10 gm/100 ml hemoglobin solution. Microcapsules of 80-100 μm diameters were prepared. The microcapsules formed were washed exhaustively with 1% Tween at 4°C and finally with 0.1 M sodium phosphate, pH 7.5, at 4°C. For immobilization of all three enzymes the same procedure was used with the addition of lactate dehydrogenase (1 mg containing 390 Units).

RESULTS AND DISCUSSION

The galactokinase and pyruvate kinase scheme for the conversion of galactose and recycling of ATP is shown in Fig. 1. 0.1 ml of microcapsules containing galactokinase and pyruvate kinase were added to 3 ml of a solution containing galactose (0.116 M), ATP (0.64 mM), phosphoenol pyruvate (0.64 mM), $MgSO_4$ (11.6 mM), and KCl (16 mM) in 0.1 M sodium phosphate buffer, pH 7.5. The conversion of galactose was followed by the formation of pyruvate, which

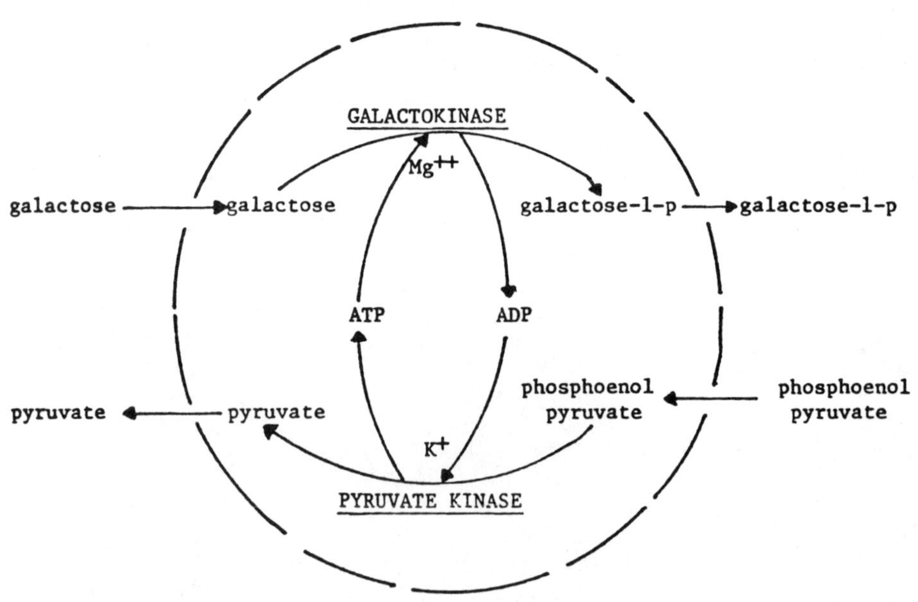

Fig. 1. Two enzyme scheme

was measured by the simultaneous presence in the solution of NADH (0.21 mM) and lactate dehydrogenase (20 μg/3 ml). The results showed that galactose was effectively converted to galactose-1-phosphate in the presence of the above reaction mixture and that ATP was efficiently recycled. In the absence of ATP recycling, by omitting the substrate phosphoenol pyruvate, galactose conversion stopped rapidly as the small amount of ATP was exhausted.

The galactokinase, pyruvate kinase and lactic dehydrogenase scheme for differentiating galactose-1-phosphate is shown in Fig. 2. The microcapsule suspension was added to a solution containing galactose in the presence of ATP (0.64 mM), phosphoenol pyruvate (0.64 mM), $MgSO_4$ (11.6 mM), KCl (16 mM) in 0.1 M Tris solution, pH 7.5. Galactose but not galactose-1-phosphate would give quantitative changes in optical density. The total galactose in the galactose and galactose-1-phosphate could be separately measured using the standard galactose oxidase method. In this way the concentration of galactose and galactose-1-phosphate could be estimated.

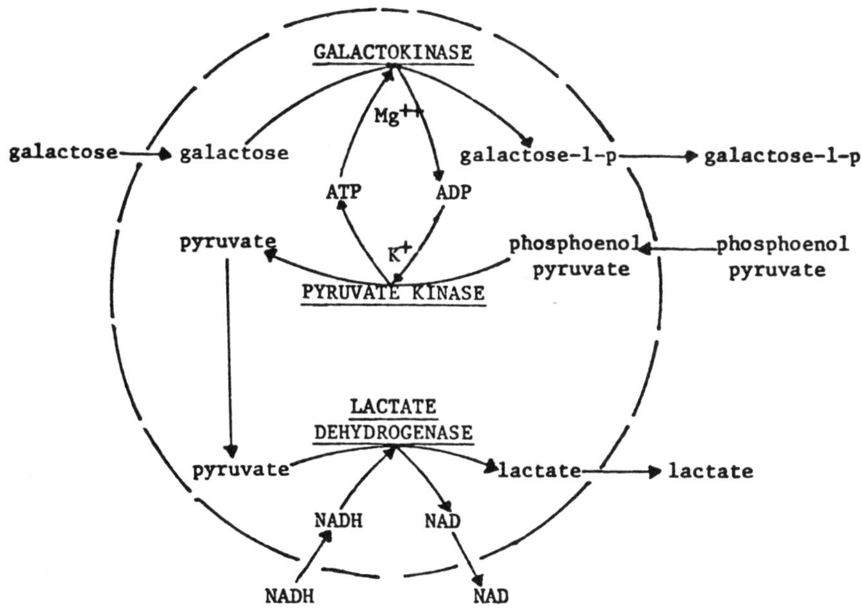

Fig. 2. Three enzyme scheme

PERSPECTIVES FOR GALACTOSEMIA

Assuming therefore that the microencapsule immobilized multienzyme system of galactokinase and pyruvate kinase efficiently converts galactose into galactose-1-phosphate with simultaneous recycling of ATP, there are implications for galactosemia due to galactokinase deficiency. The accumulated galactose can be converted to galactose-1-phosphate. Galactose-1-phosphate in the extracellular component can be excreted by the kidney; that which enters the cells can be converted by the intracellular galactose-1-P-uridyl-transferase. Unfortunately, galactosemia is more commonly due to deficiency in galactose-1-P-uridyl-transferase, which results in the accumulation of galactose-1-phosphate and galactose. Even in this latter case, the microencapsulated enzyme system could still function in converting the accumulated galactose in the extracellular fluid to galactose-1-phosphate. Galactose-1-phosphate is not transported at the same rate as galactose into cells and therefore can be excreted into the urine. In this way the continuing accumulation of galactose and galactose-1-phosphate might be diminished.

With the presence of the ATP recycling enzyme, pyruvate kinase, only a very low concentration of ATP is required to carry out the galactokinase function. Ultimately one would prefer to have cofactor immobilized within artificial cells which do not require the supplement, even of a small amount of cofactor. Work in this laboratory has demonstrated the feasibility of immobilizing cofactors with macromolecules so as to retain them within the microcapsules and allow them to recycle with the recycling enzyme (1, 5). Another approach is the microencapsulation of free cofactor and the cofactor-recycling enzymes in microcapsules, whose membrane is not permeable to the cofactor but permeable to the substrates.

ACKNOWLEDGMENT

The support of the Medical Research Council of Canada (MRC-SP-4) is gratefully acknowledged.

REFERENCES

1. CHANG, T.M.S. "Biomedical Applications of Immobilized Enzymes and Proteins," vol. 1, Plenum, New York, 1977, pp. 428, 2, 359.

2. CHANG, T.M.S. "Artificial Cells," Charles C. Thomas Publisher, Springfield, Ill., 1972.
3. CHANG, T.M.S. *Meth. in Enzymology 44*:999, 1976.
4. CAMPBELL, J. & CHANG, T.M.S. *Biochim. Biophys. Acta. 397*:101, 1975.
5. CAMPBELL, J. & CHANG, T.M.S. *Biochem. Biophys. Res. Commun. 69*:562, 1976.
6. COUSINEAU, J. & CHANG, T.M.S. in "Artificial Kidney Artificial Liver and Artificial Cells," (T.M.S. Chang, ed.), Plenum, New York, 1978, p. 302.

HOLLOW FIBER-ENTRAPPED LIVER MICROSOMES: A POTENTIAL EXTRACORPOREAL DRUG DETOXIFIER

P.R. Kastl, W.H. Baricos, R.P. Chambers* and W. Cohen

Tulane Medical School
New Orleans, Louisiana and
Auburn University*
Auburn, Alabama, USA

We are attempting to develop an extracorporeal hollow fiber drug detoxifier (EDD) based on the NADPH-dependent, multienzyme, drug hydroxylation complex present in liver endoplasmic reticulum and isolatable in particulate form as microsomes (MS). In addition to its potential use as an EDD this system serves as a model for the use of membrane-bound, cofactor-requiring, multienzyme complexes in industrial as well as medical applications. This work also may contribute toward the development of liver assist devices as well as provide for production of drug metabolites for subsequent therapeutic screening.

In the presence of NADPH and molecular oxygen this system will hydroxylate both endogenous as well as foreign compounds. This paper considers the stability of the isolated MS drug hydroxylase complex, the presence of contaminating enzymes, the supply of NADPH, and the delivery of O_2 to the reaction site.

METHODS AND MATERIALS

Phenobarbital induction of the rat liver MS drug hydroxylase system, isolation, and lyophilization were carried out as previously described (1). Microsomal p-nitroanisole (pNA) 0-demethylase activity was assayed by a modification of the method of Zannoni (2). Hexobarbital metabolism was monitored by the method of Kupfer and Rosenfeld (3) or by the following procedure utilizing ^{3}H-hexobarbital: aqueous samples were extracted with ether and unmetabolized hexobarbital separated from its more polar metabolites by TLC of the ether extract; the TLC plates were sprayed with mercurous nitrate-diphenyl carbozone; and the spots corresponding to hexobarbital and its metabolites were scraped from the plate, eluted with methanol, and counted in a liquid scintillation counter. Protein was determined by the method of Lowry (4). Glucose-6-PO_4 dehydrogenase (G6PDH), glucose dehydrogenase (GDH), and NADP were purchased from Sigma. pNA was purchased from Aldrich and recrystallized before use. Hexobarbital sodium was purchased from Winthrop Laboratories. ^{3}H-hexobarbital was purchased from New England Nuclear. The following hollow fiber devices were purchased from Amicon or Bio-Rad Laboratories: Amicon Vitafiber 50,000 nominal molecular weight cutoff, 60 cm^2 surface area, acrylic copolymer fibers; Bio-Fiber 5 Beaker 500 cm^2 surface area, silicone/polycarbonate fibers; Bio-Fiber 80/5 Cell Culture Tube 30,000 nominal molecular weight cutoff, 50 cm^2 surface area, 50% silicone/polycarbonate fibers, 50% cellulose acetate fibers.

RESULTS AND DISCUSSION

Although MS stored at -85°C were reported to be quite stable (5), we found that lyophilyzed MS while less active than fresh MS, could be stored under vacuum at -20°C or room temperature with little if any loss of the original drug detoxification activity for longer than 1 year.

Microsomes contain a variety of enzymes in addition to those for drug detoxification. Of particular significance are those which act to lower NADPH levels: NAD(P) ase, NAD(P) pyrophosphatase, and NAD(P)H oxidase. By including the competitive inhibitors pyrophosphate and nicotinamide in the reaction mixture (6,7), we were able to minimize the deleterious effects of the pyrophosphatase and NAD(P)ase, respectively. The effects of NAD(P)H oxidase were minimized by inclusion of excess amounts of an NADPH regeneration system. These factors were used in producing a working system shown in Fig. 1 for pNA. Optimal activity was obtained using the inhibitors combined with the NADPH regeneration system.

A variety of enzyme/substrate couples have been used for NADPH regeneration, including G6PDH/G6P and

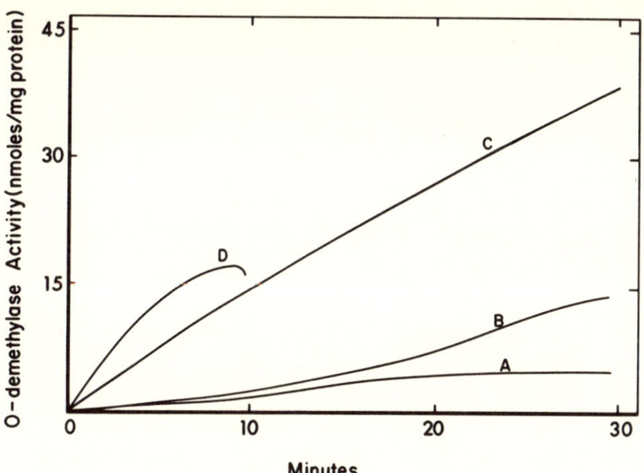

Fig. 1. Optimization of MS pNA O-demethylation. (C) complete system containing 1.2 mM pNA, 0.13 mM NADP, 50 mM pyrophosphate, 2.4 mM nicotinamide, 6 mM $MgCl_2$, 6 mM G6P, 4 units G6PDH per ml, and 0.73 mg MS protein per ml in a final volume of 2.5 ml of 0.1 M phosphate buffer, pH 7.4; (A) pyrophosphate and nicotinamide deleted; (B) pyrophosphate deleted; (D) 0.13 mM NADPH replaced NADP, also G6P, G6PDH and $MgCl_2$ deleted.

isocitrate dehydrogenase/isocitrate. However biocompatibility, delivery to the EDD, and potential untoward effects of the regeneration products led us to test the glucose dehydrogenase/glucose system. This enzyme catalyzes the NAD(P)-dependent oxidation of glucose, producing gluconic acid and NADPH. The primary advantage is that glucose can be supplied from the blood of the patient; and the gluconic acid is relatively nontoxic. We compared this NADPH regeneration couple with G6PDH/G6P in a model detoxification reaction employing pNA and found this couple slightly more effective at low enzyme levels (Fig. 2).

A major problem associated with cofactor-requiring enzyme reactors is retention of the relatively small cofactor within the reactor (8). Increasing the molecular weight of the cofactor by covalent attachment to soluble or insoluble supports is expensive and usually results in decreased enzyme activity due to steric hindrance (9, 10).

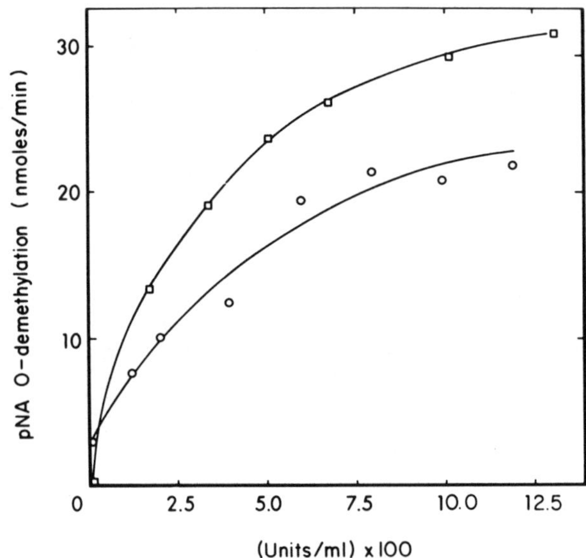

Fig. 2. Comparison of G6PDH/G6P and GDH/glucose as NADPH regeneration systems in MS pNA O-demethylation. (O) G6PDH/G6P system as in Fig. 1 (C), except 22 mM pyrophosphate; (square) GDH/glucose system containing 280 mM glucose, 1.2 mM pNA, 0.13 mM NADP, 2.4 mM nicotinamide, 22 mM pyrophosphate and 0.73 mg MS protein per ml in a final volume of 2.5 ml of 0.1 M phosphate buffer, pH 7.4.

In contrast it has been demonstrated experimentally that effective cofactor reuse can be achieved without modification of the cofactor (11-13). Preliminary calculations based on a 10,000-15,000 cm^2 hollow fiber reactor indicate that a slow infusion of as little as 26 mg/hr of NADP directly into the MS suspension would maintain an excess NADP concentration, thus assuring maximal rates of drug detoxification.

It can be calculated that at ambient temperatures and O_2 pressures, aqueous solutions contain less than 1 μmole O_2 per ml. Even for a large volume EDD (e.g. 500 ml), the dissolved O_2 cannot support the detoxification of amounts of drug commonly encountered in overdose cases. Therefore, additional O_2 will have to be supplied. Simple bubbling of either air or O_2 into a MS suspension resulted in rapid loss of drug metabolizing activity. However, we have found that O_2 could be efficiently supplied to a MS suspension by immersing in the suspension an O_2-permeable hollow fiber bundle, such as the Bio-Fiber 5 hollow fiber beaker, perfused with 5 psi O_2. As shown in Fig. 3, hexobarbital metabolism in the ambient O_2 system rapidly ceased while that of the O_2-perfused Bio-Fiber 5 system continued at a significant rate for at least 4 hr. In actual patient detoxification O_2 ideally would be obtained directly from the patient's blood.

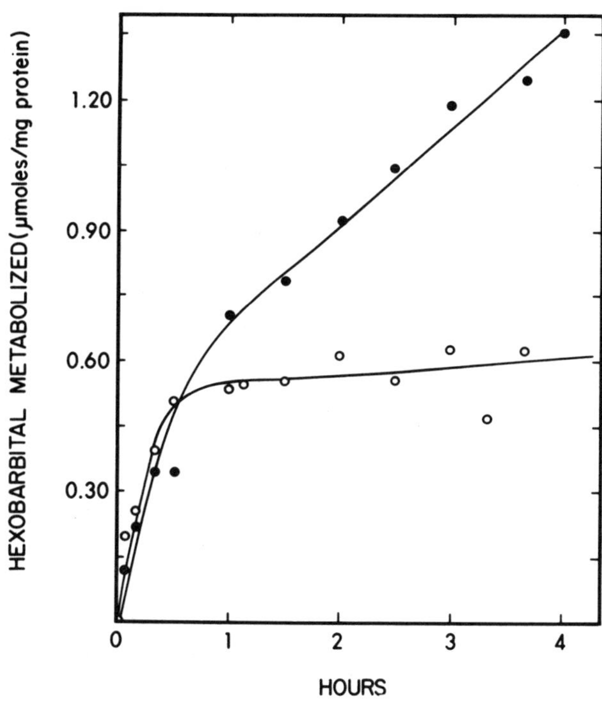

Fig. 3. Effect of oxygen infusion on MS hexobarbital metabolism. Solutions as in Fig. 1 (C) except 22 mM pyrophosphate and final volume 130 ml, containing 0.57 mg MS protein per ml, 2.34 mM ^3H-hexobarbital (1.2 x 10^6 DPM) replaced pNA, and 2 units·G6PDH per ml. (O) MS suspension with ambient oxygen; (●) MS suspension with Biofiber 5 hollow fiber beaker perfused with 5 psi oxygen.

In model EDD experiments 0.5 ml MS suspension (17 mg protein) containing 60 units of G6PDH were placed on the dialysate side of a hollow fiber. Buffered solutions of drug, simulating a patient's oxygenated blood, were circulated at 12 ml/min through the lumen of the fibers. Using this system we studied the metabolism of hexobarbital with (a) the Bio-Fiber 80/5 Culture Tube containing O_2-permeable silicone-polycarbonate as well as cellulose acetate fibers, and (b) the Amicon Vitafiber with fibers composed of an O_2-permeable acrylic copolymer. In both cases O_2 was supplied to the circulating solution (not the MS) by direct bubbling. As can be seen in Fig. 4, the rate of hexobarbital metabolism in the system utilizing the 80/5 Culture Tube was significantly lower than the initial rate observed in simple MS suspensions. This decreased rate could be attributed to mass transfer limitations associated with O_2 transfer from the circulating solution to the MS on the shell side of the fiber bundle. In the Vitafiber system however, the effects of mass transfer of both substrate and O_2 appeared to be significantly diminished; and the rate of hexobarbital metabolism approached the initial rate observed in simple MS suspensions. In addition, and most importantly, this rate was maintained for at least three hours, indicating the effective transfer of O_2 and hexobarbital from the circulating solution to the MS suspension. We repeated the Vitafiber perfusion experiment using hexobarbital in serum at ambient O_2 as the circulating solution; the rate of hexobarbital metabolism was approximately 50% of that observed in the Vitafiber system employing buffered oxygenated solutions. Presumably this decreased rate was due to mass transfer limitations resulting from binding of hexobarbital to serum proteins from lack of adequate O_2 supply, and from clogging of the fiber pores by serum proteins.

Based on the data of Fig. 4, we calculated a hexobarbital clearance for the Vitafiber-serum system of 2.7 ml/hr. Extrapolation to a 10,000 cm^2 hollow fiber yielded a theoretical clearance of 9.0 ml/min., which approaches the hemodialysis clearance of amylobarbitone and butobarbitone (30 ml/min) and phenobarbitone (60 ml/min) (14). A MS-based hollow fiber EDD probably can surpass these hemodialysis clearances through the use of more permeable membranes and improved reactor design. While it is doubtful that such a system can match the clearances of charcoal or resin adsorption devices, the MS-based hollow fiber EDD does possess advantages over these and other EDD systems.

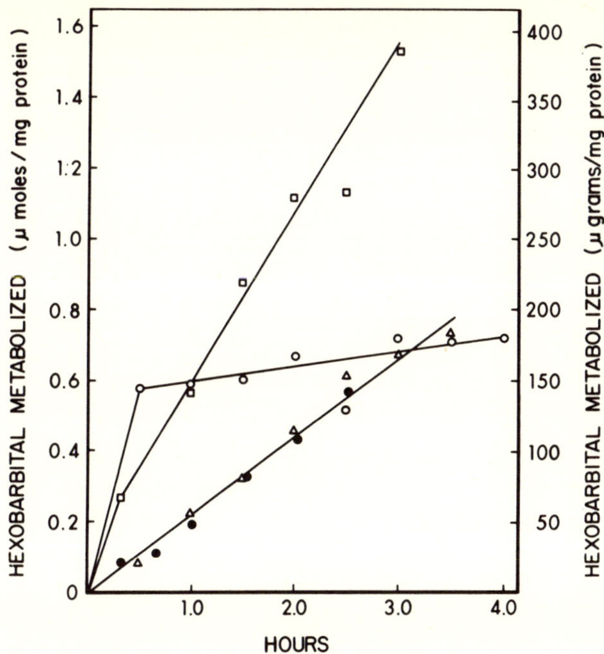

Fig. 4. Simulated extracorporeal hollow fiber drug detoxification. 0.5 ml MS were placed on the dialysate side of the hollow fiber and 30 ml of ^{3}H-hexobarbital solution (as in Fig. 3, except used 0.77 mM hexobarbital) were circulated through the lumen of the fibers at 12.0 ml/min. (O) control, MS suspension, ambient oxygen as in Fig. 3, final volume 2.5 ml; (square) Amicon Vitafiber 2.8 mg MS protein, oxygen bubbled into circulating solution; (triangle) Biofiber 80/5 17 mg MS protein, oxygen bubbled into circulating solution; (●) Amicon Vitafiber 2.8 mg MS protein, human plasma replaced buffer, ambient oxygen.

ACKNOWLEDGMENTS

This investigation was supported by grants from: the Edward G. Schlieder Educational Foundation; USPHS Postdoctoral Fellowship GM00968 to PRK; and Avondale Shipyards--Louisiana Heart Association. The expert technical assistance of P. Gonzales is acknowledged.

REFERENCES

1. COHEN, W., BARICOS, W.H., KASTL, P.R. & CHAMBERS, R.P. in "Biomedical Applications of Immobilized Enzymes and Proteins," (T.M.S. Chang ed.), Plenum, New York, 1977, p. 319.
2. ZONNONI, V.G. in "Fundamentals of Drug Metabolism and Drug Disposition," (B.N. LaDu, H.G. Mandel, and E.L. Way, eds.), Williams Wilkins Co., Baltimore, 1971, p. 566.
3. KUPFER, D. & ROSENFELD, J. *Drug Metabolism Disposition 1*:760, 1973.
4. LOWRY, O.H., ROSENBROUGH, N.J., FARR, A.L. & RANDALL, R.J. *J. Biol. Chem. 193*:265, 1951.
5. BORTON, P., CARSON, R. & REED, D.T. *Biochem. Phar. 23*:2332, 1974.
6. GILLETTE, J.R. *Prog. Drug Res. 6*:11, 1963.
7. GILLETTE, J.R., GRIED, W. & SEASAME, H. *Fed. Proc. 22*:366, 1963.
8. BARICOS, W.H., CHAMBERS, R.P. & COHEN, W. *Anal. Letters 9*:257, 1976.
9. WYKES, J.R., DUNNILL, P., LILLY, M.D. *Biochim. Biophys. Acta 286*:260, 1972.
10. LOWE, C.R. & MOSBACH, K. *Eur. J. Biochem. 49*:511, 1974.
11. BARICOS, W.H., CHAMBERS, R.P. & COHEN, W. *Enzyme Technol. Digest 4*:39, 1975.
12. CHAMBERS, R.P., WALLE, E.M., BARICOS, W.H. & COHEN, W. in "Enzyme Engineering," vol. 3. (E.K. Pye and H.H. Weetall, eds.,) Plenum, New York, in press.
13. FINK, D.J. & RODWELL, V.W. *Biotechnol. Bioeng. 17*:1029, 1975.
14. WIDDOP, B., MEDD, R.K., BAITHWAITE, R.A., REES, A.J. & GOULDING, R. *Arch. Toxicol. 34*:27, 1975.

ENZYMATIC CONVERSION OF CYCLOHEXANE TO CYCLOHEXANOL BY ISOLATED RAT LIVER MICROSOMES (MS)

W.H. Baricos, K. H. Johnson, and R.H. Steele

Department of Biochemistry
Tulane Medical School
New Orleans, Louisiana, USA

Normal rat liver MS possess little if any ability to convert medium chain length hydrocarbons (HC) to the corresponding primary alcohols *in vitro*. Pretreatment of MS donor animals with phenobarbital (75 mg/kg body wt., IP injection, daily for 5 days) resulted in MS with a slight but detectable ability to convert these HC to their primary alcohols *in vitro* (Table I). MS isolated from phenobarbital-treated rats converted cyclohexane to cycohexanol at rates 11-128 times those of the n-HC. Maximum cyclohexanol production was observed in the presence of pyrophosphate, nicotinamide, and an NADPH regenerating system. This presumably was due to the inhibition of various NADP(H) degrading enzymes such as NADP ase, NADP(H) pyrophosphatase, and NADPH oxidase present in the MS. NADH could not substitute for NADPH. Cyclohexanol production was maximal at pH 7.4 (0.1 M phosphate buffer) and at a temperature of 37°C. Yields of cyclohexanol increased with increasing MS protein up to a maximum of about 13 mg/ml of reaction mixture. The time course of cyclohexanol production was biphasic, being rapid for the first 30 min followed by a much slower but linear rate of production for as long as 3 hr. A variety of compounds were found to inhibit the production of cyclohexanol, in particular sodium dodecyl sulfate and the medium chain n-alcohols.

TABLE I

CONVERSION OF MEDIUM CHAIN HC TO ALCOHOLS BY ISOLATED RAT LIVER MS

Hydrocarbon	Untreated MS	Pb-Treated MS
	(nmoles alcohol/mg MS protein)	
Pentane	None	8.0
Hexane	none	22.0
Heptane	none	24.0
Octane	none	2.1
Cyclohexane	Trace	270

Hydrocarbon oxidation was carried out in a final volume of 2.955 ml containing: 2.5 ml of 0.1 M phosphate buffer, pH 7.4, containing 1.85 x 10^{-2}M pyrophosphate, 2.03 x 10^{-3}M Nicotinamide, 5.1 x 10^{-3}M $MgCl_2$, 1.1 x 10^{-4}M NADP, 5.1 x 10^{-3}M glucose-6-phosphate (G6P), 5 units G6P dehydrogenase, 0.4 ml MS solution in 0.25 M sucrose, and 50 μliters of the hydrocarbon. Each incubation mixture was gassed for 60 sec with 100% O_2, sealed, and incubated for 3 hr at room temperature in a Dubnoff shaking incubator. Reactions were initiated by addition of G6P dehydrogenase. Product alcohols were quantitated by gas chromatography.

ACKNOWLEDGMENTS

This research was supported by a grant from the Ethyl Corp., USA.

HYDROLYSIS OF GLUCURONIDES WITH IMMOBILIZED ENZYMES

D.J. Fink, M.K. Bean, and R.D. Falb

Battelle
Columbus Laboratories
Columbus, Ohio, USA

Metabolism of many drugs leads to the formation of conjugation products with glucuronic acid prior to excretion of urine. Thus, heroin is converted to morphine and, after conjugation with glucuronic acid, appears in the urine primarily as morphine glucuronide. Detection systems for drug glucuronides can be improved if the glucuronide is hydrolyzed first (1). In the case of morphine, a significant increase in sensitivity can be realized using enzyme-catalyzed glucuronide hydrolysis (2). With the development of techniques for enzyme immobilization, it should be possible to prepare an insoluble β-glucuronidase derivative which permits exposure of the test sample to relatively large amounts of the enzyme, making the hydrolysis amenable to instrumentation on continuous-flow analysers (3-5). Solid-phase enzymes should also be useful in other analyses such as the quantitation of steroid glucuronides (5,6).

The most active preparations of immobilized β-glucuronidase prepared in this study utilized the beef liver or *E. coli* enzymes. Immobilization was achieved by glutaraldehyde attachment to alkylamine glass or to CNBr-activated Sepharose 4B. Preparations of both enzymes had half-lives of greater than 7 days at 37° in columns through which buffer was pumped. Both soluble (7) and immobilized enzymes are substrate inhibited by p-nitrophenyl glucuronide and by phenolphthalein glucuronide. This inhibition might slow the hydrolysis of any given glucuronide in mixtures of glucuronide conjugates.

The efficacy of enzymatically hydrolyzing morphine-3-glucuronide (M3G) in urine was investigated in this study. M3G was added to urine specimens obtained from healthy donors who were not taking any medication at the time. The spiked urine was buffered and pumped through β-glucuronidase/porous glass columns. The hydrolyzed urine was extracted with isopropanol:chloroform (1:3) to remove the free morphine (8) which was then converted to the fluorescent pseudomorphine dimer with potassium ferricyanide (8). Enzymatic hydrolysis of M3G was demonstrated in urine, but the hydrolysis also resulted in the production of other compounds which interfered with the fluorescent morphine quantitation procedure. Washing the extracted morphine with borate buffer (3) reduced the interference but did not completely remove the interfering compounds.

ACKNOWLEDGMENTS

This program was supported by Contract No. DAMD 17-74-C-4091 from the U.S. Army Medical Research and Development Command. M3G used in this study was the generous gift of Dr. Robert Willette of the National Institute of Drug Abuse, Rockville, Maryland, U.S.A.

REFERENCES

1. PAYTE, J.T., WALLACE, J.E., & BLUM, K. *Curr. Therap. Res.* *13*:412, 1971.
2. FISH, F. & HAYES, T.S., *J. Forensic Sci.* *19*:676, 1974.
3. BLACKMORE, D.J., CURRY, A.S., HAYES, T.S. & RUTTER, E.R. *Clin. Chem.* *17*:896, 1971.
4. MULE, S.J. & HUSHIN, P.L. *Anal. Chem.* *43*:708, 1971.
5. VELA, B.A., ACEVEDO, H.F. & CAMPBELL, E.A. *Clin. Chem.* *14*:837, 1968.
6. GRAEF, V., FURUYA, E., & NISHIKAZE, O. *Clin. Chem.* *23*:532, 1977.
7. SZAZZ, G. *Clin. Chem.* *13*:752, 1967.
8. KUPFERBERG, H., BURKHALTER, A., & WAY, E.L. *J. Pharmacol. Exp. Ther.* *145*:247, 1964.

IMMOBILIZATION OF NEURAMINIDASE FOR THE TREATMENT OF TUMOR CELLS

Edward R. Bazarian and Lemuel B. Wingard, Jr.

Departments of Pharmacology and Chemical Engineering
University of Pittsburgh
Pittsburgh, Pennsylvania, USA

A promising approach to tumor immunotherapy is based on using neuraminidase treated cells to stimulate the immune response of the patient (1). In this technique tumor cells are incubated *in vitro* with *Vibrio cholerae* neuraminidase (VCN) and then injected into a patient or test animal possessing the same type of tumor. Remission has been observed with several types of tumor (1, 2).

VCN cleaves the terminal N-acetyl-neuraminic acid (sialic acid) residues from cell surface glycoproteins. The cells remain viable in that their ability to synthetize proteins or exclude dye macromolecules is not compromised. One theory for the enhanced immunogenicity of the VCN-treated cells postulates that the removal of sialic acid exposes cell surface antigens, which cause the enhanced immune response. A second theory postulates that it is adsorbed VCN on the re-injected cells that is the causative agent. We decided to immobilize VCN to test the validity of the second theory as well as to provide easy separation of the VCN from the treated cells prior to re-injection.

VCN from Behring Diagnostics, 500 U/ml (E.C.3.2.1.18) was covalently attached to the inside wall of 1.0 mm inner diameter nylon-6 tubing from Portex Ltd. (3). The tubing was activated with dimethyl sulfate to form the imidate, to which a lysine spacer was attached. The lysine was activated with either glutaraldehyde or dimethyl adipimidate, followed by attachment of the enzyme. The enzyme activity was measured by incubation with 1% mucin (Sigma) as substrate in 0.05 M acetate buffer, pH 5.5, 0.9% NaCl, and 0.1% $CaCl_2$. With the immobilized enzyme the incubation was for 2 hr at 37°C; for the soluble enzyme only 15 min at 37°C was required.

With adipimidate coupling the immobilized VCN released 10 to 18 μg of diacetylneuraminic acid (DANA) per ft of tubing; with glutaraldehyde coupling 3 to 11 μg/ft were released. No DANA was released when nylon tubes free of enzyme were incubated for 2 hr at 37°C with the substrate solution. The soluble enzyme showed a gradual loss of activity at pH values above about 8.8, with 60% activity loss at pH 10. The immobilized VCN was quite stable for up to 3 hr at 37°C. An unexplained apparent increase in activity was noted upon repeated use of the enzyme-tubes at 37°C.

From the results so far, we concluded that VCN was successfully immobilized on nylon tubes, and that the immobilized enzyme retained a significant portion of its activity.

REFERENCES

1. SEDLACEK, H.H., SEILER, F.R., & SCHWICK, H.G. *Klin. Wschr.* *55*:199, 1977.
2. HOLLAND, J.F. & BEKESI, J.G. *Med. clin. North America* *60*:539, 1976.
3. CAMPBELL, J., HORNBY, W.E. & MORRIS, D.L. *Biochim. Biophys. Acta* *384*:307, 1975.
4. WARREN, L. *J. Biol. Chem.* *234*:1971, 1959.

ENZYME THERMISTOR ANALYSIS IN CLINICAL CHEMISTRY AND ENVIRONMENTAL AND PROCESS CONTROL

B. Mattiasson, B. Danielsson and K. Mosbach

Chemical Center
University of Lund
Lund, Sweden

The combination of enthalpimetry and immobilized enzymes makes a pair with, in many respects, unique properties: *generality* in detection and assay of any reaction evolving or consuming heat and *specificity* by the immobilized enzyme chosen.

The enzyme thermistor consists of a small insulated column filled with immobilized enzyme. Heat changes due to enzymic reaction in the enzyme bed are registered as a changed resistance of the thermistor mounted at the top of the column. Disturbing unspecific heat effects can be compensated for by using an inert reference column (1,7). A continuous flow of buffer is pumped through the enzyme thermistor unit; and substrate and sample are introduced either as pulses (15-60 sec) or as a continuous flow.

The metabolites of clinical interest which have been assayed with the enzyme thermistor unit both in pure solution and in serum samples are shown in Table 1.

Environmental analysis may be carried out by three basically different ways: (a) use of enzymes to which the substance to be analyzed is a substrate (7); (b) analysis of substances based on their inhibitory properties on a specific enzyme reaction (8); (c) assay of integrated effects on living cells immobilized within the thermistor column (9).

TABLE 1

SUBSTANCES ANALYZED WITH THE ENZYME THERMISTOR

Substance	Enzyme used	Concentration* Range (mmol/l)	Ref.
Glucose	Glucose oxidase + catalase	0.002 - 0.5	(1)
Urea	Urease	0.01 - 500	(2)
Penicillin G	Penicillinase	0.1 - 150	(3)
Cholesterol	Cholesterol oxidase	0.03 - 0.15	(4)
Cholesterol esters	Cholesterol oxidase + cholesterol esterase	0.03 - 0.15	(4)
Lactose	Lactase + glucose oxidase	0.05 - 10	(4)
Hydrogen peroxide	Catalase	0.005 - 10	(5)
Oxalic acid	Oxalate decarboxylase	0.1 - 3	(6)

*It is likely that the concentration ranges given can be extended by proper optimization.

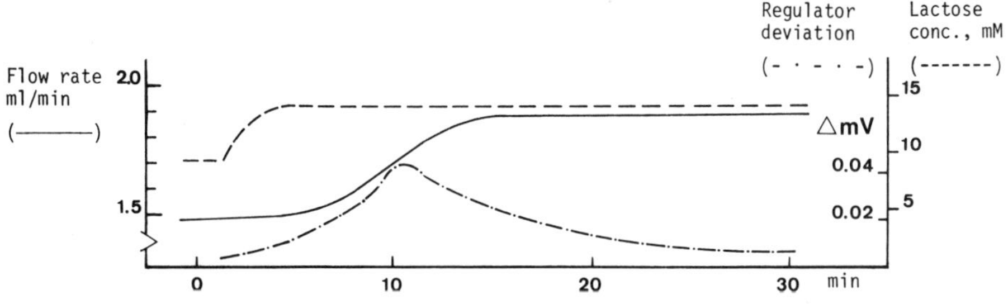

Fig. 1. Control of flow reactor.

Cyanide was assayed in blast furnace waste water using immobilized rhodanese (7), whereas the inhibitory properties of heavy metal ions on urease activity were exploited when assaying Hg^{2+}, Cu^{2+} and Ag^{+} (8). By subsequent biospecific elution the heavy metal ions could be washed off individually, thus making it possible to assay more than one metal ion at the same time. Hg^{2+} was assayed in concentrations down to 0.2 ppb.

The enzyme thermistor unit also has been used to assay continuously the effluent from an enzyme reactor in order to regulate the flow through the reactor, thus allowing control of the product concentration (10). A plug flow reactor with immobilized β-galactosidase was fed with lactose of a constant concentration; and the eluent was assayed for glucose present. As long as the glucose level was constant, the thermistor gave a constant Δt. If, however, the lactose concentration or the performance of the enzymic reactor varied, the glucose concentration in the effluent would be affected, leading to a change in the thermistor signal. Via a regulator the flow rate was adjusted to give the initial product concentration. Thus, as can be seen from Figure 1, the thermistor could continuously register and control the degree of production in a bioreactor, a technique that should be applicable to both enzyme and fermentation reactors.

REFERENCES

1. DANIELSSON, B., GADD, K., MATTIASSON, B. & MOSBACH, K. *Clin. Chim. Acta*, in press.
2. DANIELSSON, B., GADD, K., MATTIASSON, B. & MOSBACH, K. *Anal. Lett.* 9:987, 1976.

3. MOSBACH, K., DANIELSSON, B., BORGERUD, A. & SCOTT, M. *Biochim. Biophys. Acta 403*:256, 1975.
4. MATTIASSON, B., DANIELSSON, B. & MOSBACH, K. *Anal. Lett. 9*:217, 1976.
5. MATTIASSON, B. *FEBS Lett. 77*:107, 1977.
6. MATTIASSON, B., DANIELSSON, B. & MOSBACH, K., in "Enzyme Engineering", vol. 3 (E.K. Pye & H.H. Weetall, eds.), Plenum Press, New York, in press.
7. MATTIASSON, B., MOSBACH, K. & SVENSSON, A. *Biotechnol. Bioeng.*, in press.
8. MATTIASSON, B., DANIELSSON, B., HERMANSSON, C. & MOSBACH, K. submitted for publication.
9. MATTIASSON, B., LARSSON, P.O. & MOSBACH, K. *Nature 268*:519, 1977.
10. DANIELSSON, B., MATTIASSON, B., KARLSSON, R. & WINKVIST, B.F. submitted for publication.

NYLON TUBE BOUND ENZYMES IN SYSTEMS FOR THE DETERMINATION OF SUBSTRATES IN SERUM AND URINE BY ELECTROCHEMICAL AND CALORIMETRIC DETECTORS

P. Kirch, J. Danzer, G. Krisam, and H.-L. Schmidt

Institut for Chemie Weihenstephan
Technical University of Munchen
Freising-Weihenstephan, Federal Republic Germany

By combination of enzyme reactors with different sensors specific and sensitive substrate detection systems have been obtained. Nylon tube bound enzymes (1) have been used for the analysis of substrates in biological samples. The measuring devices consisted of a piston or peristaltic pump, an injection part, and the specific detector; calibration curves were obtained from recorder responses.

With a 10-cm nylon tube reactor (1 U glucose oxidase and 15 U catalase) in a LKB 2107 heat flux calorimeter, glucose in serum and urine (20 μl samples) was determined in the range of 0.05 - 0.6 μmoles $\pm$ 0.01 μmoles (2). Because of the low reaction enthalpy of the urea hydrolysis (1.7 kcal/mole) the determination of urea in urine (0.1 - 1.5 μmoles) with a urease nylon tube reactor was less precise. By combination of the same enzyme reactors with a thermistor sensor (3) and by monitoring the temperature difference in the buffer caused by the reaction heat, a detector was obtained, which had about half the sensitivity but also half of the response time, in comparison to the calorimeter device. With the enzyme thermistor glucose in serum and urine, urea in urine, saccharose in a yeast suspension (0.05 - 0.8 μmole/ml), and lactose in skimmed milk (0.05 - 1.0 μmole/ml) have been determined.

In another flow system for the determination of urea in urine and serum a urease nylon tube (0.8 U/10 cm) was combined with an Orion potentiometric ammonia gas electrode (4). After passage through the enzyme tube the buffer was alkalized in order to operate at optimal sensitivity and stability of the electrode. 50 μl samples of 0.0001 to 0.01 M urea solutions were needed; and deviations of the results by a reference method were not more than 2% for urine and less than 5% for native serum. The time for one analysis was 5 min.

Finally, the glucose oxidase-amperometric oxygen electrode (5) was tested using a nylon tube enzyme reactor (1 U glucose oxidase and 15 U catalase per 10 cm tube). Glucose was determined in solutions of 0.05 - 15 mmol/l within 2 min (error < 5%).

Yields of enzyme immobilization on the nylon tubes were not higher than 2 - 5%; however, the activities of the reactors did not decrease more than 20% within 3 months. As the nylon tube enzyme reactors were not largely affected by native biological liquids, the reactors seem to be suitable for automated devices for routine substrate assay of such samples.

REFERENCES

1. MORRIS, D.L., CAMPBELL, J. & HORNBY, W.E. *Biochem. J.* *147*:593, 1974.
2. SCHMIDT, H.-L. & KRISAM, G. *Biochim. Biophys. Acta* *429*:283, 1976.
3. KRISAM, G. & SCHMIDT, H.-L. in "Application of Calorimetry in Life Sciences," (J. Lamprecht and B. Schaarschmidt, eds.), W. de Gruyter, Berlin, 1977, p. 39.
4. KIRCH, P. & SCHMIDT, H.-L unpublished
5. UPDIKE, S.J. & HICKS, G.P. *Nature* *214*:986, 1967.

ANALYTICAL APPLICATION OF COLLAGEN ENZYMATIC MEMBRANES FOR A GLUCOSE SENSOR

D.R. Thevenot,* P.R. Coulet,** and D.C. Gautheron**

Universite Paris Val de Marne*
Creteil, France and
Universite Claude Bernard (Lyon I)**
Villeurbanne, France

The aim of this work was to test the analytical potentialities of our enzymatic collagen membranes combined with electrochemical detectors. Our glucose sensor is based on glucose oxidation by dissolved oxygen, as catalyzed by β-D-glucose oxidase (GOD, EC 1.1.3.4), to produce gluconic acid and hydrogen peroxide. The GOD is bound to the collagen film by the acyl-azide procedure (1-3). The hydrogen peroxide is detected by anodic oxidation on a platinum disk electrode at +650mV versus a Ag/AgCl reference electrode. A differential method is used by comparison of the current output of two electrodes, one containing GOD(E_1) and the other a non-enzymic membrane (E_2).

Centre Technique du Cuir, Lyon, France, provided industrial collagen membranes. After activation and coupling, the thickness was about 0.3 mm and the GOD activity was 50-100 mU/membrane (1 cm diameter). This activity was controlled by (a) a pH-stat in which the rate of gluconic acid formation was monitored by KOH addition; (b) peroxidase (EC 1.11.1.7) and 2.2'-azino-di-[3-ethylbenzthiazoline sulfonate (6)] (ABTS) with colorimetric detection of H_2O_2 either in batch or in a flow-through cell; or (c) electrochemical detection of H_2O_2 by the E_2 electrode. The GOD membranes were stored in 0.1 M acetate buffer at pH 4.5; after 45 days the activity decrease was 15-20% at 20°C and 50-80% at 30°C.

Two glucose sensor devices were used: dipped or flow-through electrodes. After glucose injection into stirred or circulating 0.1 M acetate buffer pH 5.6, the following data were recorded, where I represents current:

$$(I_1-I_2),\ \frac{d}{dt}(I_1-I_2),\ \frac{d^2}{dt^2}(I_1-I_2)\quad \text{and } I_2.$$

The response time was 30 to 50 sec for the first derivative dynamics and 2 to 3 min for a stationary response of (I_1-I_2). About 2 to 3 min were needed for washing the electrode. The lowest detection limit was 0.1 μM and 0.5 μM for dipped and flow-through electrodes, respectively, with linear calibration curves from 0.1 μM to 3-4 mM glucose.

The selectivity of the sensor was related to the enzymatic properties, with a selectivity coefficient larger than 2,000 for fructose, lactose and sucrose. The use of the differential method widely improved the selectivity for H_2O_2, ascorbate, uric acid, and tyrosine. The variation of the sensor was estimated by 15 successive assays as S.D. lower than 5% in the 10 μM to 3 mM range. The calibration curves remained linear even after 40 hr operation at 30°C and 250 days storage at 4°C, allowing accurate repeated determinations for 160 micro assays tested. Calibration was necessary because of a slight decrease in the calibration curve slope of about 20% after 20 hr operation (cummulated time of measurements) and 20-50% after 40 hr. Sensor responses were temperature dependent (4 to 5%/°C around 30°C) and could be used in the 15-40°C range.

In conclusion the collagen film was suitable for GOD immobilization, with GOD activity maintained and stability enhanced. A membrane loading of 50-100 mU/membrane (1cm diameter) was sufficient to obtain accurate measurements. Using these enzymic membranes and appropriate electronics, we obtained a very high sensitivity over a large concentration range for stationary and dynamic responses proportional to glucose concentrations of 0.1 μM to 3-4 mM. This glucose sensor is suitable for whole blood samples and adaptable for automated analysis with sample processing and data printing.

REFERENCES

1. COULET, P.R., JULLIARD, J.H. & GAUTHERON, D.C. French Patent (ANVAR) No. 2,235,133; 1973.
2. COULET, P.R., JULLIARD, J.H. & GAUTHERON, D.C. *Biotechnol. Bioeng.* *16*:1055, 1974.
3. GAUTHERON, D.C. & COULET, P.R., this volume.

IMMOBILIZATION OF ENZYME ON PMG AND ITS APPLICATION TO THE UREA MONITORING APPARATUS

Y. Yugari, Y. Minamoto, K. Komiya, K. Mitsugi, and N. Mimura*

Central Research Laboratories
Ajinomoto Co. Inc., Kawasaki and
Toranomon Hospital*, Tokyo
Japan

Poly-γ-methyl-L-glutamate (PMG) has a protein-like character (1,2) and easily gives various types of fiber or film. Also, it coats various surfaces of solid materials. PMG has the proper mechanical strength and hydrophilic features to be suitable as a matrix for enzyme immobilization. Entrapment of enzyme in PMG fiber has been tried (3); however, the chemical immobilization of enzyme on polyglutamate (PG) has not been reported yet.

We have succeeded in covalently binding the enzymes to the azidized PG (4,5) and have developed a urea-monitoring apparatus using the immobilized urease column (4-6).

Glass beads (800-100 mesh) were coated with PMG MW $1\text{-}2 \times 10^5$, Ajinomoto Co., B type) and binder, dissolved in chloroform. After drying, PMG was derivatized to its hydrazide with hydrazine hydrate and then to its azide with $NaNO_3$ and HCl. The azidized PG-glass beads were mixed with various enzymes at pH 7.5-9.0 and 4°C for 12-18 hours.

During the process of immobilization little or no enzyme activity was lost; and the amounts of enzyme immobilized on PG were high (Table 1). The optimal pH and Km values of the immobilized enzymes were almost the same as those of the native enzymes. Heat stabilities also

TABLE 1

IMMOBILIZATION OF ENZYMES

ENZYME (origin)	Bound Enzyme (mg/100mg of azidized PG)	Enzyme Act. (%)	
		On Azidized PG	In Washings PG
Urease (Jack Beans)	5.7	95	5
Amino Acylase (Fungus)	1.8	40	59
Glucose Oxydase (Fungus)	1.1	66	33
Uricase (Yeast)	6.3	93	4
Trypsin (Bovine Pancreas)	30	98	1

TABLE 2

HEAT STABILITY OF IMMOBILIZED AND NATIVE ENZYMES

Enzyme	Heat Treatment	Residual act. (%)	
		Immobilized	Native
Urease	75°C,30 min. pH 7.0	67	12
Amino Acylase	70°C,30 min. pH 7.2	73	11
Uricase	45°C, 10 min. pH 8.5	75	5
Glucose Oxidase	65°C,30 min. pH 6.0	65	10
Trypsin	55°C, 30 min. pH 8.0	82	6

were markedly improved (Table 2). Immobilized urease, for example, lost its activity not more than 20%, even after one year preservation at 25°C.

On the application of the immobilized enzymes in clinical fields, we have devised a convenient apparatus which is useful for the determination of the urea content in blood. The apparatus consists of an immobilized urease column, an ammonia electrode, and a built-in computing circuit. The urea nitrogen concentration in blood BUN′ was estimated (6) from the urea nitrogen concentration in the dialyzate (DUN) by the following equation:

$$BUN' = (F_D)\ (F_{QB})\ (F_{QD})\ (DUN)$$

where F_D was an efficiency factor for dialysis, F_{QB} was a factor for blood flow rate, and F_{QD} was a factor for dialyzate flow rate. These three factors were experimentally determined in advance and stored in the computer. The correlation between the BUN value of serum, which was directly determined using an autoanalyzer, and the BUN' value measured by the monitor was linear with the following regression coefficient:

$$BUN = 1.00(BUN') + 0.17 \quad (r = 0.991)$$

By connecting this monitor with an artificial kidney BUN can be determined continuously without any loss of blood from the patient; and the course of dialysis can be recorded on the chart. During dialysis malfunction of the artificial kidney can be detected immediately; and the end point of the dialysis can be foretold, even at the very beginning of the therapy. The details of this work will be published soon (5).

REFERENCES

1. BRESLOW, E., BEYCHOK, S., HARDMAN, K.D. & GURD, F.R.N. *J. Biol. Chem* *240*:304, 1965.
2. TAKASHIMA, S. *Biopolymers* *1*:171, 1963.
3. DINELLI, D. *Process Biochem.* *7(8)*:9, 1972.
4. YUGARI, Y. & MINAMOTO, Y. US Patent No. 3,985,617; 1976.
5. YUGARI, Y. *et al.*, to be published.
6. MINAMOTO, Y., KATO, N., KOMIYA, K., YUGARI, Y., NASUNO, S., SUZUKI, Y., & MIMURA, N. *J. Japanese Medical Instruments* *46*:suppl. 176, 1976.

Session V
INDUSTRIAL APPLICATIONS FOR ENZYMES: STATE OF THE ART

Chairmen: W. Marconi and H. Weetall

INDUSTRIAL APPLICATIONS OF IMMOBILIZED ENZYMES: STATE OF THE ART

R.D. Sweigart

Corning Glass Works
Corning, New York, USA

What is the state of the art worldwide in immobilized enzymes today? What technology is available for sale or licensing? Where are immobilized enzymes being used commercially? Why have some systems succeeded commercially while others have failed? How about the future outlook for immobilized enzymes in industrial applications? The purpose of this paper is to step back from the research bench and attempt to suggest answers to some of these questions to better guide the thinking and planning of those involved in immobilized enzyme development work. Emphasis will be placed on commercial aspects and concerns rather than indepth technical comparisons of available technologies and will be limited to technologies available for general sale or licensing. No attempt will be made to cover medical or analytical applications.

COMMERCIALLY AVAILABLE IMMOBILIZED ENZYME TECHNOLOGY

Table 1 presents a composite of the more visible immobilized enzyme systems available commercially through sales and/or licensing and the companies providing these technologies. Two reasons appear most significant for the small number of immobilized enzyme systems in commercial use. First is the fact that the development and optimization of immobilized enzyme systems can be expensive. Because there is no single general purpose immobilization technique applicable to all enzymes, the return from any application must be great enough to justify the develop-

TABLE 1

COMMERCIALLY AVAILABLE
IMMOBILIZED ENZYME TECHNOLOGY
(SALES/LICENSING)

COMPANY	ENZYMES								
	GI	GA	L	I	PA	AA	A	F	H
Car-Mi	*								
Clinton	*								
Corning	*(a)	0	0	0		0			
Diamond Shamrock			0						
Gist - Brocades	*	0	0	0					
ICI	*								
Lehigh University			0						
Novo	*								
Snam Progetti	0	0	*	0	*	0		0	*
Tanabe Seiyaku					*	X	X	X	

* industrial scale
0 lab or pilot scale
X industrial scale but not presently available for licensing
(a) licensed to CPC International

GI = glucose isomerase
GA = glucoamylase
L = lactase
I = invertase
PA = penicillin amidase
AA = amino acylase
A = aspartase
F = fumarase
H = hydantoinase

ment investment. Such high return applications are not that prevalent at this point in time. Second is the fact that the economics of enzyme immobilization for whatever the reason (enzyme cost, immobilization cost, enzyme life, etc.) are not yet attractive enough to challenge such established process methodologies as soluble enzymes and chemical synthesis. Nor has the technology, with the exception of glucose isomerase, suggested important new products that take real advantage of the economic benefits of enzyme immobilization. However, as more enzymes become commercially available and with expanded research and development in such areas as multiple enzyme systems and the regeneration and reuse of co-factors, the long term outlook should improve significantly. There is one other factor that is encouraging continued research in immobilization. Since this is an expanding technology area, it is very probable that important non-enzymatic applications of immobilization technologies will be developed. An example of this would be the use of immobilized antibodies and antigens on porous inorganic support materials for solid phase radioimmunoassay kits used in medical diagnostics.

GLUCOSE ISOMERASE

Table 2 lists commercially available glucose isomerase technologies. Key patents are listed under each company name. For each system the table indicates the enzyme source together with the basic immobilization process, the reactor configuration normally used, and where available, basic performance data. Although each system listed employs a different enzyme source, a different level of enzyme purity, and a different immobilization technique, there appears to be relatively little difference in reported performance of the systems for which data are available. Column reactors operating in the downflow mode appear to be most popular. Heights up to five meters appear to be feasible. Productivity in the range of 2000 lbs of 42% product (dsb) per pound of enzyme composite is reported for several of the systems. Although the operating conditions are not listed, the recommended temperature and pH are almost identical.

GLUCOSE ISOMERASE VS. OTHER IMMOBILIZED SYSTEMS

Let us now consider more subtle characteristics of the glucose isomerase systems that contributed to its commercial success. To highlight these characteristics Table

TABLE 2

GLUCOSE ISOMERASE SYSTEMS

Company and Patents	Enzyme Source	Immobilization Technique and Performance
CAR-MI U.S. 3,625,828 3,654,081 3,779,869	*Streptomyces olivaceus*	Granular particles containing whole cells; cross-linked with glutaraldehyde. Column reactor; 2,000 lb 42% syrup/lb enzyme mix.
CLINTON U.S. 3,788,945 3,623,953	*Streptomyces wedmorensis*	Enzyme immobilized on DEAE cellulose via ion exchange. Shallow bed reactor
CORNING* U.S. 3,847,740 3,850,751 3,868,304 3,982,997 3,992,329	CPC patented strains of *Streptomyces olivochromogenes.*	Enzyme adsorbed on controlled pore alumina carrier (.46 mm dia; APD 140-220A°). Column reactor; carrier can be regenerated.
GIST BROCADES U.S. 3,834,988 3,838,007	*Actinoplanes missouriensis* (Licensed from Anheuser-Busch)	Gel entrapment of crude mycelium fixed enzyme followed by crosslinking with glutaraldehyde. Columns to 5 m bed depth; t 1/2 500 hrs. Column; 1.5 BVH initial.
ICI U.S. 3,645,848 3,821,086 3,935,068	*Arthrobachter* (Licensed from R. J. Reynolds Tobacco Co.)	Flocculating agent used to immobilize within cells; solid paste then extruded and dried into cylindrical pellets. Columns to 5 m bed depth; 2,000 lb 42% syrup/lb enzyme mix 1.5 BVH initial.
NOVO BR. 1,361,387 1,362,365	*Bacillus coagulans*	Enzyme mixed with inorganic diluent and formed into solid spheres. Columns to 4.5 m bed depth; 1,000-1,360 lb 45%/lb enzyme mix.

*Exclusively licensed to CPC International, Englewood Cliffs N.J. Licensing rights includes sub-licensing rights.

3 compares glucose isomerase to a potentially important but commercially unsuccessful system, glucoamylase, on which extensive work was done in the late 60's and early 70's (1-3).

The cost of the isomerase enzyme is high compared to glucoamylase. This suggests that immobilization costs for glucoamylase must be low and system performance high to compete with the soluble enzyme process. Secondly, there is a significant difference in enzyme stability. For continuous food processing a temperature range of 60°C is strongly preferred to minimize microbial growth prob-

TABLE 3

GLUCOSE ISOMERASE VS. GLUCOAMYLASE VS. LACTASE

	Immobilized System		
BASIC CHARACTERISTICS	GI	GA	LACTASE
ENZYME COST	HIGH	LOW	HIGHER
ENZYME STABILITY	HIGH*	LOW*	HIGH
CONVERSION EFFICIENCY	HIGH	LOW	HIGH
END PRODUCT	NEW	OLD	NEW

* Compared at 60°C

lems. This is an acceptable temperature range for glucose isomerase. Because of the poor thermal stability of commercially available glucoamylase, however, it is necessary to operate immobilized reactors at about 40°C to prolong enzyme life and achieve favorable economics. Pilot scale-up work at Iowa State University (4) has suggested that such operation is feasible if the total system is maintained in a sterile condition. Corn sweetener producers, however, show concern about being able to maintain such conditions in large-scale production.

Conversion efficiency indicates the ability of the immobilized enzyme system to produce the required end product at attractive costs. Again, glucose isomerase performs well in this area, while glucoamylase struggles to produce a 93 dextrose equivalent (D.E.) product when a 97 D.E. product is needed for commercial viability.

The last characteristic listed in Table 3 compares the nature of the end products. Immobilized glucose isomerase made possible the production of an important new product, high fructose corn syrup, while the glucoamylase system was aimed at displacing a well-established, highly optimized soluble enzyme process for high volume production of glucose. In the latter situation, system performance and economics must be dramatically superior to get serious commercial attention. At present, the potential economic benefits of immobilized glucoamylase appear marginal except for possible capacity expansion or new product applications involving low temperature processing (e.g., light beer).

In hindsight, it would appear that the enthusiasm for immobilized glucoamylase may have been prompted not only by a potentially attractive market opportunity but also by the early attitude that immobilization had to be economically superior to the use of soluble enzymes. Hopefully, the industry is now more mature in selecting applications for commercial development.

Table 3 also extends the comparison to include a potentially important new immobilized enzyme, lactase, used to hydrolyze lactose in milk or cheese whey into sweeter and more soluble sugars. This system appears similar to glucose isomerase with an even higher enzyme cost, good enzyme stability (particularly with lactase derived from *A. Niger*), and conversion efficiency. Most important, however, is the fact that hydrolyzed lactose represents a new product opportunity. Several scale-up evaluations are now in progress to define the long term performance/economics of various immobilization techniques and to consider markets for the hydrolyzed products.

ECONOMIC CONSIDERATIONS: IMMOBILIZED ENZYME SYSTEMS

All data presented thus far reinforce the importance of economics to the commercial success of any immobilized enzyme system. We have already discussed the importance of enzyme cost. Lab-scale work can be carried out with experimental quantities of enzyme, but for commercial scale-up, it must be determined that the required preparation will be available in the volumes needed and at acceptable costs. Immobilization costs include the costs of the carrier material and the enzyme attachment process. If the carrier can be repeatedly regenerated and reused, as is the case with porous inorganic supports, initial carrier cost becomes less of a factor. It is critical that the lab scale work reflects to the extent possible the operating conditions that will be encountered in scale-up.

Capital investment can be another key economic consideration. If high substrate volumes are involved, an immobilized enzyme reactor capable of handling high flow rates with minimal pressure drops can significantly reduce equipment size and investment. This suggests carriers that do not deform under pressure. Cleanup costs might well include such things as substrate pre-treatment to minimize enzyme life. Be sure such costs are included in the overall economic assessment of the total system. If a food product is involved, daily sanitization of the reactor may be required. This is not only a cost consideration in itself, but may also be a potential cause of enzyme deactivation or carrier deterioration, especially if harsh sanitizing solutions are required. Also, if any component of the immobilized enzyme system could be detrimental to the functionality of the end product, the necessary cleanup costs must be included in total economic evaluations.

OPTIONS AVAILABLE FOR IMMOBILIZED ENZYME DEVELOPMENT

It is obvious that no one immobilized enzyme technology will best meet all the needs of each industrial application. For certain conversions it has been shown that soluble enzymes provide best performance and economics. Therefore, the researcher interested in applying immobilized enzymes must take a hard look at the options available to him.

An intermediate alternative is the collaborative effort with a qualified partner. This can be appropriate if one is strong in one facet of the technology but requires outside expertise to develop a total solution. The fastest, lowest initial cost and minimal risk approach is represented by purchasing of available technology. An excellent example of this strategy has been the purchase of glucose isomerase technology by a number of corn sweetener manufacturers. This has enabled them to quickly establish a position in an expanding market with little or no investment in research and development. This alternative can be especially appropriate for companies with limited technology in the particular application of interest, but with the market resources and position required to gain share of the market quickly.

REFERENCES

1. WILSON, R.J.H. & LILLY, M.D. *Biotechnol. Bioeng.* *11*: 349, 1969.
2. BACHLER, M.J., STRONDBERG, G.W. & SMILEY, K.L. *Biotechnol. Bioeng.* *12*:85, 1970.
3. WEETALL, H.H. & HAVEWALA, N.B. in "Enzyme Engineering," (L.B. Wingard, ed.), J. Wiley, New York, 1972, p. 241.
4. WEETALL, H.H., VANN, W.P., PITCHER, JR., W.H., LEE, D.D., LEE, Y.Y. & TSAO, G.T. in "Immobilized Enzyme Technology Research and Applications," Plenum Press, New York, 1975, p. 269.

REGENERATION OF ATP BY IMMOBILIZED MICROBIAL CELLS AND ITS UTILIZATION FOR THE SYNTHESIS OF NUCLEOTIDES

H. Samejima, K. Kimura, Y. Ado, Y. Suzuki* and T. Tadokoro.*

Tokyo Research Laboratory
Kyowa Hakko Kogyo Co., Ltd.
Tokyo, Japan and
Kyowa Petro-chemicals Co., Ltd.*
Yokkaichi-shi, Mie-Prefecture, Japan

A number of biosynthetic methods for the production of various important nucleotides from their precursors using microbial cells have been developed in Japan (1); and some of these nucleotides have been produced industrially for more than a decade. For the production of those nucleotides an energy-rich phosphate compound, namely ATP, usually is required; and it is supplied to the reaction media either exogenously or by the naturally occurring ATP regeneration system involved in microbial cells. In the present paper the authors describe the immobilization of microbial cells and the utilization of such immobilized cells as the ATP regenerator for the continuous production of various nucleotides.

MODEL REACTIONS FOR NUCLEOTIDE BIOSYNTHESIS

Three examples were chosen as model reactions for the nucleotide biosynthesis. The first example was the production of ATP from AMP by *Saccharomyces cerevisiae* cells (2). In this process 2 moles of ATP were regenerated from ADP in order to produce 1 mole of ATP from AMP and the glycolytic pathway of the yeast cells. The second example was the production of cytidine di-phosphate-choline (CDP-choline) from CMP and choline chloride by *S. cerevisiae* (3,4). CDP-choline is a precursor of the phospholipid, lecithin, and is used as a therapeutic agent for brain nerve injury. For the synthesis of CDP-choline from CMP and choline chloride the enzymes choline kinase, CMP-kinase, and CDP-choline pyrophosphorylase were involved; and 3 moles of ATP were regenerated from equimolar ADP in order to synthesize 1 mole of CDP-choline. The third example was the production of coenzyme A (CoA) from pantothenate, cysteine, and ATP by *Brevibacterium ammoniagenes* cells (5). These cells had been immobilized already by others (6,7). Five enzymes were involved in this process. In order to produce 1 mole of CoA 3 moles of ATP were regenerated and another 1 mole of ATP was supplied as a raw material for building up the CoA molecule.

IMMOBILIZATION OF MICROBIAL CELLS

The microbial cells mentioned above were immobilized by polyacrylamide gel and by microencapsulation in ethylcellulose.

For the acrylamide entrapment cyclohexane, as dispersion medium, and sorbitan monolaurate, as dispersing agent, were mixed and stirred vigorously under nitrogen gas bubbling at 10 to 15°C for 10 min. To this mixture was added a suspension of microbial cells in 0.1 M phosphate buffer, pH 6.8, containing 25% mixed monomers of acrylamide and N,N'-methylene-bis-acylamide (95: 5, w/w), 5% N,N,N',N'-tetramethylethylenediamine (as polymerization stabilizer) and 2.5% ammonium persulfate (as polymerization initiator). The mixture was stirred vigorously under nitrogen gas bubbling at 10 to 15°C for 30 min. Then, another 2.5% ammonium persulfate was added; and stirring was continued for another 30 min. After the reaction the resultant polymerized gel was recovered by filtration and washed with 10 times the volume of 0.1 M phosphate buffer, pH 6.8. The immobilized cell preparations were spherical beads with diameters of 100 to 500 microns.

The micro-encapsulation of microbial cells in ethylcellulose, with or without chitosan pretreatment, involved mixing 10 g ethylcellulose powder, 145 g benzene (as dispersion medium), 55 g n-hexane (as dispersion aid), and 4 g sorbitan monolaurate (as dispersing agent) with vigorous stirring at 5°C for 10 min. To this mixture was added a suspension of microbial cells in 55 ml water, containing 1% NaCl (and 1% chitosan). The mixture was stirred vigorously for 10 min in order to form a water/oil (w/o) emulsion. This emulsion was then dropped into 600 ml of 1% polyethyleneglycol aqueous solution and stirred vigorously at 5°C for 30 to 40 min in order to form a water/oil/water (w/o/w') emulsion. Then 1,000 ml of n-hexane was added to the emulsion to precipitate the ethylcellulose encapsulated cells. The encapsulated cells were recovered by filtration and washed with 5 times the volume of 0.1 M phosphate buffer, pH 7.5. The encapsulated cell preparations also were spherical beads having diameters of 100 to 300 microns.

Some characteristics of immobilized cells thus prepared are summarized in Table 1. In comparison with polyacrylamide gel, microencapsulation with ethylcellulose was able to entrap 5 to 6 times more microbial cells in the unit volume of the final gel and to retain 4 to 5 times higher enzymatic activities. The mechanical strength of polyacrylamide gel was superior to that of ethylcellulose microcapsules; but the latter also could be used for long-run column operations without appreciable mechanical troubles. It is still hard to say which method would give better stability in long-run operations because there are many factors affecting the stability of immobilized cells.

TABLE 1

COMPARISON OF METHODS FOR ENTRAPMENT OF MICROBIAL CELLS

Entrapment Method	*S. cerev.* ATP			*B. ammonia.* CoA			*S. cerev.* CDP-C		
	a	b	c	a	b	c	a	b	c
Polyacrylamide-Gel Entrapment	40	14	>5	60	37	14	40	14	>2
Microencapsulation with Ethylcellulose	200	62	10	--	--	--	200	62	--
Microencapsulation with Ethylcellulose & Chitosan	300	71	>5	250	52	>5	220	62	10

a: Entrapped Cells (mg-cell/mg-Gel)
b: Retention of Activity (%)
c: Stability in Operation (Days)

CONTINUOUS COLUMN OPERATIONS OF IMMOBILIZED CELLS

The typical apparatus for continuous column operation of immobilized cells consisted of two column reactors in series, with each reactor equipped with a gas purge tube on the top to eliminate CO_2 gas evolved during the reaction. The substrate solutions were sterilized, either by autoclaving or by millipore-filtration, and were held aseptically in a feed tank before charging to the reactors. The substrate solutions were fed to the reactors at a constant flow rate by a rate-controlled pump, as ascending flow. When sterile conditions were needed the entire process line of apparatus was sterilized by passing lysozyme solution through the system or immobilized lysozyme columns were placed before and after the main reactor.

In Fig. 1 the results of the continuous production of ATP from AMP using ethylcellulose encapsulated *S. cerevisiae* cells are shown. In these experiments the substrate solution was composed of 166 mM glucose, 43 mM 5'-AMP (15 mg/ml), 4 mM $MgSO_4$, and 1 M phosphate buffer of pH 8.0. The solution was passed through the column reactor at a space velocity of 0.2 hr^{-1} and at 35°C. At the beginning the molar conversion from AMP to ATP was more than 70%. Though the enzyme activity decreased gradually, still a conversion of more than 50% was maintained during 10 days of operation.

In a similar manner the continuous production of CDP-choline was carried out using ethylcellulose-chitosan encapsulated *S. cerevisiae* cells. The substrate solution

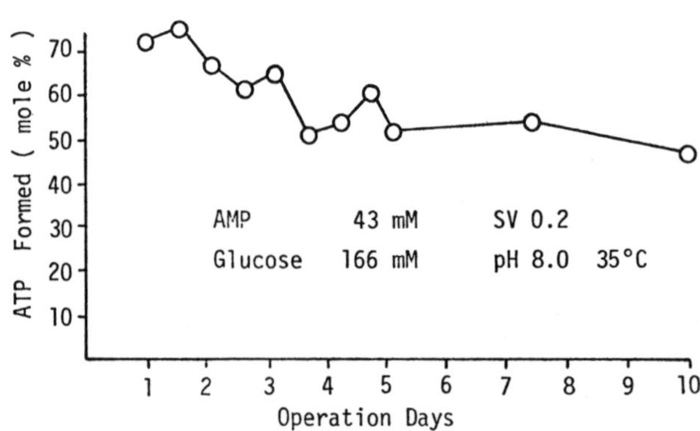

Fig. 1. Long Run Test of Ethylcellulose Encapsulated *S. cerevisiae* Cells for Production of ATP from AMP

was composed of 400 mM glucose, 20 mM 5'-CMP (6.5 mg/ml), 100 mM choline chloride, 50 mM $MgSO_4$ and 0.4 M phosphate buffer of pH 7.5. The solution was passed through the reactor at a space velocity of 0.25 and at 35°C. During the first eight days the molar conversion from 5'-CMP to CDP-choline was maintained at 50 to 60%.

From the above experimental data the approximate turnover numbers of endogenous ATP in immobilized yeast cells were calculated for both ATP and CDP-choline productions (Table 2). The concentrations of endogenous ATP in the immobilized yeast cells used for ATP and CDP-choline productions were determined to be 4 μM and 4.3 μM, respectively. Conversion in the early stages of operation from AMP to ATP and from CMP to CDP-choline were assumed to be 70% and 50%, respectively. Under such assumptions the turnover numbers of endogenous ATP, for ATP production and CDP-choline production, were calculated to be 3,010 and about 1,400 cycles/hr, respectively. These turnover numbers of ATP suggest that yeast cells themselves might be useful practical tools for ATP regeneration.

TABLE 2

ESTIMATION OF TURN OVER NUMBER OF ATP

Item	ATP		CDP-C	
Key Substrate (mM)	AMP	43	CMP	20
Product	ATP	30.1	CDP-C	10
Conversion (%)		70		50
ADP to ATP (mole/mole product)		2		3
Endogenous ATP (mM)	4×10^{-3}		4.3×10^{-3}	
Reaction Time (hr)		5		5
T.O.N. of ATP (cycles/hr)*		3010		about 1400

$$\text{*T.O.N. of ATP} = \frac{(\text{mM of Product})\ (\text{No. of ATP required})}{(\text{Conc. of endo. ATP (mM)})\ (\text{Reaction Time (hr)})}$$

The production of CoA using polyacrylamide gel entrapped cells of *Brevibacterium ammoniagenes* is shown in Fig. 2. In this experiment immobilized lysozyme columns were placed just before and after the main reactor. The substrate solution was composed of 10 mM sodium pantothenate (2.4 mg/ml), 15 mM disodium ATP, 10 mM cysteine, 10 mM $MgSO_4$, and 0.15 M phosphate buffer of pH 6.8. The solution was passed through the reactors at a space velocity of 0.04 hr^{-1} and at 37°C. The molar conversion from pantothenate to CoA was about 17% (CoA 0.9 mg/ml) at the beginning of the reaction and 12.5% after 9 days. For the production of CoA 1 mole of ATP was incorporated into each mole of CoA produced. Therefore, part of the ATP required for the reaction was added in the substrate solution; but the conversion was not as good as for the production of ATP or CDP-choline. Further improvement of the CoA production was tried; and the results are shown in Fig. 3. The reactions were carried out in stirred vessels for 16 hours. The yield of CoA was much higher with both cells than with *B. ammoniagenes* cells alone. Evidently the *S. cerevisiae* cells contributed to the ATP regeneration required for the total reaction of CoA synthesis.

DISCUSSION

As hitherto mentioned, the naturally occurring ATP regeneration system in microbial cells seems a very useful tool of ATP regeneration required for the synthesis of energy-rich compounds by immobilized enzyme or cell systems because such natural ATP regenerating systems are

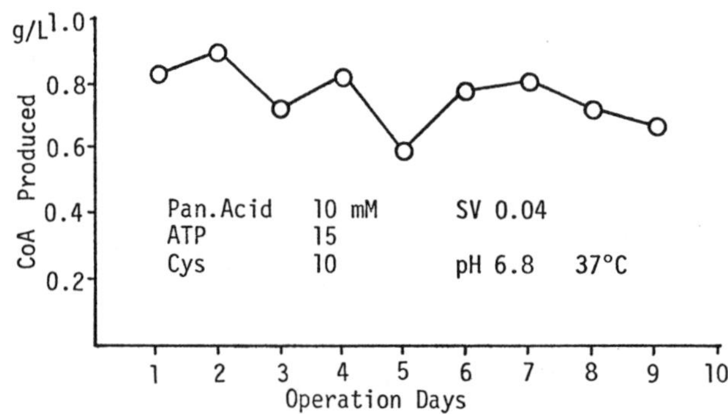

Fig. 2. Long Run Test of Polyacrylamide Gel Entrapped Cells of *B. ammoniagenes* for Production of CoA

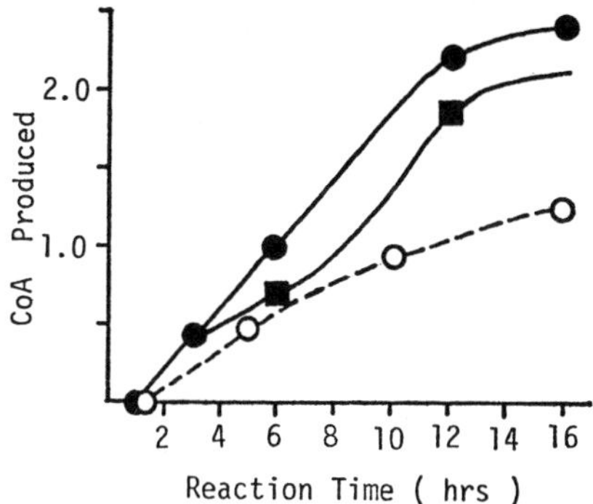

Fig. 3. Synthesis of CoA by Co-immobilized *B. ammoniagenes* and *S. cerevisiae*. (●) Native cells, both types, with the same solution as in Fig. 2 except 155 mM glucose, 400 mM inorganic phosphate, and 15 mM AMP in place of the ATP; (square) both type cells co-immobilized; (o) *B. ammoniagenes* alone with a solution of 10 mM pantothenate, 15 mM ATP, and 10 mM cysteine. The vertical axis is in g/l.

usually rather cheap and stable, and also inexpensive energy donors like glucose can be used in such systems. Furthermore, co-immobilization of strong ATP regenerator cells with other cells can broaden the variety of enzyme reactions with economic values.

Whitesides *et al.* (8) suggested that immobilized adenyl kinase and acetate kinase was the most economical system then available for ATP regeneration. However, we believe that in some cases naturally assembled ATP regenerating systems must be more economical if suitable targets and combinations can be selected.

REFERENCES

1. OGATA, K. in "Microbial Production of Nucleic Acid-Related Substances," (K. Ogata, S. Kinoshita, T. Tsunoda, and K. Aida, eds.), Kodansha, Tokyo, 1976.
2. TOCHIKURA, T., KUWAHARA, M., YAGI, S., OKAMOTO, H., TOMINAGA, Y., KONO, T., & OGATA, K. *J. Ferment. Technol.* 45:511, 1967.

3. NAKAYAMA, K. & HAGINO, H. Japanese Patent No. 48-40757; 1973.
4. TOCHIKURA, T., KIMURA, A., KAWAI, H., TACHIKI, T., & GOTAN, T. *J. Ferment. Technol.* *48*:763, 1970.
5. OGATA, K. Japanese Patent No. 49-13994; 1974.
6. SHIMIZU, S., MORIOKA, H., TANI, Y., & OGATA, K. *J. Ferment. Technol.* *53*:77, 1975.
7. CHIBATA, I., KAKIMOTO, T., NISHIMURA, N., & NABE, K. Japanese Open Patent No. 50-126884; 1975.
8. WHITESIDES, G.M., LAMOTTE, A., ADALSTEINSSON, O., & COLTON, C.K. *Meth. in Enzymology* *44*:887, 1976.

PRODUCTION OF 6APA IN THE PENICILLIN G FERMENTATION PLANT BY USING FIBER-ENTRAPPED PENICILLIN AMIDASE

F. Giacobbe,* A. Iasonna,* and F. Cecer**

Biochem Design S.p.A.* and
SNAM Progetti Microbiological Laboratory**
Roma* and Monterotondo**, Italy

In the last few years improvements in enzyme immobilization techniques have enabled enzymatic hydrolysis of penicillin G for the production of 6APA to become a viable alternative to chemical hydrolysis. This paper presents a novel process method for producing 6APA by enzymatic hydrolysis. In the overall process penicillin G is separated from the fermented broth by solvent extraction; the end product of the solvent extraction is generally a crude concentrated solution of penicillin G,K from which the penicillin is crystallized by azeotropic distillation with butanol. In the conventional 6APA enzymatic process the dried crystals of penicillin G are dissolved in water, and sent to the hydrolysis stage to yield 6APA and phenyl acetic acid. It would appear that if the 6APA plant was located adjacent to the penicillin fermentation plant, then it would be possible to hydrolyze directly the crude solution of penicillin G obtained from the solvent extraction, by-passing the crystallization stage. In this stage about 8-10% of the penicillin G is lost either in the mother liquor or through degradation during the azeotropic distillation. In addition it should also be possible to recover the phenyl acetic acid, which could be used as a precursor in the penicillin fermentation.

Research work was started at SNAM Progetti to investigate this possibility. A crude solution of penicillin was obtained from an industrial plant of a leading penicillin G manufacturer and was hydrolyzed directly using fiber-entrapped penicillin amidase. The results initially were very disappointing. The residual activity of the fiber-entrapped enzyme was drastically reduced when the solution of penicillin G, prepared from the crystallized product, was replaced with a crude penicillin solution. In spite of the fact that the crude solutions had been stored at low temperature, they showed signs of product degradation. So, initially it was believed the loss of activity was due to these degradation products of penicillin G. Subsequent research, however, has shown that the probable reason for the loss of activity was the presence of compounds which were transferred from the fermented broth to the crude solution during the extraction stage and subsequently removed during the crystallization. Hence, the pure crystals did not show this loss of activity. Finally, we discovered a method to remove these inactivating agents from the crude penicillin solution. We have now succeeded in hydrolyzing the treated crude penicillin solution without a sensible loss in residual activity of the fiber-entrapped amidase (Fig. 1). When a pretreated crude solution was hydrolyzed, the residual activity was reduced by about 10%, compared to pure penicillin solution. In addition the initial residual activity could be restored any time if the crude treated solution was replaced by the pure solution of penicillin G.

The kinetics of the hydrolysis of the pure and pretreated penicillin solutions are compared in Fig. 2. It is seen that there were no substantial differences when the consumption of caustic soda during the hydrolysis was plotted against time for both solutions. The recovery of 6APA from the treated solution of penicillin G was then performed with only minor modification for the usual operating procedure used in the conventional process. The characteristics of the 6APA produced from the treated solution were identical to those of 6APA obtained from the pure penicillin solution and at about the same degree of purity. The yield of 6APA obtained through by-passing the crystallization is shown in Table 1, where the conventional process utilizing a pure penicillin solution is compared with the integrated process utilizing a pretreated solution of crude penicillin. From this table it can be seen that the hydrolysis yield remained unchanged, and the integrated process presented a yield about 5% higher than that of the conventional process.

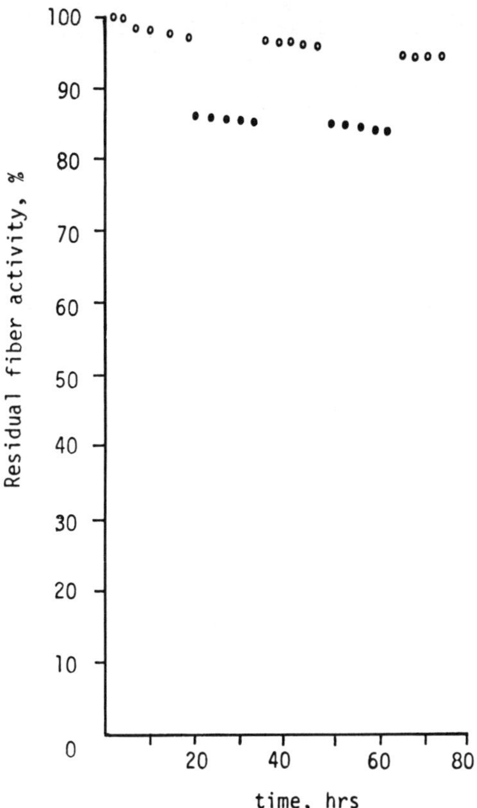

Fig. 1. Hydrolysis of Penicillin G,K residual activity with fiber-entrapped penicillin amidase. (o) Pure penicillin solution; (●) treated penicillin solution.

If we now translate these results into economics and compare the conventional, the integrated enzymatic process, and the chemical process, we can observe the following:

(a). The yield of the chemical process was slightly higher than that of the enzymatic process, which was due to higher losses of 6APA in the mother liquor during the recovery of 6APA from the enzymatically hydrolized solution. This solution had a lower concentration of 6APA because it is not possible to hydrolize enzymatically a concentrated solution of penicillin G since it causes inhibition of the enzyme.

(b). Less expensive raw materials were involved in the enzymatic process due to its very simple mechanism of reaction.

(c). The cost of utilities was lower in the enzymatic hydrolysis because the hydrolysis was performed at ambient temperature and refrigeration was not required.

(d). The investment cost was lower in the enzymatic hydrolysis due to a simpler processing scheme.

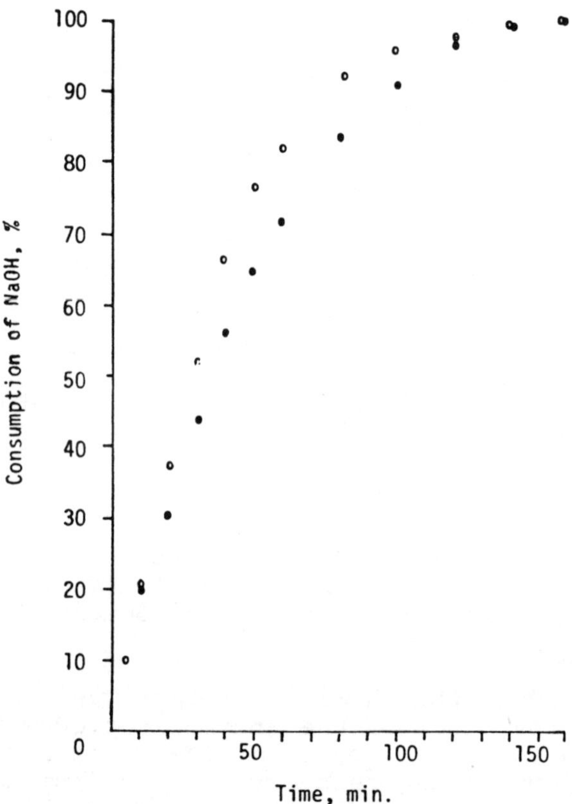

Fig. 2. Consumption of NaOH during the hydrolysis. (o) Pure penicillin G solution; (●) treated penicillin G solution.

The above observations have been quantified in Table 2, which indicates a breakdown of the manufacturing cost of 6APA for the three processing alternates. These manufacturing costs are totally dependant on the price of the penicillin G. The price of penicillin G is cyclical in the market and subject to the fluctuation. Fig. 3 shows the manufacturing cost of the 6APA as a function of the penicillin price. For the integrated process, the cost of the penicillin G can be discounted by the cost of the crystallization stage, which has been estimated at about 2.0\$/BU. The enzymatic processes also show lower manufacturing costs of 6APA, both in the low and high penicillin G price ranges. This means that in the small scale production of 6APA, where higher prices have to be paid for

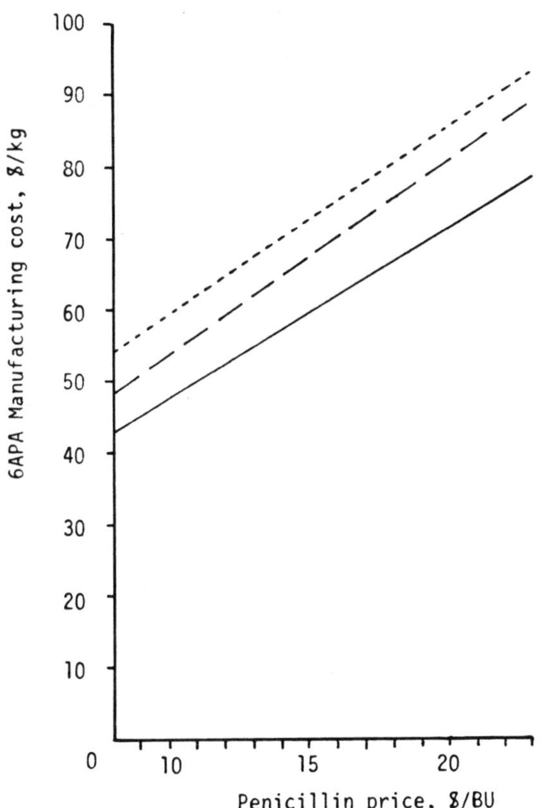

Fig. 3. 6APA manufacturing cost versus penicillin G,K price based on a plant capacity of 40,000 kg/yr. (--) Chemical process; (— —) enzymatic conventional process; (—) enzymatic integrated process.

TABLE 1

COMPARISON OF THE YIELDS OF 6APA PRODUCED BY CONVENTIONAL AND INTEGRATED PROCESSES

Item	Conventional process (moles)	Integrated process (moles)
Crude penicillin G solution	100	100
Crystallization losses	7.3	-
Treating losses	-	0.5
Penicillin G feed to the hydrolysis	92.7	99.5
6APA residue in the fiber	8.3	8.32
6APA in the hydrolyzed solution	79.8	85.80
6APA losses in the mother liquor	7.4	8.78
Total 6APA produced	80.7	85.34
Hydrolysis yield, %	95.0	94.6
Crystallization yield, %	91.6	90.0

TABLE 2

6APA MANUFACTURING COST*

Item	Chemical process	Conventional enzymatic process	Integrated enzymatic process
Penicillin G	43.72	46.01	39.65
Chemicals	7.49	0.78	1.39
Fiber-entrapped amidase	-	2.67	2.67
Utilities	4.64	0.91	0.91
Labor	3.12	3.95	3.95
Operating supplies	0.29	0.36	0.36
Depreciation	6.16	4.99	3.85
Other overhead	2.99	2.82	2.54
Total	68.41	62.49	55.28

*Figures in $/kg; basis of 40,000 kg/yr plant capacity and penicillin G price of $14.35/BU

the penicillin G raw material, the enzymatic hydrolysis can be competitive with the chemical hydrolysis. The integrated process in particular appears to be very attractive for those manufacturers of penicillin G who are able to integrate the production of penicillin G with that of 6APA. In addition, the values shown in Fig. 3 have not included the possibility of recycling the phenyl acetic acid. This could contribute an estimated saving of 10% tawards the cost of the raw materials needed for the production of penicillin G. This assumes that 60% of the phenyl acetic acid is recovered and recycled. The chemical route has been well developed over the years and is nearly at a practical maximum; while the enzymatic routes have considerable room for improvement. Crystallization losses can be reduced by higher fiber loading, thereby allowing more concentrated penicillin G solutions to be used. We have concluded that the integrated process presents a large potential. The economics of this integrated process are sufficiently attractive to study the potential for revamping existing chemical 6APA plants located adjacent to penicillin fermentation facilities or to allow a penicillin manufacturer, which is not yet producing 6APA, to consider entering the market. For some time now, penicillin G has been considered a bulk commodity. With this process it is now conceivable that 6APA will achieve commodity status.

EXPERIMENTAL PROCEDURES USED

The standard procedure described by Dinelli (1) was used to immobilize the penicillin acylase. 10 kg of a solution of penicillin acylase containing glycerol, which had been phosphate buffered at pH 8, was added to a solution of 5 kg cellulose triacetate in 71.4 l methylene chloride at 4°C with stirring. The emulsion, which was formed, was extruded through a spinnerette into a coagulation bath, containing toluene. The fibers formed were then dried to remove all organic solvents.

The fibers containing the entrapped enzymes were divided into five equal portions, each corresponding to 1 kg of cellulose acetate. A portion was placed, parallel to the longitudinal axis, in a jacketed column (43 x 14 cm), and then attached at the bottom. 20 l of 6% w/v penicillin G (potassium salt), buffered at pH 8.2 by 0.02 M phosphate buffer, was then continuously pumped through the jacketed column until a 97% or more conversion had been achieved. The pH was maintained at 8.2 by the automatic addition of

NaOH solution. Solutions of the potassium salt of penicillin G and the crude solutions, extracted from the fermentation broth, were alternatively hydrolyzed. In order to determine the residual activity of the penicillin acylase, the times necessary to achieve 50% conversion of penicillin G to 6APA, were measured and compared to the times when the fiber had not been used previously.

The amount of 6APA in the hydrolized penicillin G solutions was measured by the hydroxilamine method (2). The purity of the 6APA extracted from the final reaction mixture was determined by the same method. Iodometric (3) and p-dimethylaminobenzaldehyde (4)(5) methods were also used.

REFERENCES

1. DINELLI, D. *Process Biochem.* 7:9, 1972.
2. BATCHELOR, F.R., CHAIN, E.B., HARDY, T.L. MANSFORD, K.R.L. & ROLINSON, G.H. *Proc. Roy Soc.* B154:498, 1961.
3. ALICINO, J.F. *Anal. Chem.* 33:648, 1961.
4. BOMSTEIN, J. *Anal. Chem.* 37:576, 1965.
5. BALASINGHAM, K., WARBURTON, D., DUNNILL, P. & LILLY, M.D. *Bioch. Bioph. Acta* 276:250, 1972.

ENZYMATIC DEACETYLATION OF CEPHALOSPORINS

J. Konecny

CIBA-GEIGY LIMITED
Basel, Switzerland

The acetyl esterase from *B. subtilis* ATCC 6633 is a stable, nonspecific and relatively costly enzyme with a molecular weight of about 150,000 and an activity maximum at pH 8. The V_{max} for all substrates examined are typically 110 to 130 μmole/min/mg pure protein; and the values of K_m range from 7 to 15 mM for various cephalosporins. Nearly pure enzyme preparations are obtained in high yields from the cell extract (1,2). The enzyme has been immobilized on several carriers, including controlled pore glass (CPG) and brick powder (2,3). The present work describes the analytical and preparative uses of this esterase as a catalyst for deacetyling cephalosporins, with emphasis on the choice of reaction conditions, the carrier, and the reactor for specific applications.

Cephalosporin antibiotics and intermediates are costly labile compounds, which are hydrolyzed selectively at various positions by appropriate enzymes (4-6) to give the corresponding carboxylic acids. Non-enzymatic hydrolysis of the materials is acid and base catalyzed (7) and leads to a mixture of products (7,8) the rates being as high as 2%/hr at pH 8 and 25°C.

EXPERIMENTAL

All experiments were carried out at 25°C. Activities, defined as 1 unit = 1 μmole/min, were measured with 85 mM 2-methoxyethyl acetate substrate at pH 8 (9,10). A differential recirculation reactor (11) was employed in all the work except in the continuous runs, where a 3 mm diameter column containing 2 ml of catalyst was used. In the analytical titrations a solution of substrate in 40 ml 0.1 M KCl/10 mM phosphate buffer of pH 8.0 was circulated at 10 l/hr through 2.4 ml of CPG-esterase. The Radiometer titration assembly with 1 N NaOH was used. Microtitrations with soluble enzymes were carried out in 5 ml of the same buffer with 0.1 N NaOH in the Radiometer burette. The present work was carried out with zirconia-clad CPG of 0.2-0.8 mm particles and brick powder of 0.4-0.5 mm particles, packing density 0.95 g/ml, internal surface 2 m^2/g, internal volume 0.6 ml/g, pores mostly > 550 Å, and binding capacity for pure esterase about 4 mg/g (2). Polyethylene imine-treated (12) brick powder and alkylamino CPG were derivatized with glutaraldehyde (13,2). The free esterase had a specific activity of 107 units/mg protein, lactamase P-99 (14), as assayed with 15 mM cephalosporin C at pH 7, a specific activity of 620 units/mg. About 700 units of the lactamase were coupled by the same procedure to 1 g of alkylamino CPG.

RESULTS

Yield losses due to side reactions depend on the employed levels of enzymatic activity (residence time) and also on the pH, as shown in Fig. 1. The activity of the two supports measured at low buffer concentrations, is shown as a function of the enzyme coupled in Fig. 2. The costs of enzyme enrichment from 1 to 100 units/mg protein are low enough to permit the use of the cheap carrier having a low binding capacity.

Alkalimetric titrations were developed for the assay of samples which were used as analytical standards in the liquid chromatography of cephalosporin C, cephacetril, and their deacetyl derivatives (7). By using the esterase to assay the parent compounds and to prepare analytically defind solutions of the deacetyl derivatives, it was established that the lactamase-catalyzed reactions were stoichiometric and generated exactly 1 equivalent acid/mole deacetyl cephalosporins and 2 equivalent/mole of the parent compounds.

The curves shown in Fig. 3 illustrate the operational stability of the CPG-esterase when used in a recirculation reactor at low buffer concentrations and high recirculation rates. The runs were carried out in uninterrupted succession with 5 g cephacetril in 100 ml 20 mM phosphate.

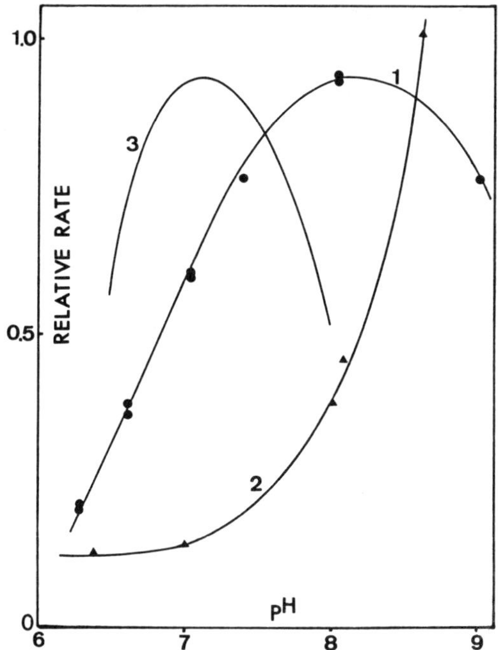

Fig. 1. Effect of pH on the rate of 1) enzymatic deacetylation of cephacetril, 2) non-enzymatic decomposition of the substrate, and 3) on the ratio of the two rates.

The bed had a cross sectional area of 3.8 cm^2 and contained 6.7 g catalyst, prepared by coupling 250 units esterase/g. The reaction time for 98% conversion was initially about 0.5 hr at pH 8 and 1 hr at pH 7. The thoroughly washed catalyst was stored under 0.1 M phosphate pH 7 at room temperature over weekends and over night. The used catalysts were subsequently stored for 60 days at 4°C and 30 days at room temperature without loss of activity. After an initial activity loss of 20% brick powder-esterase displayed comparable operational stability.

Continuous deacetylation runs were carried out with 2 g brick powder. The quantity of esterase coupled was 114 units/g. The curves in Fig. 4 show the conversion of two substrates, with different K_m values (2), as a function of space velocity (bed volumes/hr). The pH of the feed,

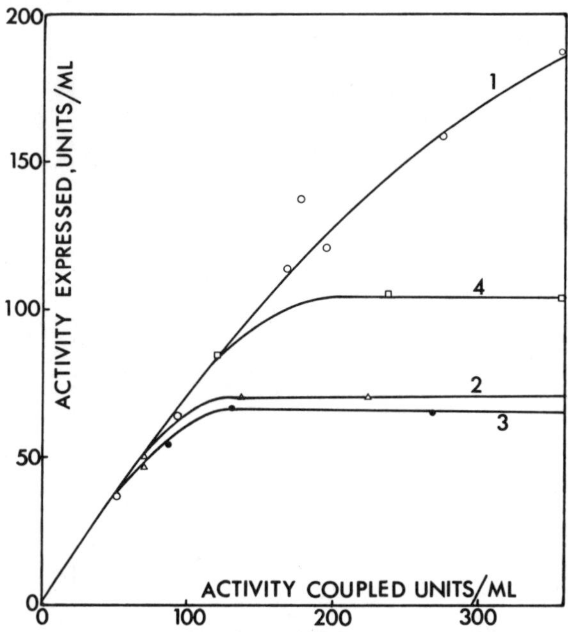

Fig. 2. Relation between expressed activity and the quantity of enzyme consumed in coupling to 1 ml of carrier: 1) CPG (0.2-0.8 mm particles) and pure enzyme (spec. activity 120 units/mg protein), 2) same with crude enzyme (spec. act. 13), 3) 0.4-0.5 mm brick particles and nearly pure enzyme (spec. act. 107), 4) same but 0.2-0.3 mm brick particles. Measurements in 0.1 M NaCl containing 10 mM phosphate (CPG) or 20 mM phosphate (brick powder) at pH 8. Packing density of brick powder 0.95 g/ml, of CPG 0.46 g/ml.

containing 200 mM phosphate, was 8.0; that of the effluent at 99% conversion was about 6.8. The data in Fig. 5 show the substrate conversion at constant flow rates as a function of operational time. Saturation of the feed with toluene failed to inhibit microbial growth. Operations had to be interupted three times to wash the catalyst and repack the columns.

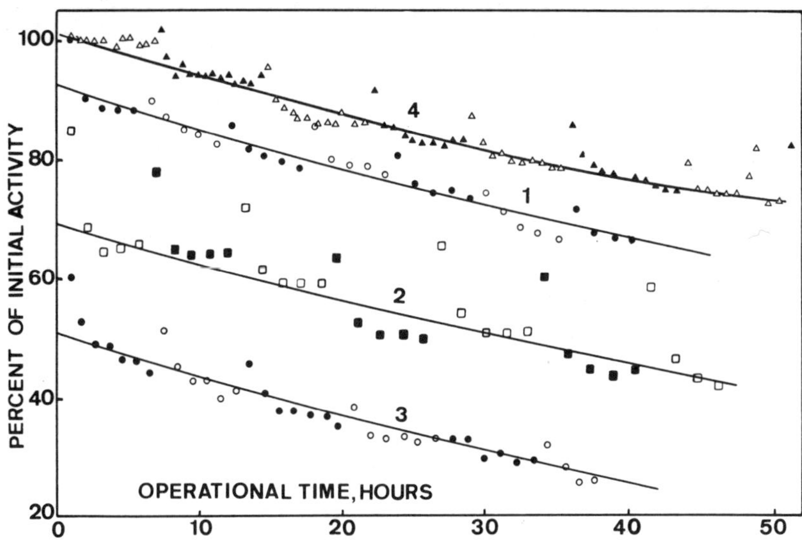

Fig. 3. Operational stability of CPG-esterase in the recirculation batch reactor at the following recirculation rates: 1) 12, 2) 37, 3) 25 l/hr at pH 7, and 4) 25 l/h at pH 8; 25°C. Contrasting points indicate successive days. Curves displaced vertically from origin for clarity.

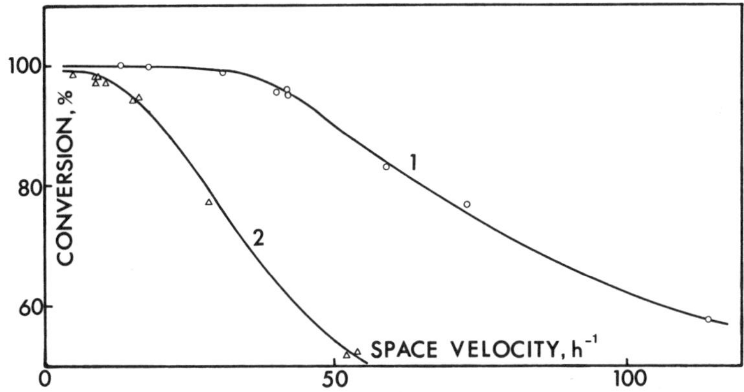

Fig. 4. Conversion of 1) 85 mM 2-methoxyethyl acetate and 2) 76 mM cephalosporin C as a function of space velocity (bed volumes/hr). Feed: Substrates in 200 mM K-phosphate pH 8; 25°C.

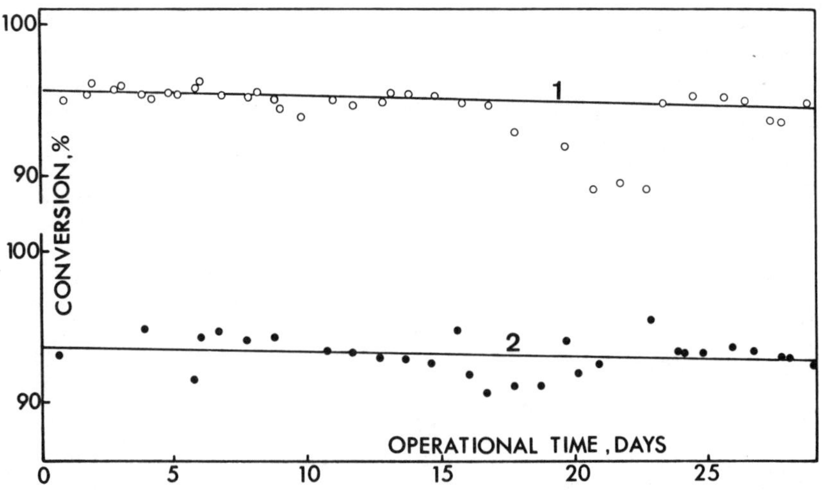

Fig. 5. Operational stability of brick powder-esterase in a continuous reactor. Substrates and space velocities: 1) 85 mM 2-methoxyethyl acetate, 40 hr^{-1}, 2) 76 mM cephalosporin C, 16 hr^{-1}. Buffer and temperature as in Fig. 4.

DISCUSSION

A recirculation reactor is convenient for intermittent preparation of g to kg quantities of products and has the advantage that low buffer/substrate ratios may be used. However, this mode of operation is laborious on a technical scale, with adverse effects on catalyst life and catalyst costs. A continuous reactor consisting of a single packed bed has one drawback, a stoichiometric excess of buffer must be used. On the other hand, the high buffer capacity and low flow rates increase catalyst life.

REFERENCES

1. KONECNY, J. in "Enzyme Engineering" vol. 3 (E.K. Pye and H.H. Weetall, eds.) Plenum, New York, in press.

2. KONECNY, J. & VOSER, W. *Biochim. Biophys. Acta 485*: 367, 1977.
3. KONECNY, J. in "Survey of Progress in Chemistry," vol. 8 (A.F. Scott, ed.), Academic Press, New York, 1977, p. 195.
4. VANDAMME, E.J. & VOETS, J.P. *Rev. Applied Microbiol. 17*:311, 1974.
5. FLYNN, E.H. "Penicillins and Cephalosporins," Academic Press, New York, 1972.
6. O'CALLAGHAN, C.H., KIRBY, S.M., MORRIS, A., WALLER, R.E. & DUNCOMBE, R.E. *J. Bacteriology 110*:988, 1972.
7. KONECNY, J., FELBER, E. & GRUNER, J., *J. Antibiotics 26*:135, 1973.
8. HAMILTON-MILLER, J.M.T., RICHARDS, E. & ABRAHAM, E.P. *Biochem. J. 116*:385, 1970.
9. KONECNY, J. & SLANICKA, J. *Biochim. Biophys. Acta 403*:573, 1975.
10. KONECNY, J. *Biochim. Biophys. Acta 481*:759, 1977.
11. FORD, J.R. LAMBERT, J.H. COHEN, W. & CHAMBERS, R.P. in "Enzyme Engineering" (L.B. Wingard, Jr., ed.), Wiley, New York, 1972, p. 267.
12. ROYER, G.P., GREEN, G.M. & SINHA, B.K. *J. Macromol. Scie., Chem. A 10(1-2)*:289, 1976.
13. WEETALL, H.H. & FILBERT, A.M. *Meth. in Enzymology 34*:59, 1972.
14. ROSS, G.W. *Meth. in Enzymology 43*:678, 1975.

ENZYMATIC PRODUCTION OF SUGARS FROM HEMICELLULOSE

G.B. Oguntimein, J.R. Frederick, A.R. Fratzke,
M.M. Frederick, and P.J. Reilly

Department of Chemical Engineering and Nuclear
Engineering
Iowa State University
Ames, Iowa, USA

For many years there has been an intense research effort to unlock the secrets of the enzymatic hydrolysis of cellulose. More recently the breakdown of lignin by enzymes has drawn interest. Comparatively neglected has been the enzymatic hydrolysis of the third major component of cellulosic materials, hemicellulose. Yet no breakdown of native cellulose is likely to be economically feasible without incorporating a means to hydrolyze the hemicelluloses that may comprise up to 30% of the total dry weight.

We have been engaged in a program to elucidate the functions and characteristics of the xylanase family, which break down xylan, the major hemicellulose of hardwoods and cereal grains. With the exception of β-xylosidase, little is known of these enzymes. This search has gained new emphasis with the sudden interest in xylitol. This sugar alcohol is the hydrogenation product of xylose, the major breakdown product of xylan, and is reported to be noncariogenic and even anticariogenic. Very sweet, with a cooling effect upon dissolution, xylitol has been-finding increasing use in chewing gum. A better method for the production of xylose could significantly reduce both its cost and that of xylitol.

We report here (a) two methods to purify β-xylosidase without electrophoresis, one of which removes virtually all the interfering β-glucosidase, (b) properties of purified β-xylosidase, and (c) methods to partially separate the remainder of the xylanases into approximately 15 enzymes active on xylan.

MATERIALS AND METHODS

The β-xylosidase and other xylanases were Rohm and Haas Rhozyme HP-150 Concentrate from *Aspergillus niger*, reportedly precipitated from the fermentation broth by alcohol. Two types of xylan were employed. The first, larchwood xylan-lot 1, was prepared in approximately 1% concentration by adding 2% Sigma Lot 62C-2820 xylan to water and centrifuging out the undissolved solids. The second, larchwood xylan-lot 2, was prepared in much the same manner and was composed of approximately 1% Sigma Lot 125C-00582. There were major differences between the two in their susceptibility to attack by different xylanase fractions. In addition, lot 2 contained an appreciable amount of starch.

β-Xylosidase was assayed by adding 0.1 ml samples to 1 ml of 4 mM o-nitrophenyl-β-D-xylopyranoside and 0.9 ml of 0.05 M acetate buffer at pH 4.0. After 15 min incubation at 40°C, the reaction was stopped by the addition of 2 ml 20% Na_2CO_3 and the absorbance determined at 400 nm. Xylanase activity was determined by adding 0.1-1.0 ml xylanase solution and 0.5 ml 1% xylan to 1.5 ml 0.075 M acetate buffer at pH 4.8. After incubation for 20 min at 40°C, 2 ml of Somogyi's reagent (1) was added, the solution heated at 100°C for 20 min, and cooled. After adding 2 ml of Nelson's reagent (2), the solution was allowed to stand for approximately 1 hr before measuring the absorbance at 500 nm. Protein was determined by Lowry's method (3). With column eluates, the absorbance at 280 mm was measured.

PURIFICATION OF β-XYLOSIDASE

In the first method the crude enzyme was dissolved in water (15%); and ammonium sulfate was added to 95% of saturation. The resulting precipitate was dissolved in 0.05 M acetate buffer, pH 4.8, dialyzed with an Amicon XM-100A ultrafiltration membrane to remove most of the xylanases and other small enzymes, and concentrated with the same membrane. The retentate was chromatographed on Sephadex G-150. The β-xylosidase fraction was dialyzed with 0.02 M phosphate buffer, pH 6.8, passed through DEAE-Sephadex A-25, and eluted with NaCl. The eluate was dialyzed with 0.05 M citrate buffer, pH 3.1, passed through SP-Sephadex, and eluted with NaCl. Only the last step could separate β-xylosidase from β-glucosidase. Purification factors were 30- to 100-fold with a final specific activity of about 60 U/mg in 28% yield.

The second method to purify β-xylosidase employed differential $(NH_4)_2SO_4$ precipitation. To approximately 20% crude enzyme in 0.075 M citrate buffer, pH 4.55, was added enough $(NH_4)_2SO_4$ to reach 69% saturation. This removed over 95% of the xylanases and β-glucosidase, while leaving approximately 40% of the original activity of the β-xylosidase in the supernatant. One portion of the supernatant was applied to a column containing Sephacryl S-200 Superfine gel permeation resin and eluted with 0.015 M citrate buffer. The overall purification was 15.7-fold, giving a specific activity of 15.6 U/mg in 32.3% yield. The other portion was dialyzed in cellulose tubing against 0.05 M citrate buffer at pH 3.1. The resulting solution was applied to a SP-Sephadex C-25 column and eluated to give a 27-fold purification and a final specific activity of 26.7 U/mg at an overall yield of 19.4%.

β-XYLOSIDASE PROPERTIES

The optimum pH for activity in 0.037 M Tris-citrate buffer at 40°C was 3.75. There was high activity from pH 2.5 to 5.0. The Arrhenius relationship was obeyed by β-xylosidase in 0.05 M acetate buffer at pH 4.0 between 30 and 70°C, with an energy of activation of 11.9 kcal/mol. In 0.05 M acetate buffer at pH 4.0, half-lives of β-xylosidase increased from 0.14 hr at 75°C to 113 hr at 62°C, yielding the exceptionally high activation energy for decay of 121 kcal/mol. While β-xylosidase had a rather wide pH range for activity, its range for stability was very narrow. At 65°C in 0.037 M Tris-citrate buffer of 0.4 ionic strength, the optimum was between pH 4.0 and 4.5, with stabilities being very low at both pH 3.5 and 5.5. β-Xylosidase was immobilized on Corning alkylamine porous silica that had been activated with 10% glutaraldehyde. Approximately 50% of the enzyme activity was taken up by the carrier.

SEPARATION OF XYLANASES

A 15% solution of crude enzyme in 0.025 M acetate buffer, pH 4.5, was treated with Ultrogel AcA 54. The four eluate peaks were labelled A through D in reverse order of apparent molecular weight. While Fraction A was effective only on larchwood xylan-lot 1, Fractions B and C were more effective on lot 2. Fraction D attacked lots 1 and 2 equally. Isoelectric focusing of the four fractions showed that A had one or two xylanases with pI's of approximately 3.5, and D just one xylanase with a pI of about 3.5. Fractions B and C each had many enzymes active on xylan with pI's of 3 to 7.

Fractions B and C were each separated on SP-Sephadex C-25 at pH 4.5 using 0.025 M acetate buffer and NaCl as eluent to give four peaks (B) and six partially resolved peaks (C), respectively. The nonretarded eluates from the SP-Sephadex columns were titrated to pH 5.4 with concentrated NaOH and applied to DEAE-Sephadex C-25 columns. The enzymes were eluated with 0.025 M acetate buffer and NaCl to divide Fraction B into three peaks and C into four.

Samples of all the peaks and many of the shoulders of the eluates from the two types of column were incubated for several days with 2% larchwood xylan-lot 2 in 0.075 M acetate at pH 4.8, until hydrolysis neared completion. Samples of the supernatant contained various mixtures of xylose, arabinose, and at least eight oligosaccharides, as determined by TLC on silica plates using a 6:3:2 mixture of ethyl acetate, acetic acid and water (4). While most of the oligosaccharides were present in each sample, a few of the enzymes did not produce xylase or arabinose.

DISCUSSION

The procedure described here that purifies β-xylosidase by fractional precipitation with ammonium sulfate, followed by gel permeation or ion exchange chromatography, achieves satisfactory yield and activity. In addition, it removes interfering β-glucosidase without resorting to electrophoresis. Because of its simplicity, it can be employed to purify β-xylosidase in larger scale processes than can other methods.

β-Xylosidase in soluble form has been shown to be an extremely stable enzyme, even in the absence of its substrate. It is most active and stable at pH's well below neutrality, which should reduce its susceptibility to microbial contamination. The effects of temperature on both the activity and the stability of β-xylosidase follow an Arrenius relationship. The enzyme in preliminary experiments was successfully immobilized to alkylamine porous silica. If β-xylosidase were highly stable, it could find use in large-scale systems.

The xylanase complex has been partially separated by gel permeation and ion exchange chromatography into fifteen or more enzymes, all of which produce a mixture of oligosaccharides of varying sizes, and nearly all of which produce xylose and arabinose upon incubation with xylan. It is premature at this point to classify all of these enzymes as endo-hydrolases, since the numerous branch points of xylan could have prevented more complete enzymic attack leading to the production of much smaller oligosaccharides.

Why so many xylanases are present is something of a puzzle that awaits more complete separation and characterization. While many seem to have similar action patterns on xylan, their pI's range from 3 to 7, which is sufficient variation to allow separation by ion exchange chromatography. In addition, there are sufficient differences in molecular weight, from 17,000 to roughly 50,000 daltons, to permit separation into four groups by gel permeation chromatography. It is likely, however, that some of the enzymes that have been detected in this work are closely related isozymes that differ only slightly in charge or molecular weight.

ACKNOWLEDGMENT

This project was supported by National Science Foundation Grants APR74-2011 and AER77-00198 and by the Engineering Research Institute, Iowa State University.

REFERENCES

1. SHAFFER, P.A. & SOMOGYI, M. *J. Biol. Chem.* *100*:695, 1933.
2. NELSON, N. *J. Biol. Chem.* *153*:375, 1944.

3. LOWRY, O.H., ROSEBROUGH, N.J., FARR, A.L. & RANDALL, F.J. *J. Biol. Chem.* *193*:265, 1951.
4. OVODOV, Y.S., EVTUSHENKO, E.V., VASKOVSKY, V.E., OVODOVA, R.G., & SOLOV'EVA, T.F. *J. Chromotog.* *26*:111, 1967.

ENZYMES AND SEPARATION PROCESSES

J.A. Howell, J.S. Knapp and O. Velicangil

Department of Chemical Engineering
University College of Swansea
Wales, UK

Two difficult physical separations are the dewatering of sewage sludge and the ultrafiltration of cheese whey. The difficulties arise from the physical structure of macromolecules of biological origin: and in both cases the use of enzymes to modify the structure and hence separation characteristics has been investigated.

It has been shown that treatment of primary sewage sludge (PSS) with the cellulase enzymes from *Trichoderma viride* will degrade much of the cellulose present (4 to 22% of the solids), causing changes in the physical properties of the sludge (1,2). It was suggested that these changes might aid further treatment. The cellulase treatment also caused a rapid improvement in the ultrafilterability of diluted PSS, greatly reducing filter blinding. During hydrolysis up to 75% of the cellulose was converted in 24 hours, releasing reducing sugars; 7 g/l of reducing sugar was achieved during hydrolysis of autoclaved PSS in the presence of mertiolate (biostat). Others also have proposed cellulolysis as a treatment for sewage sludge (3), and *Candida utilis* and other species have been grown on media derived from human or animal wastes (4-6).

In the ultrafiltration of cheese whey the sharp decrease in flux of ultrafiltration membranes to less than half of their initial value within the first few hours has been attributed to irreversible consolidation of the macromolecular gel layer on the membrane with time (7).

The investigation described here is concerned with the preparation of a self-cleaning membrane, by attaching a proteolytic enzyme to the membrane surface. In this way clogging proteins should be hydrolyzed as they are deposited, thus increasing the permeability of the gel layer. The system has been-operated continuously for periods of up to 72 hr, which compares with the normal operating period of 20-24 hr in the dairy industry.

METHODS AND MATERIALS

Raw primary domestic sewage sludge was obtained from treatment works at Pwll and Bishopston. The filterability of the sludge was measured as the specific resistance to filtration (8) using a pressure bomb 15 lb/in^2 pressure and 100 ml of sample. Standard analytical techniques were used. The cellulase preparation was produced by *T. viride* QM 9123 (ATCC 24, 449) on 2% Solka Floc SW40 (Brown Co., Berlin, New Hampshire).

For the cheese whey studies papain (E.C. No. 3.4.4.10) was purchased from Sigma. Ultrafiltration membranes (type PM-10, 150 mm diameter) were obtained from Amicon. Cheddar cheese whey was donated by Unigate Ltd. from local dairies.

The papain was attached to the membrane by methods described elsewhere (9, 10); and the amount of bound protein was determined by difference without sacrificing the membranes (11). The esterase activity of free and immobilized papain was assayed by the standard pH-stat method using BAEE as a substrate. The esterase activity of the immobilized enzyme membranes was determined before they were incorporated into the cell and used for ultrafiltration studies. Between 12 and 40% by weight of the papain originally in solution was immobilized and about 15% of the free solution activity was retained, depending upon the enzyme concentration and the degree of prior hydrolysis. In the hydrolysis step an increase in HCl concentration from 3 to 10 N increased the yield of bound enzyme from 17.2 to 40%, but retained only low activity (3.7% of the initial solution activity in comparison with 15.8% for the lower protein loading).

To investigate the flux decay of modified membranes with time, cheddar cheese whey was continuously ultrafiltered in a thin-channel cell with total recycle. Two experiments were performed over two different time periods and employing two distinctive enzyme loadings. Pasteurized whey (pH 5.65-5.2) was pressurized to 2 atm. The temperature was kept at 37°C.

EXPERIMENTAL RESULTS

Sludge hydrolysis was carried out at 50°C and at the prevailing pH of the sludge, which was 5.5 to 5.9 and usually 5.6. The cellulase preparation was a 20 times concentrated *T. viride* culture filtrate and was mixed with PSS in a ratio of 1 ml enzyme to 39 ml PSS.

The effect of cellulase treatment on the specific resistance to filtration (SRTF) was determined. There was an initial sharp increase in SRTF during the first few hours of treatment followed by a decline. It was shown that sludge left to stand at room temperature (about 25°C) underwent negligible change in SRTF over one day. Further, it was demonstrated that the increase in SRTF was not due to shear caused by agitation of the sludge during treatment. Sludge incubated at 50°C in the absence of cellulase also showed a marked increase in SRTF.

The effect of conditioner on the filterability of cellulase treated sludge also was examined. $Al_2(SO_4)_3 \cdot 16 H_2O$ was added to the sludge to give a final concentration of 0.5% (0.081% as Al_2O_3). Aluminum sulphate was not able to overcome the deleterious effect of cellulase treatment on filterability.

It was shown, using a stirrable pressure bomb filter at 10 lb/in,^{2} that cellulase treatment for 7 hr decreased the filterability of PSS considerably in unstirred filtrations. However, when the sludge was stirred at about 200 rpm by a propeller situated about 7 mm above the filter surface (Whatman No. 1), cellulase treated sludge had very similar filtration characteristics to untreated sludge.

Sludge diluted with distilled water was subjected to varying periods of cellulase treatment after which settleability was assessed. The cellulase treatment greatly affected the rate of settling of very dilute (6.0 g/l suspended solids) PSS. Enzyme treated sludge initially settled at about 30 times the rate of untreated sludge. However, untreated PSS of 16.5 g/l suspended solids settled only about 2.5 times faster after cellulase treatment; no effect was noticed with sludge of 24 or

32 g/l suspended solids. Two, 6, and 24 hr of treatment increased the initial settling rate, compared with untreated sludge, by 5, 7.9, and 9 times, respectively. The suspended solids content of the sludge zone was concentrated by 50, 120, and 170%, respectively in a one-hour settling period compared with a 7% concentration of untreated sludge in the same time. However, more concentrated (16, 24 and 32 g/l) sludges showed no significant settling in a one hour period, even after 24 hr cellulase treatment.

The filtrate produced after cellulolysis of sludge has been used successfully as a growth medium for microorganisms. This sewage sludge extract (SSE) has been shown to support growth of *C. utilis*, *S. lipolytica*, *Rh. glutinis*, and *Tr. cutaneum* yeasts as well as *P. putida* bacterium. *C. utilis* was grown on SSE, supplemented by 2 g/l KH_2PO_4 and 1 g/l $(NH_4)_2SO_4$. The initial pH of the medium was 4.2, due to the aluminium sulphate flocculant, and increased during growth to 7.0. The final growth yield was 4.85 g/l; and the mean generation time was about 4 hr. During growth the BOD_5 was reduced from about 13,000 to 3,000 mg/l; and the protein concentration was reduced from 2.4 to 1.25 g/l.

In the ultrafiltration experiments the initial flux losses with immobilized papain amounted to 35% for an enzyme loading of 0.088 mg/cm^2 and 49% for 0.068 mg/cm^2. However, the flux with the immobilized enzyme membranes exceeded that of the control runs (no enzyme) after about 8-16 hr. The pH versus flux relationship of the enzyme membrane was quite different from that of the control (no enzyme) membrane at 37°C and 25-32 lb/in^2 pressure. It exhibited a shift of two pH units in its minimum toward lower values and no maximum in the tested range.

DISCUSSION

A new potential application of immobilized enzyme systems has been introduced by this investigation. The integrals of flux-versus-time curves gave higher overall yields of ultrafiltration for the protease-coupled membranes with less sensitivity to pH changes.

Cellulase treatment had no beneficial effect on the filtration of primary sewage sludge through conventional filter media (filter papers). In fact in systems in which the filter cake was allowed to build up, short periods of cellulase treatment increased the SRTF. Cellulase treatment could improve greatly the settling of very dilute sewage sludges; but would show very little effect on more concentrated material.

Overall it seems likely that cellulase treatment will be of no value as a pretreatment for physical methods of primary sewage sludge dewatering; but it could have application to more dilute sludges.

REFERENCES

1. STUCK, J.D. & HOWELL, J.A. *Chem. Eng. Prog. Symp. Ser. 70* (No. 136):337, 1973.
2. CING-MARS, G.V. & HOWELL, J.A. *Biotechnol. Bioeng. 19*: 377, 1977.
3. OSE, H. & MORISHITA, S. *Chem. Abs. 84*:21768f, 1976.
4. SIMARD, R.E. & THANH, N.C. *Chem. Eng. Prog. Symp. Ser. 70* (No. 136):309, 1973.
5. ADDYMAN, C., M.Sc. Thesis, University of Newcastle-upon-Tyne, 1973.
6. THOMAS, K. M.Sc. Thesis, University of Newcastle-upon-Tyne, 1975.
7. BLATT, W.F., DRAVID, A., MICHAELS, A.S. & NELSEN, L. in "Membrane Science and Technology," (J.E. Flinn, ed.,) Plenum, New York, 1970, p. 47.
8. Department of the Environment, "Analysis of Raw, Potable and Waste Waters," H.M.S.O., London, 1972.
9. VELICANGIL, O. M.Sc. Thesis, University College of Swansea, 1976.
10. INMAN, D.J. & HORNBY, W.E. *Biochem. J. 129*:255, 1972.
11. LOWRY, O.H., ROSEBROUGH, N.J., FARR, A.L., & RANDALL, R.J. *J. Biol. Chem. 193*:265, 1951.

ENZYMATIC HYDROLYSIS AND PROTEIN ULTRAFILTRATION FOR CONVERTING ACID WHEY TO VALUABLE PRODUCTS

Robert W. Coughlin and Marvin Charles*

University of Connecticut
Storrs, Connecticut and
Lehigh University*
Bethlehem, Pennsylvania, USA

The paper covers preliminary bench scale data and design and flow diagrams of our pilot plant for hydrolysis of the lactose in cheese whey, using fungal lactase immobilized on porous alumina. The reactors operate in the fluidized bed mode. The performance of 3-inch pilot scale reactors is compared with that of a 1-inch bench scale unit. The larger reactors appear to provide a higher conversion of lactose at the same residence time. This may be due to better distribution of the liquid in the feed to the larger reactors.

The paper also provides an approximate economic analysis and return-on-investment calculation for a 100,000 lb/day whey hydrolysis unit alone and in combination with various related processes for (a) protein removal and recovery, (b) demineralization, and (c) concentration of the liquid products. In most cases prospects appear favorable for reasonable rates of return on invested capital, especially in those instances wherein whey protein is also recovered and sold.

The full text of this paper will be published in early 1978 in the American Institute of Chemical Engineers (AIChE) Symposium Series: Food, Pharmaceutical and Bioengineering 1976-1977, Volume 74.

MODEL PROTEOLYSIS OF β-CASEIN WITH IMMOBILIZED TRYPSIN AND ITS TECHNOLOGICAL SIGNIFICANCE

E.H. Reimerdes

Bundesanstalt fur Milchforschung
Kiel, Federal Republic Germany

Immobilized enzymes were used as a model system to study proteolytic processes which occur in milk during processing. The properties of milk depend on the protein composition, especially on the hydrophobic caseins. β-casein is about 30% and γ-casein up to 10% of the total casein. The γ-caseins are fragments of β-casein, as derived from sequence analysis (1). During storage of raw milk at 4°C the γ-casein fraction increases steadily, resulting in irreversible changes of technological properties. During our research we isolated and characterized two serine proteinases, associated with the casein micelles, showed the high specifity of these enzymes for β-casein degradation (2), and did further characterization using immobilized trypsins.

Electrophoretically pure β-casein was isolated by urea/pH-fractionation and ion exchange chromatography. The absence of milk proteinases was tested with benzoyl-arginine-4-nitroanilide. The trypsin immobilization was performed with CNBr activation, anhydrides, oxirane groups, and photoalkylation. In the last three preparations acrylic beads (Rohm) were used. Anhydride coupling caused an increase of negative charges due to carboxylic groups. By photoalkylation the reactive group were covalently bound to glycine residues on the trypsin surface (3). Thus there was no change of charges at the enzyme surface; and any changes in the kinetics might be due to steric hindrance by the coupling spacer to the enzyme surface. All preparations showed good activity with small substrate molecules; but entirely different results were obtained with pure β-casein.

The immobilized trypsins showed large differences during proteolysis. The fresh preparations were highly active against β-casein but lost activity rapidly, except when photoalkylated. Activity was only partially recovered by drastic purification. The photoalkylated enzyme could be used for several experiments. The proteolysis very much depended on temperature and time. Under special operating conditions we were able to get a similar proteolytic pattern as during cold storage of raw milk. Two γ-casein and one phosphoprotein fragment were obtained by ion exchange chromatography and identified with native homologues by electrophoresis and dansylation. The very hydrophobic γ-caseins were extractable with 1-propanol/diethylether. This has been used for the quantitation of the γ-casein formation during cooling of raw milk. Thus, protein degradation in milk could be simulated with immobilized trypsins under special conditions. The enzyme preparations showed unspecific binding of the γ-caseins, which depended on the coupling system. The fragments of β-casein have been purified and are used as standards for an electrophoresis-densitometry procedure for the quantitative determination of proteolysis during milk processing.

REFERENCES

1. MERCIER, J.C., GROSCLAUDE, F. & DUMAS, B.R. *Milchwiss* *27*:402, 1972.
2. REIMERDES, E.H. & KLOSTERMEYER, H. *Milchwiss* *29*:517, 1974.
3. KRAMER, D.M., LEHMANN, K., PENNEWISS, H., & PLAINER, H. *Coll. Protides of Biol. Fluids, Brugge* *23*: 505, 1975.

DETOXIFICATION OF ORGANOPHOSPHATE PESTICIDES USING IMMOBILIZED ENZYMES

Douglas M. Munnecke

Institut fur Bodenbiologie
Bundesforschungsanstalt fur Landwirtschaft
Braunschweig, Federal Republic Germany

Microbial enzymes which have the ability to degrade organophosphate insecticides are being examined to determine if such enzymes, either free or immobilized, may be used industrially for pesticide detoxification, and secondly, to establish if their economics will allow their use in environmental pollution control systems (degradative processes). In previous research (1) a mixed microbial culture was adapted to growth on parathion (0,0-diethyl-O-nitrophenyl phosphorothioate) as the sole carbon and energy source. This gram negative, predominately pseudomonad culture (1,5) contained an enzyme(s) which had the ability to hydrolyze a wide range of organosphosphate insecticides (2) without the need for cofactors or special salts. Therefore, immobilization studies were started (3) to determine the feasibility of using immobilized parathion hydrolase for the detoxification of pesticides. A crude enzyme extract of specific activity 3-5 U/mg protein was obtained by disrupting cells by sonification and centrifuging at 10,000 x g for 15 min. 60% of the parathion hydrolase activity remained in the supernate after this treatment.

This crude enzyme extract was bound to porous glass (120-250 μm diameter) by azide coupling (4). Up to 59 mg of protein/g glass could be bound with over 95% of the initial protein bound. The bound parathion hydrolase activity ranged up to 50 U/g of glass. An optimal temperature of 35°C was found for the bound enzyme; and temperatures above 50°C caused enzyme inactivation. Maximal enzymatic activity occurred between pH 8.5 and 9.5, while the enzyme was stable up to pH 10.5. Organic solvents such as xylene, benzene, hexane, acetone, ethanol, methanol, and ethyl ether at 0.1% concentration inhibited the enzymatic activity by 10-50%. Salts at 50 g/l had relatively little influence on the enzyme activity, while industrial wastewater inhibited the enzyme by only 20%. The enzyme-glass retained 95% of its initial activity after 40 days in wastewater at room temperature.

The kinetics of continuous parathion hydrolysis were examined using a fluidized bed reactor (7.5 cm x 70 cm) containing 40 g of parathion hydrolase-glass of 5 U/g. Parathion (10 ppm in phosphate buffer, pH 8.5) was pumped through the reactor at up to 20 l/hr. Greater than 95% of the parathion was degraded; while at the maximum tested flow rate of 45 l/hr, 86% of the parathion was hydrolyzed.

Preliminary economic calculations showed that for parathion hydrolase-glass to be more economical than continuous addition of soluble enzyme to wastewater, the enzyme-glass system must have an operational stability of more than 6 weeks. With this operational stability enzymatic pretreatment would cost approximately 2% of the cost for operating a tertiary waste treatment plant. In return, by increasing the biodegradability of the pesticide waste, and decreasing the toxicity of the waste, an economical advantage might be attainable.

REFERENCES

1. MUNNECKE, D.M., & HSIEH, D.P.H. *Appl. Microbiol. 28*: 212, 1974.
2. MUNNECKE, D.M. *Appl. Environ. Microbiol. 32*:7, 1976.
3. MUNNECKE, D.M. *Appl. Environ. Microbiol. 33*:503, 1977.
4. MASON, R.D. & WEETALL, H.H. *Biotechnol. Bioeng. 14*: 637, 1972.
5. BAUMGARTEN, Personal communications, Bayer AG, Wuppertal, 1977.

PRODUCTION OF L-TRYPTOPHAN

F. Bartoli, G.E. Bianchi and D. Zaccardelli

Snamprogetti S.p.A.
Laboratories for Biochemical Processes
Monterotondo (Roma), Italy

N-acetyl-DL-tryptophan is produced (1) by room temperature condensation of DL-serine and indole in the presence of acetic acid and acetic anhydride at 62% yield. A part of the unreacted indole is easily extracted with ether and recycled. The overall conversion of indole is 85%. The DL-serine is the product of a method developed in Snamprogetti laboratories (2). The acetyl-DL-tryptophan in turn is asymmetrically hydrolyzed by acylase entrapped in cellulose triacetate fibers to produce L-tryptophan and acetyl-D-tryptophan.

In the acylase process the feed tank contains 9.85 kg acetyl-DL-tryptophan, 23.8 g $CoCl_2.6H_2O$, 1.6 kg NaOH (for pH 7.0), and 200 kg water at 45°C. This solution is recycled for 5.5 hr through the enzyme hydrolysis reactor at 350 l/hr. The reactor is 770 mm high by 190 mm dia with 4 kg of dry fiber, containing 0.28 kg protein/kg dry fiber (1,600,000 I.U./kg dry fiber). The hydrolyzed material is evaporated to 15 l and then separated, based on solubility differences, to give 3.9 kg of L-tryptophan (95% yield) and 15 l of acetyl D-tryptophan solution. The later is mixed with 3 l of acetic anhydride and held 5 hr at 45°C to give 4.7 kg of racemized and precipitated product (yield 96%). The racemized acetyl-D-tryptophan is then recycled to the feed tank.

The same reactor is able to hydrolyze the N-acetyl derivatives of different amino acids (Fig. 1).

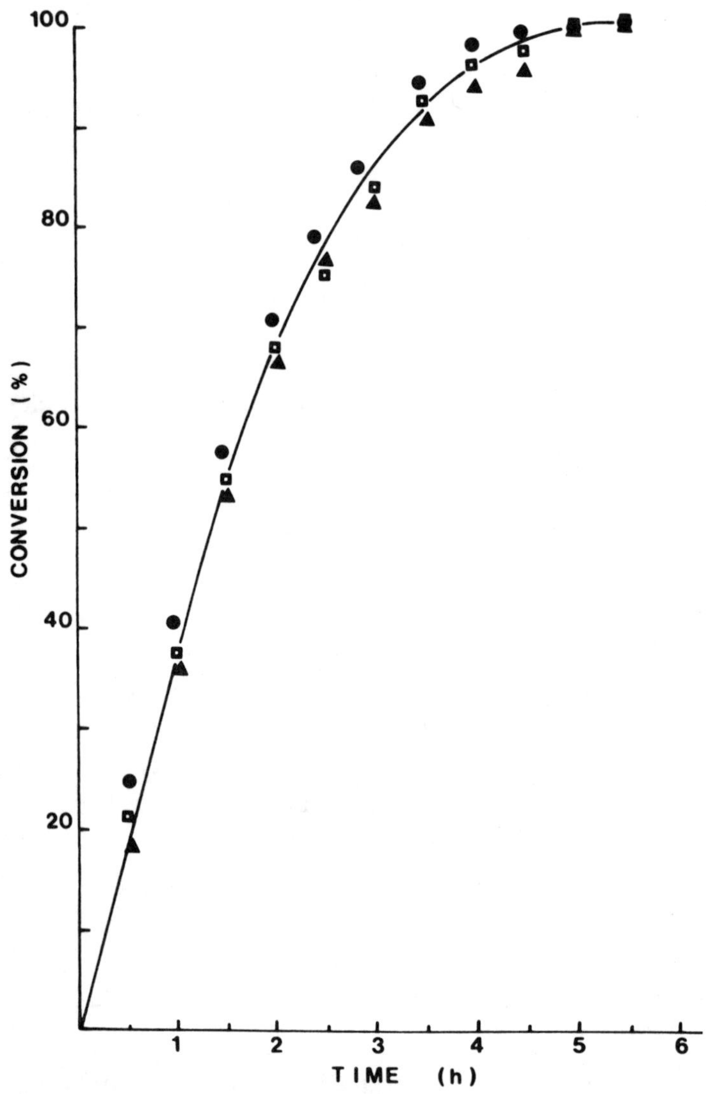

Fig. 1. Hydrolysis for tryptophan (square), methionine (circle), and valine (triangle).

REFERENCES

1. GHOSAL, S. *J. Indian Chem. Soc.* *41*:496, 1964.
2. German Patent No. 2,515,622; (*Chem. Abs.* *84*:31496e, 1976).

ELECTROCHEMICAL ASSAY OF OXIDOREDUCTASE ENZYMES

Robert W. Coughlin and Thomas C. Wallace

University of Connecticut
Storrs, Connecticut and
Clinton Corn Processing
Clinton, Iowa, USA

Because many coenzymes are electroactive this property can be used to provide an electrical signal related to the activity of a corresponding oxidoreductase enzyme. An example of the coupling of an enzymatic reaction and an electrochemical reaction involves the coenzyme NAD (1). The NADH is oxidized to NAD^+ at the anode and within the surrounding electrolyte solution. The enzyme alcohol dehydrogenase (ADH) catalyzes the reduction of NAD^+ to NADH by ethanol. If the electrolyte solution contains at first only NAD^+ and ethanol but no enzyme, then neither enzymatic nor electrochemical reaction will occur until enzyme is added. It is this electrochemical reaction which starts only after addition of enzyme, which is the basis for the assay methods to be described.

A steady-background current can be established at a Pt anode immersed in an electrolyte containing NAD^+ and ethanol. This current undergoes an abrupt increase caused by the addition of ADH. We have found that the slope of the current vs. time curve is proportional to the amount of enzyme added. Similar proportionality characteristics have been shown for the assay of lactate dehydrogenase (LDH) and malate dehydrogenase (MDH). The assay of each of these enzymes in blood serum is of importance for clinical diagnosis. In the case of MDH it is really the enzyme glutamic oxaloacetic transaminase (GOT) which is of clinical interest; the latter enzyme is usually assayed by coupling its reaction to the MDH-catalyzed reaction of oxaloacetic acid. The details are given elsewhere (2,3).

The foregoing method of enzyme assay is essentially a batch technique, e.e., the anode compartment must be flushed, washed, and refilled for each subsequent determination of enzyme. To improve the method and to make it more adaptable to easy automation, we developed a flow system. When a pulse of enzyme-containing sample is injected into a flowing stream of appropriate substrates and reagents, a current pulse is produced when the electrodes encounter the flowing parcel of liquid containing enzyme, reacting substrates, and reaction products. The magnitude of the pulse height is proportional to the amount of enzyme injected in the respective pulse, for ADH, LDH, and MDH. If NAD^+ is produced from NADH by enzymatic reaction in the flowing reagent solution, then the result will be a decrease in pulse height caused by enzymatic reaction; such a result was observed.

REFERENCES

1. COUGHLIN, R.W. & ALEXANDER, B.F. *Biotechnol. Bioeng.* *17*:1379, 1975.
2. WALLACE, T.C. & COUGHLIN, R.W. *Anal. Biochem.* *80*:133, 1977.
3. WALLACE, T.C., LEH, M.B. & COUGHLIN, R.W. *Biotechnol. Bioeng.* *19*:901, 1977.

DESIGN CONSIDERATIONS AND OPERATIONAL STRATEGIES OF A COMMERCIAL GLUCOSE ISOMERASE REACTOR SYSTEM

K. Venkatasubramanian

H.J. Heing Company and Rutgers University
Pittsburgh, Pennsylvania USA

Relatively few reports in the literature address the problems of large-scale industrial reactor design and operation. Several important design considerations and process engineering problems are discussed in this communication. A commercial high fructose corn syrup (HFCS) production facility, including an immobilized glucose isomerase reactor, forms the basis of this discussion.

The first major design consideration is selection of the catalyst system. Immobilized glucose isomerase is available commercially from a number of suppliers including Novo Industri A/S, Denmark; ICI Americas, USA; and Gist Brocades NV, Holland. An objective comparison of the performance of the different catalysts is required. Table I summarizes several design parameters to be evaluated in this connection.

From an evaluation of this type plus catalyst cost data--the optimal enzyme system is chosen. The reactor feed and effluent specifications are then arrived at easily. From a knowledge of catalyst characteristics listed in Table I, the isomerase reactor vessel size is calculated. Since the activity of the bound enzyme decreases continuously, it is preferrable to have a number of columns in contrast to one gigantic column. This would minimize production fluctuations with respect to capacity and conversion level. For a given plant capacity, the optimum number of columns can be estimated theoretically. Both series and parallel operations of the reactor system are possible. While the former has some theoretical kinetic advantages, parallel operation is preferred since it provides the greatest operational flexibility.

While glucose isomerase is a fairly uniform granule and has low pressure drop characteristics, it is dimensionally less stable than rigid granules, e.g. activated carbon; and it is compressible. Control of flow to avoid surges is of particular importance. It is characteristic of a compressible granule that a high flow-high pressure drop condition of even short duration will compress and shrink the bed. This shrinkage is not reversible unless specific actions are taken (e.g. backwash to re-expand the bed at minimum or physical removal and reloading at maximum).

Compression of the bed of granules to reduced volume has severe effects on pressure drop. Under normal downflow operating mode, compression is insignificant and behavior is close to that for a rigid granule. On the other hand, if flow is such that compression is experienced, pressure drop will progressively increase until one approaches the elastic limit of the particles; and the rate of bed deformation is reduced. This latter condition is undesirable, both from the standpoint of the flow problems created and the reduction of the apparent activity of the enzyme. The latter effect becomes evident through reduction of interstitial volume and sharp reduction of true residence time in the enzyme bed. Column height-to-diameter ratio should be so chosen-that pressure drop problems are minimized.

Adequate provisions must be made to complete swelling of particles prior to regular column operation. Uniformity of bed packing is also mandatory. In downflow operation, the substrate percolates through the packed bed by gravity; to facilitate this, a sufficient liquid head (above the bed) must be maintained at all times. Feed inlet and outlet are so designed to ensure that the packed bed is never allowed to become dry.

It is necessary to operate the reactor system to obtain a finished product of uniform quality and constant fructose content (typically 42% dry solids basis). By adjusting the flow rate through the column periodically, nearly constant conversion can be maintained. By properly sequencing the order in which the reactors are brought into stream, it is possible to operate the entire system in such a way that the composite stream from the system meets capacity and conversion constraints. Modest variations in operating temperature and conversion level (followed by back-blending) are additional process control strategies to optimize the reactor battery performance. It is possible to control this entire operation on-line through a mini-computer or a microprocessor.

TABLE 1

LIST OF PARAMETERS IMPORTANT IN REACTOR DESIGN

Biochemical Characteristics

1. Activity
2. Operational Stability (half-life) and Activity Decay Profile
3. Productivity in Usage Life Time
4. Optimal Substrate Concentration
5. Effect of Oligosaccharides Concentration
6. Effect of Dissolved Oxygen
7. Minimum and Maximum Residence Times
8. By-Product Formation
9. pH and Temperature Sensitivity
10. Storage Stability
11. Protein/Enzyme Elution
12. Microbial Growth, If Any
13. Reactor Effluent Quality (Composition, Color, Odor, Protein Content, pH, etc.)

Mechanical Characteristics

1. Particle Size, Shape and Size Distribution
2. Density (Dry Bulk Density and Wet Density)
3. Swelling Behavior
4. Compressibility
5. Cohesion
6. Particle Attrition

Hydraulic Characteristics

1. Pressure Drop
2. Mode of Flow (Upflow versus Downflow)
3. Bed Compaction-
4. Axial Dispersion and Channeling
5. Residence Time Distribution
6. Stratification
7. Length-to-Diameter Ratio
8. Minimum Velocity for Onset of Fluidization

ACKNOWLEDGMENT

The author is indebted to Lee Harrow, Corporate Technical Director, H.J. Heinz Company, and J.W. Connolly, President and staff of the Hubinger Company for their encouragement in the preparation of this paper. The assistance of Diane Otto is gratefully acknowledged.

Session VI
MULTIENZYME SYSTEMS, IMMOBILIZED WHOLE CELLS, AND ORGANELLES

Chairmen: I. Chibata and K. Mosbach

ENZYME SEQUENCES IN THE LIVING CELL

G. Rickey Welch

Biochemistry and Molecular Biology
The University of Texas Medical School
Houston, Texas, USA

In this article I explore briefly views on enzyme sequences in the living cell and some implication regarding practical applications.

There is a growing feeling that a significant portion of the protein structure participates in energy transduction, in order to achieve a high free energy event at the active center. From a thermodynamic viewpoint (1) individual protein molecules consist of relatively few discrete particles in comparison to familiar macroscopic objects, and statistical fluctuations in thermodynamic properties assume much greater importance.

The basic issue is that of localizing (for a specified time period) at the active center some of the internal fluctuating energy of the protein structure. Maybe enzyme structures have been programmed in evolution to collimate or funnel some of the internal energy to the active center (2-7). However, it is not yet clear whether the protein can be regarded as a harmonic solid, involving a system of coupled oscillators (8,9), or as a dense hard-sphere fluid composed of particles that are connected by flexible links (10). Also, the environment of a given enzyme *in vivo* may be far different from that *in vitro* and may, in fact, be quite heterogeneous. Even for single soluble enzyme *in vivo*, the bulk cytoplasm might generate viscosity effects not realized *in vitro* (11).

Careri *et al.* (12) recently reviewed critically the statistical time events in enzymes. They suggested that such events may be coupled statistically and that the macromolecule and its thermal environment are likely to be on a common frequency such that some free energy can be exchanged between the protein molecule and its environment through the coupling of events at the protein surface. More specifically, Careri and Gratton (13) proposed that fluctuation of the bound water density at the enzyme surface can be coupled with spontaneous conformation fluctuation at the active center. In this regard the effects of inorganic salts on hydration of protein surface groups can influence markedly the activation volume (and activation free energy) of enzyme reactions (6). Fluctuation in other types of surface conditions (e.g., ionic environment) might be equally important for enzyme catalysis (12,14).

THE FABRIC OF INTERMEDIARY METABOLISM

On the basis of facts and logic, it can be maintained that most (if not all) enzymes of intermediary metabolism operate *in vivo* in some sort of structured state, as multienzyme aggregates and/or membrane-associated systems (15). A direct approach is to construct structural rate equations, which would reflect the geometry of the enzyme molecule immersed in an organized setting. In general, an enzyme situated in a physically structured state *in vivo* might be subject to a defined spectrum of external influences. Hence, we might write the rate constant, k, for a given chemical process involving the enzyme as follows:

$$k = k^* \left(\prod_{r=1}^{n}\right)\left(\alpha_r\right)^{i_r} \qquad \text{(Eq. 1)}$$

The intrinsic constant, k*, represents the rate in the absence of external effects, as developed from the theory of absolute reaction rates. k* depends on ΔG_i, the intrinsic activation free energy (per molecule) for the given enzymatic step when there are no external influences. The α_r are dimensionless interaction coefficients (16) defined by

$$\alpha_r = \exp\left(-\frac{\Delta G_r}{k_B T}\right) \qquad \text{(Eq. 2)}$$

and correspond to contributions from n types of structural interactions. The exponent i_r is an integer expressing the number of interactions of type r. The functional nature of the α_r in organized multienzyme systems has been analyzed elsewhere (15).

For cellular metabolism the input and output takes a set of concentrations and respective spatial fluxes of a group of metabolites and localizes them and then transforms them into local pools and fluxes specific to given metabolic processes. Normally, one characterizes mathematically the flow of matter within the organized multienzyme system by describing macroscopically the spatiotemporal behavior of the concentration S_i of an intermediate substrate. This is done by setting up the following kind of mass balance relation:

$$\frac{\partial}{\partial t}[S_i(r,t)] = D_i \nabla^2 S_i(r,t) + \frac{D_i}{k_B T} \underset{\rightarrow}{\nabla} [S_i(r,t) \nabla U_i(r)] + X_{i-1}[S_{i-1}(r,t)] - X_i[S_i(r,t)] \qquad \text{(Eq. 3)}$$

where D_i reflects the diffusive properties of S_i in the structured matrix, and U_i is the interaction potential between S_i and the matrix (15). The X_i and X_{i-1} are reaction rates for the respective steps in the process $S_o \rightarrow S_1 \rightarrow S_2 \rightarrow \ldots S_i \ldots \rightarrow S_n \rightarrow P$. However, such a formalism usually is limited to steady states and linear relations; also the difficulty is great in providing a macroscopic description when the system exhibits inhomogeneity and anisotrophy (15, 17). In my opinion the theory of network thermodynamics (17, 18) is better suited to date for analyzing the above interrelationships between dynamic metabolic processes and organizational complexity in the living cell. This theory is not subject to the above limitations. Moreover, its approach is based on system topology, both spatial and functional. Bunow and Aris (19) have presented a matrix method, formally related to the network theory, which may be of particular value in picturing the operation of structured enzyme systems.

SOME ECONOMIC BENEFITS IN THE DESIGN OF THE METABOLIC MACHINERY

Since viable cells continually overcome the random field in their environment, enzymes may be programmed to correlate efficiently the chemical processes at the active center with the stochastic properties of the ambient medium. A recent stochastic model applied to transmembrane transport involving a two-enzyme sequence (20) considers a finite number of individual molecular enzyme channels distributed uniformly in the total cellular membrane. It is found that the coefficient of variation of the total transport rate is nonzero, even for a large exterior concentration of the first substrate. Therefore, local fluctuations in product-supply would be expected; and the variances might not be so small as to be neglected. Such local fluctuations might affect adversely the overall cellular economy.

As an alternate the mosaic model seems more workable. Here, topographically localized regions of metabolic activity are viewed as patches on intracellular particulate structures (Fig. 1). Spatial juxtaposition of a number of individual molecular channels related to a specific metabolic process would constitute collectively a given-patch. Thus, pools of end-product may be generated locally at sites for immediate utilization (15).

The economy of cellular metabolism also can be seen in the reduction of transient time in steady state transitions for a naturally-occurring five-enzyme sequence (22). The free-energy cost of transition is directly proportional to the transient time (23).

At the level of the component enzymes, catalytic facilitation in the clustered state may contribute to the free energy of activation through the following factors: (a) steric and enthalpic factors arising from activation-effects in aggregated systems; (b) structural effects, such as stabilization in an optimally open configuration that obviates some post-catalytic refolding (24); (c) stabilization in a conformation poised for catalytic action in the absence of substrate (25); (d) freezing each component polypeptide into a single (optimal) configuration for catalysis (26); (e) retaining the energy in excited internal vibrational states within the protein structure for specific utilization in subsequent catalytic events (27); and (f) electrostatic contributions (24). For reactions in ionic solution, the ΔG^* for the formation

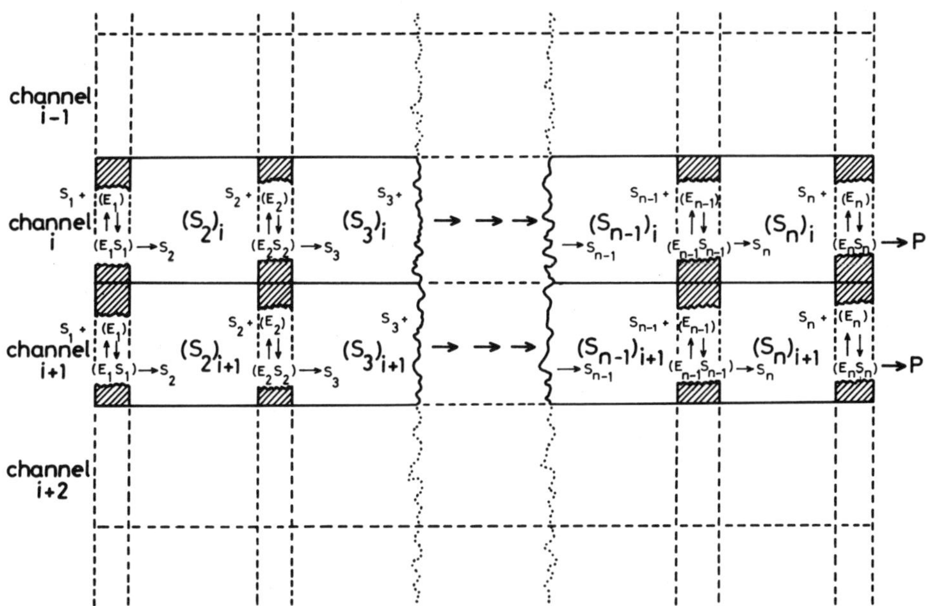

Fig. 1. Schematic representation of a mosaid structured multienzyme system catalyzing the reaction sequence

$$S_1 \xrightarrow{E_1} S_2 \xrightarrow{E_2} \cdots \xrightarrow{En-1} S_n \xrightarrow{En} P$$

Each subunit contains one enzyme molecule (hatched region) plus an associated volume compartment (analogous to lipoprotein subunits in membrane-bound enzyme schemes). The diagram also illustrates the quantization phenomenon, whereby each channel maintains its own respective concentration of a given substrate, $(S_j)_i$ (21).

of an activated complex from two ions will contain a part ΔG_{es}*, due to the free-energy change associated with the electrostatic forces between the two reactants as they are brought together. This has the form

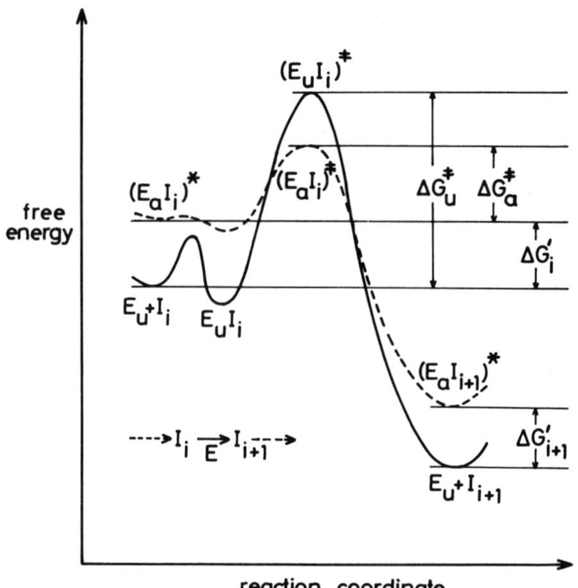

Fig. 2. Potential effects of enzyme clustering on the free-energy profile of the reactions of intermediary metabolism. As discussed in the text, catalytic facilitation by structured multienzyme systems may entail a smoothing effect on the overall profile, resulting from a lowering of energy barriers and/or a raising of energy valleys. Enzyme E (structured a, or unstructured u) catalyzes the reaction $I_i \rightarrow I_{i+1}$ among metabolic intermediates.

$$\Delta G_{es}^* = \frac{Z_1 Z_2 e^2}{\varepsilon r^*}$$

where Z_1 and Z_2 are the number of charges on the respective ions, e the electronic charge, ε the dielectric constant of the medium, and r* some critical distance. For the formation of an enzyme-substrate complex in solution, the net charge of the protein will affect the overall interaction potential. Hence, in general, ΔG_{es}^* will not be negative (rate-enhancing). Within the confinement of a multienzyme cluster, a nascent intermediate-substrate molecule might not see the overall charge of the protein (or matrix structure), but only that in the vicinity of the active center. Also, the value of ε might be significantly lower (than for water) in structured regimes. Then a negative (rate-enhancing) value of ΔG_{es}^* might be a built-in feature of the organized state. The potential influence on ΔG^* of these combined factors, resulting from aggregation, is illustrated in Fig. 2. Some features, e.g., (a), (d), and (f), might lower ΔG^*, while others, e.g., (b), (c) and (e), might elevate the free-energy valley (see $\Delta G_i'$ and ΔG_{i+1} in Fig. 2).

PRACTICAL ASPECTS

There has been some interest in creating functional multienzyme aggregates by covalently linking the respective protein moieties (28). It seems doubtful that this approach will become a general technique for producing aggregates which possess the functional properties of naturally occurring systems. For single enzyme species it is a usual procedure to optimize the enzymatic activity with respect to the corresponding enzyme free in solution; we may need to redefine this reference state. For example, the α-spectrum presupposes differences between the enzyme structure and function *in vitro* and *in vivo*. The relationship of the environment and the time-lag hysteresis effect exhibited by a host of single enzyme activities (and which pervades the entire multienzyme system of the polyaromatic pathway in *Neurospora crassa*) may be a further indication that enzyme conformation *in vitro* can be deceptive. Hence, when preparing immobilized enzyme systems, it may be advantageous in some cases to introduce minor conformation alterations into the protein structure, so as to optimize the enzyme prior to immobilization.

ACKNOWLEDGMENTS

Most of this work was completed while the author was a Research Fellow of the Instituts Internationaux de Physique et de Chimie at the Service de Chimie Physique 2, Universite Libre de Bruxelles, Belgique.

REFERENCES

1. COOPER, A. *Proc. Natl. Acad. Sci. USA 73*:2740, 1976.
2. LUMRY, R. & BILTONEN, R. in "Structure and Stability of Biological Macromolecules," (S.N. Timasheff and G.D. Fasman, eds.). Marcel Dekker, New York, 1969, p. 65.
3. SOMOGYI, B. & DAMJANOVICH, S. *J. Theor. Biol. 51*:393, 1975.
4. FROHLICH, H. *Proc. Natl. Acad. Sci. USA 72*:4211, 1975.
5. GREEN, D.E. *Ann. N.Y. Acad. Sci. 227*:6, 1974.
6. LOW, P.S. & SOMERO, G.N. *Proc. Natl. Acad. Sci. USA 72*:3305, 1975.
7. CARERI, G. in "Quantum Statistical Mechanics in the Natural Sciences," (B. Kursunoglu, S.L. Mintz, and S.M. Widmayer, eds.), Plenum, New York, 1974, p. 15.
8. KEMENY, G. *J. Theor. Biol. 48*:231, 1974.
9. SHOHET, J.L. & REIBLE, S.A. *Ann. N.Y. Acad. Sci. 227*:641, 1974.
10. MCCAMMON, J.A., GELIN, B.R., & KARPLUS, M. *Nature 267*:585, 1977.
11. DAMJANOVICH, S. & SOMOGYI, B. in "Proceedings First European Biophysics Congress," vol. 6 (E. Broda, A. Locker and H. Springer-Lederer, eds.), Verlag der Wiener Medizinischen Akademie, Vienna, 1971, p. 133.
12. CARERI, G., FASELLA, P., & GRATTON, E. *Critical Reviews Biochem. 1975*:141.
13. CARERI, G. & GRATTON, E. *Bio Systems 8*:185, 1977.
14. IKEGAMI, A. *Biophys. Chem. 6*:117, 1977.
15. WELCH, G.R. *Prog. Biophys. Mol. Biol. 32*:103, 1977.
16. RICARD, J., MOUTTET, C., & NARI, J. *Eur. J. Biochem. 41*:479, 1974.
17. OSTER, G., PERELSON, A., & KATCHALSKY, A. *Nature 234*:393, 1971.
18. OSTER, G., PERELSON, A.S., & KATCHALSKY, A. *Quart. Rev. Biophys. 6*:1, 1973.

19. BUNOW, B. & ARIS, R. *Math. Biosci.* 26:157, 1975.
20. SMEACH, S.C. & GOLD, H.J. *J. Theor. Biol.* 51:59, 79, 1975.
21. SOLS, A. & MARCO, R. *Curr. Top. Cell. Reg.* 2:227, 1970.
22. WELCH, G.R. & GAERTNER, F.H. *Proc. Natl. Acad. Sci. USA* 72:4218, 1975.
23. WELCH, G.R. *J. Theor. Biol.*, in press.
24. LAIDLER, K.J. & BUNTING, P.S. "The Chemical Kinetics of Enzyme Action," 2nd Edit., Oxford University Press, London, 1973.
25. VALLEE, B.L. & WILLIAMS, R.J.P. *Proc. Natl. Acad. Sci. USA* 59:498, 1968.
26. KARUSH, F. *J. Amer. Chem. Soc.* 72:2705, 1950.
27. RABINOVITCH, B.S. & FLOWERS, M.C. *Quart. Rev. Chem. Soc. (London)* 118:122, 1964.
28. MOSBACH, K. & MATTIASSON, B. *Meth. in Enzymology* 44:453, 1976.

IMMOBILIZATION OF ENZYMES, MICROBIAL CELLS, AND ORGANELLES BY INCLUSION WITH PHOTO-CROSSLINKABLE RESINS

Saburo Fukui, Atsuo Tanaka, and Gerald Gellf

Laboratory of Industrial Biochemistry
Faculty of Engineering, Kyoto University
Kyoto, Japan

This communication deals with the application of photo-crosslinkable resins to inclusion of enzymes, microbial cells and organelles. The method consists of mixing (a) liquid oligomers of suitable photo-crosslinkable resin(s) containing photo-sensitive functional groups, (b) an appropriate initiator, and (c) an enzyme solution or suspension of cells or organelles, followed by illumination with near-ultraviolet light for only a few minutes. This simple and convenient procedure produces tailor-made matrices in which biologically active macromolecules are successfully entrapped.

MATERIALS AND METHODS

The photo-crosslinkable resin oligomers used in this study were poly(ethylene glycol) dimethacrylate (PEGM) and an analogue (ENT) of different chain lengths.

PEGM was prepared by refluxing commercially available poly(ethylene glycol) and excess methacrylate in toluene in the presence of small amounts of p-toluene sulfonic acid. When poly(ethylene glycol) 2000 was used as the starting material, the oligomer containing photo-sensitive methacrylate groups at both terminals was referred to as PEGM-2000. A typical preparation procedure for ENT consisted of reacting equimolar hydroxyethylacrylate and isophorone diisocyanate at 70°C in the presence of a suitable catalyst, such as organic tin compounds or ternary amine compounds (for example, a mixture of 0.5% tributylamine and 0.5% hydroquinone). After 2 hr was added a half molar ratio of poly(ethylene glycol); and the reaction was continued for 5 hr at 70°C. The resulting product was used as a photo-crosslinkable resin oligomer (ENT) without further purification. When poly (ethylene glycol)-2000 was used as the starting material, the resin oligomer was named ENT-2000. In some experiments, analogues of ENT containing a poly(propylene glycol) chain instead of a poly(ethylene glycol) chain were used. This type of resin was referred as to ENTP.

Entrapment was done by mixing one part of PEGM or ENT with 0.01 part of an initiator, benzoin ethyl ether, and melting at 50°C. To the mixture was added 1.5 parts of a solution of enzyme or a suspension of microbial cells or organelles. The homogeneous liquid mixture was layered (0.4 mm thick) on a sheet of transparent polyester film, covered with another sheet of the same film, and illuminated with a Toshiba Chemical Lamp FL 20BL (wavelength range 300-400 nm; maximum intensity at 360 nm) for 3 min. The resin film thus formed, was cut into small pieces (5 X 5 mm) and used as test samples. For comparison, the conventional polyacrylamide method and a method which entrapped enzymes or cells inside proteinic gels formed by crosslinking of albumin molecules with glutaraldehyde (1) were tested.

RESULTS AND DISCUSSIONS

Single enzymes, such as invertase and glucose isomerase were successfully entrapped by using photo-crosslinkable resins. With yeast invertase the activity of the enzyme entrapped was 40 - 50% that of the native enzyme. The relative activity was affected by the proportion of water, enzyme, and oligomers in the immobilization system and by the chain length of the oligomers. The entrapped invertase showed almost the same V_{max} and about a 5 times higher K_m, compared with the native counterpart. The pH and temperature dependencies of the enzyme were changed somewhat by the immobilization. The details of these experiments have been published (2, 3).

CDP-choline is not only an important intermediate in the biosynthesis of phospholipids but also is considered to be an effective drug for brain injuries. A practical procedure to convert CMP into CDP-choline or its analogues has been described (4). The most unique part of this method was the continuous supply of energy (ATP) from glucose by glycolysis of yeast. They chose the yeast *Hansenula jadinii* because of its strong activity to phosphorylate various free aminoethanols, including choline. One of the important matters in this process is that yeast cells should be dried or treated with detergents, because CMP is not phosphorylated by intact cells but instead is decomposed to uracil. Entrapment of these dried or detergent-treated cells of *Hansenula jadinii* inside photo-crosslinked ENT-1000 resin yielded an excellent biocatalyst which could be used for long-term repeated reactions. On the other hand, inclusion of the cells by polymerization of acrylamide monomers resulted in an almost complete loss in catalytic activity.

Conversion of steroids, such as 1,2-dehydrogenation of hydrocortisone to yield prednisolone, is one of the most interesting projects in enzyme technology. Application of gel-entrapped cells as a biocatalyst in these processes has been extensively studied. We examined the influence of the hydrophobic character of the gel matrices of entrapped *Arthrobacter simplex* cells on the reaction rate of 1,2-dehydrogenation of hydrocortisone by changing the ratio of ENTP to ENT. Fig. 1 shows the effect of the ratio of ENTP-2000 to ENT-4000 on the formation of prednisolone.

Peroxisomes and mitochondria, isolated from methanol-grown cells of *Kloeckera* sp. No. 2201 (6-8), also were immobilized by entrapment. The immobilization was carried out using PEGM-2000 or ENT-4000 and entrapping the microbodies inside proteinic gels formed by crosslinking albumin with glutaraldehyde (1). The relative activities of catalase, alcohol oxidase, and D-amino acid oxidase for the gel-entrapped microbodies measured in the presence of sucrose are shown in Table 1. Almost similar relative activities for catalase (46 - 48%), D-amino acid oxidase (49%), and alcohol oxidase (73 - 80%) were observed with both ENT-4000-entrapped and albumin-entrapped microbodies. The apparent 2-fold higher activity for alcohol oxidase, as compared with catalase and D-amino acid oxidase, could be explained by the fact that 2 moles of methanol are oxidized by the combined action of alcohol oxidase and the hydrogen peroxide catalase system. The apparent K_m of the D-amino acid oxidase of the ENT resin-entrapped microbodies was somewhat higher than that of free microbodies. The V_{max} of the former was about one-fifth that of the latter. However, no significant loss in the enzyme activities appeared during the immobilization process, be-

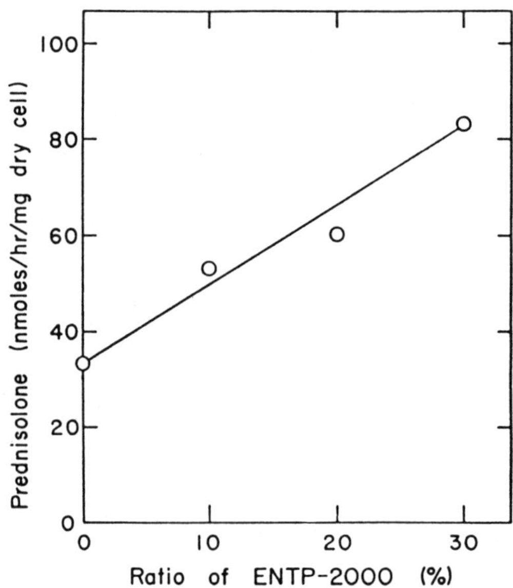

Fig. 1. Effect of mixing ratio of ENTP and ENT on prednisolone formation.

cause the activities of the microbody enzymes entrapped in the gels approached those of the free microbodies, when the microbody resin films were ground to fine powders. Only little differences were observed between the parameters assayed in the buffer solution containing sucrose (microbodies in gels should be intact) and those assayed in the absence of sucrose (microbodies were burst inside gel matrices). This result suggested that a transportation barrier due to the microbody membrane would not be significant but that diffusion limitations through the gel matrices had an effect.

Gel-entrapped microbodies can be useful as a model system to study the *in situ* function of organelles in living cells. Table 2 shows the results of stoichiometric experiments carried out in the presence and absence of sucrose and with added 3-amino-1,2,4-triazole, an inhibitor of catalase. Approximately 2 moles of formaldehyde were formed by the action of gel-entrapped microbodies with consumption of one mole of oxygen, irrespec-

TABLE 1

COMPARISON OF THE ACTIVITIES OF ENZYMES IN GEL-ENTRAPPED MICROBODIES

Microbody	Specific enzyme activity (nmoles/min/mg protein)*		
	Catalase	Alcohol oxidase	D-amino acid oxidase
Free	4310×10^3	791	15.9
	(100)	(100)	(100)
ENT-4000-entrapped	1980×10^3 (46)	630 (80)	7.82 (49)
Albumin- entrapped	2080×10^3 (48)	580 (73)	--

*Figures in ( ) represent % of free.

tive of the presence or absence of sucrose. 3-Amino-1,2,4-triazole decreased the formation of formaldehyde to one half without any change in the amount of oxygen consumed. This result would offer confirmatory evidence for the synergistic action of alcohol oxidase and catalase on the initial oxidation step of methanol (9).

The gel-entrapped microbodies should be useful as biocatalysts for assay or treatment of several primary alcohols, hydrogen peroxide, and D-amino acids. Figure 2 (left) shows the stabilities of catalase and alcohol oxidase in the microbodies entrapped in ENT-resin matrices and in albumin gels in comparison with those in free microbodies during storage at 4°C in 0.05 M potassium phosphate buffer, pH 7.2, containing sucrose. Entrapment of microbodies in gels, especially by the albumin-glutaraldehyde method, enhanced the stability of catalase to a moderate degree; whereas the catalase in free microbodies was rather unstable under the experimental conditions. On the other hand, no significant difference was observed

TABLE 2

STOICHIOMETRIC STUDIES OF METHANOL OXIDATION BY GEL-ENTRAPPED MICROBODIES IMMOBILIZED WITH ENT-4000

Protein (μg)	Addition to buffer used in the assay	O_2 consumed (A)	Formaldehyde formed (B)	B/A
		(nmoles/min/mg/protein)		
390	0.65 M Sucrose	259	526	2.03
317	None	215	401	1.87
317	0.15 M 3-Amino-1,2,4-triazole	203	207	1.02

between the stability of alcohol oxidase in free microbodies and gel-entrapped ones. Figure 2 (right) depicts the comparison of the stabilities of catalase and alcohol oxidase in ENT-entrapped and albumin-entrapped microbodies in repeated use for 5 mM hydrogen peroxide decomposition and methanol oxidation. Although the gel-entrapped microbodies lost 20 to 30% of the original activity of catalase during the initial 5 repeated reactions, the enzyme activity became stable thereafter. When the hydrogen peroxide concentration was reduced to 1 mM, catalase in gel-entrapped microbodies retained the original activity even after 12 repeated reactions. In the case of alcohol oxidase a marked difference was observed between the ENT resin-entrapped microbodies and albumin-entrapped ones. The higher stability of the latter may have been ascribable, at least in part, to partial crosslinking of the enzyme molecule with the protein gels.

Hence, it would be concluded that albumin-glutaraldehyde is superior to photo-crosslinkable resin with respect to the immobilization of yeast microbodies. However, the albumin-glutaraldehyde method was not useful for the immobilization of yeast mitochondria, because the respiratory activity of the organelles was completely lost during entrapment by this method. On the other hand the photo-crosslinkable resin method was applicable to the immobilization of mitochondria (data, not shown).

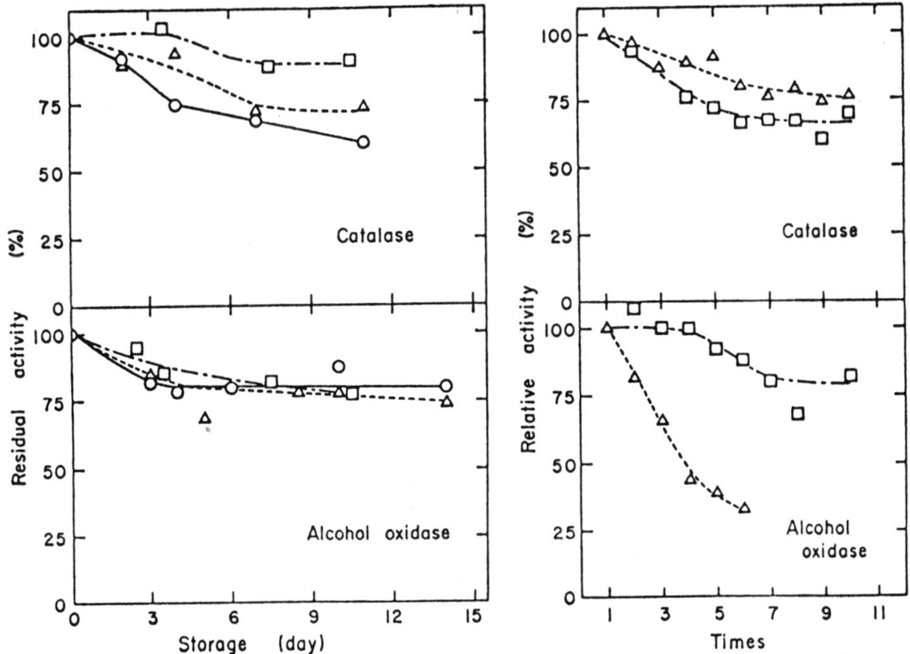

Fig. 2. Time-course changes in the activities of catalase and alcohol oxidase in free microbodies (circle) and in microbodies entrapped in ENT-resin (triangle) and albumin-glutaraldehyde protein gel (square) during storage and repeated reactions. Left shows effect of storage at 4°C; right shows effect of repeated use.

ACKNOWLEDGMENT

The authors are indebted to the late K. Ogata and his coworkers for the generous gift of a strain of *Kloeckera* sp. No. 2201 and to T. Yamamoto and T. Iida, Kansai Paint Co., Ltd., for kindly supplying the photo-crosslinkable resin oligomers. G. Gellf is on leave from the University de Technologie, Compiegne, France.

REFERENCES

1. BROUN, G., THOMAS, D., GELLF, G., DOMURADO, D., BERJONNEAU, A.M., & GUILLON, C. *Biotechnol. Bioeng.* *15*:359, 1973.
2. FUKUI, S., TANAKA, A., IIDA, T. & HASEGAWA, E. *FEBS Lett.* *66*:179, 1976.
3. TANAKA, A., YASUHARA, S., FUKUI, S., IIDA, T. & HASEGAWA, E. *J. Ferment. Technol.* *55*:71, 1977.
4. KIMURA, A. & ARIMA, K. *Abst. 4th Internat. Specialized Symp. Yeasts* and *5th Internat. Ferment. Symp.* 498, 1976.
5. FUKUI, S., TANAKA, A., KAWAMOTO, S., YASUHARA, S., TERANISHI, Y. & OSUMI, M. *J. Bact.* *123*:317, 1976.
6. SAHM, H., ROGGENKAMP, R., WAGNER, F. & HINKELMANN, W. *J. Gen. Microbiol.* *88*:218, 1975.
7. VAN DIJKEN, J.P., VEENHUIS, M., KREGER-VAN RIJ, N. J.W. & HARDER, W. *Arch. Microbiol.* *102*:41, 1975.
8. FUKUI, S., KAWAMOTO, S., YASUHARA, S., TANAKA, A., OSUMI, M. & IMAIZUMI, F. *Eur. J. Biochem.* *59*: 561, 1975.
9. ROGGENKAMP, R., SAHM, H. & WAGNER, F. *FEBS Lett.* *41*: 283, 1974.

IMMOBILIZED CELL SYSTEMS

W.R. Vieth and K. Venkatasubramanian

Department of Chemical and Biochemical Engineering
Rutgers University
New Brunswick, New Jersey, USA

Our work in this area began with simple systems (1-3), but we are now examining more complex reaction sequences involving entire metabolic pathways. This basic understanding is essential to place classical fermentations on a heterogeneous catalysis basis, which is, perhaps, the ultimate goal of bound cell systems. Some of our results have been published (4, 5). Presented here are a few of our recent findings on immobilized cell-mediated processes currently under investigation in our laboratory (Table 1). We have focused our attention on the first two items of Table 1, citric acid synthesis and biophotolysis of water.

EXPERIMENTAL

Reconstituted bovine hide collagen was used as the carrier matrix in all of our immobilized whole cell and organelle work. Methods of attaching whole cells to collagen have been described in detail elsewhere (2, 6). Briefly, they involve mixing cells with a collagen dispersion at an appropriate pH, casting, drying a membrane, and tanning with glutaraldehyde to a desired level of mechanical strength. Organelle (chloroplast) immobilization was carried out by first isolating stripped chloroplasts from intact spinach (*Spinacia oleracea*) leaves by a special procedure (7) which preserves the integrity of the photosynthetic electron transport system (PETS); stripped chloroplasts are then mixed with a 4% (w/v)

TABLE 1

COLLAGEN-IMMOBILIZED WHOLE CELL SYSTEMS

Microorganism Organelle	Substrate (Product)	Comments
Aspergillus niger	Sucrose (Citric acid)	Primary metabolite
Chloroplast	Water (Oxygen)	Immobilized organelle; First step to biophotolysis of water
Anacystis nidulans	Water (Oxygen)	Immobilized algae cells
Anacystis nidulans	Nitrate (Ammonia)	Biological nitrogen fixation
Streptomyces griseus	Glucose (Candicidin)	Antibiotic synthesis; secondary metabolite
Pseudomonas aerguginosa	--	Concentration of plutonium from waste waters
Mycobacterium rhodochorus	Cholesterol (Δ^4-cholestenone)	--

collagen dispersion at pH 8.5 and cast to form a thin membrane after drying. Other variations of this procedure have also been employed.

Activity of bound chloroplast preparations was monitored by the Hill assay which essentially involves measuring oxygen evolution polarographically (8). Citric acid was determined colorimetrically (9); reaction products were also analyzed chromatographically (10). Unreacted sucrose substrate in citric acid production was determined by converting it to glucose quantitatively and measuring its concentration by the Glucostat procedure (11).

RESULTS AND DISCUSSION

Typical activities of free and immobilized chloroplasts and algae cells (*A. nidulans*) are shown in Table 2. Immobilized chloroplasts retained about half of the activity of free chloroplasts; while the activity retention

TABLE 2

TYPICAL ACTIVITIES OF COLLAGEN-CHLOROPLAST AND COLLAGEN *A. NIDULANS* MEMBRANES

Sample**	Chlorophyll Content (mg/ml* or mg/g)	Specific Activity (μmoles O_2/ mg, hr)
Stripped chloroplast preparation	2.5*	35.4
Collagen-chloroplast membrane	11.8	18.5
Algae cells	11.9	29.1
Collagen-algae cells membrane	3.6	7.68

** Assay conditions: 25°C; light intensity 40,000 Lux; about 4 mil thick membranes; 70% of dry weight of the membrane was collagen; assay medium consisted of pH 7.4, 0.34 M sucrose or sorbitol, 0.01 M NaCl, 0.6 mM phenylenediamine and 1.5 to 3.0 mM potassium ferricyanide. All assays were short term assays (3 to 8 mins.).

was about 27% in the case of bound algae cells. The normal photosynthetic rate in an actively growing plant is about 450 μmoles O_2/mg chlorophyll/hr. The observed activities thus represented an energy recovery efficiency of 1.5 to 3.0%, when compared with a growing plant. The storage stability at 4°C of a stripped chloroplast suspension is compared with that of a collagen-chloroplast membrane in Fig. 1. Intermittant batch assays conducted at 25°C using potassium ferricyanide as electron acceptor indicated that the fixed chloroplast preparation was active after 15 days of storage. Immobilized *Anacystis nidulans* exhibited a rather sharp decay of activity on repeated batch contacts; in one experiment 17% of the initial activity was retained after four successive batch assays conducted over 30 min.

Preliminary data show that *Aspergillus niger* cells attached to collagen exhibit good activity retention and yield 8 to 40% of the final citric acid concentration

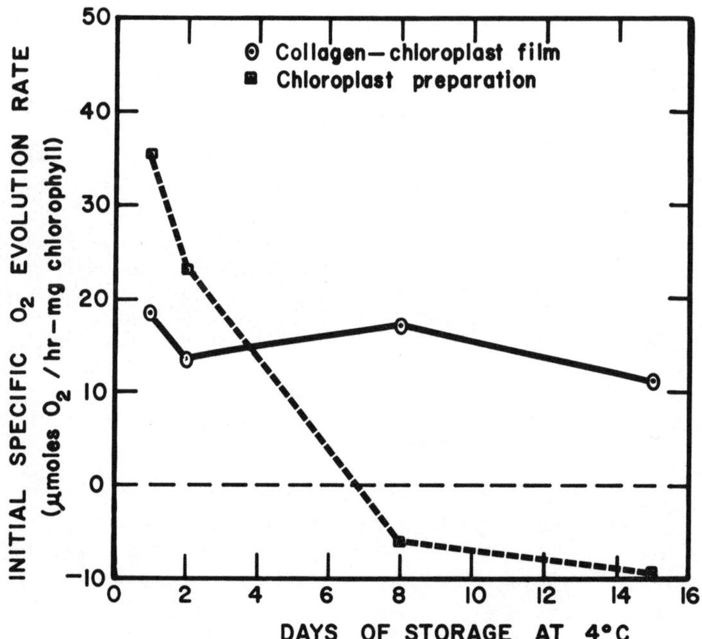

Fig. 1. Stabilization of stripped chloroplast-collagen membrane.

obtainable in fermentation starting with sucrose. The optimum pH for both immobilization and glutaraldehyde tanning has a range of 4.0 and 6.0. Considering the acidic nature of the product, this is not surprising. A portion of the data on tanning conditions are summarized in Fig. 2; 5% aldehyde for one minute appears to yield the best combination of activity and mechanical strength. There is also an apparent relationship between the degree of tanning and the viability of the organism in the immobilized state. Based on repeated batch experiments and continuous reactor experiments, the half-life of the catalyst has been estimated at 150 to 180 hr. The activity of the bound cell preparation increased linearly with increasing cells-to-carrier ratio. Although membranes containing as high as 70% cells (dry weight basis) were

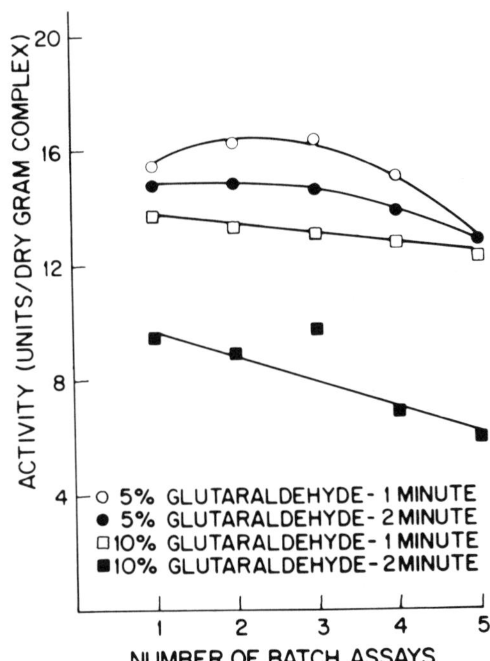

Fig. 2. Effect of glutaraldehyde tanning conditions on activity retention of immobilized *Aspergillus niger*.

prepared, their mechanical strength sharply declined above a cell/collagen weight ratio of 50%. The observed linear dependence implied that the matrix structure was sufficiently below the point of saturation, i.e., not all the available binding sites were occupied.

Another factor of importance was dehydration of the matrix. It was found best to keep the membranes hydrated when not in use, following the initial drying of the matrix. Repeated hydration-dehydration-lead to significant activity loss. Furthermore, at least a fraction of the activity decay during immobilization appeared due to dehydration. When stored under dry conditions prior to use, the membrane activity declined slowly.

There was a definite correlation between culture age and the activity of the membrane (Fig. 3). If the

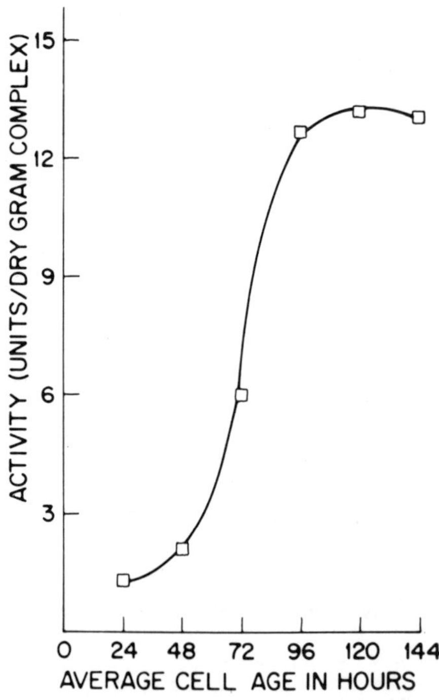

Fig. 3. Influence of cell age (prior to immobilization) on catalyst activity.

time course of citric acid formation in a fermentor was superimposed on this curve, it closely followed Fig. 3, signifying that the physiological state (in aggregate) of the organism very likely remained the same in the free and the immobilized forms. It is worth noting that the cells appeared to be viable and fully capable of reproduction within and on the surfaces of the carrier matrix. Newly formed pellets of the mold were clearly identifiable some 36 to 48 hr after the immobilized membrane chips were incubated in a medium identical to that used in the fermentor. Microscopic examination and reculturing of the system provided additional credence to our thesis that the observed growth was not due to contaminants in the system. Reproduction within the matrix, in fact, led to increased acid production. When one of the nutrients (nitrogen) was deliberately limited in the medium, reproduction, and hence additional acid formation, was curbed. Based on several other sets of experiments, it can be said that nitrogen starvation arrested cellular proliferation within the matrix. One might invoke the possibility that the observed growths occurred in the medium rather than in the matrix, triggered by the elution of the organism from the carrier into the medium. Neither centrifugation nor subculturing of the medium showed any evidence to support this postulation. In addition, the membrane showed an increase in its dry weight after 36 to 75 hr incubation in whole fermentor medium. This increase in weight was also dependent on the extent of tanning. For instance, membranes tanned with 1% and 3% glutaraldehyde for two min registered 80% and 35% increase in weight, respectively, after 72 hr of incubation. All these findings reinforce the observation that it is possible to immobilize the cells on collagen and maintain them in a viable state, thus preserving their internal cofactor regenerating machinery.

Turning to process engineering aspects, we have investigated external and internal transport constraints, overall process kinetics, oxygen transfer, and competing kinetic phenomena involving side reactions. In a very simplified way, the observed kinetics of citrate synthesis by immobilized cells can be described by an apparent first-order behavior up to a substrate concentration of 15 g/l. Presented in Fig. 4 and 5 are data relating to external and internal mass transfer. The effect of linear velocity on the observed reaction rate (Fig. 4) shows the presence of a significant boundary layer resistance below a flow rate of 235 ml/min. The existence of non-negligible pore diffusional resistance is deducible from

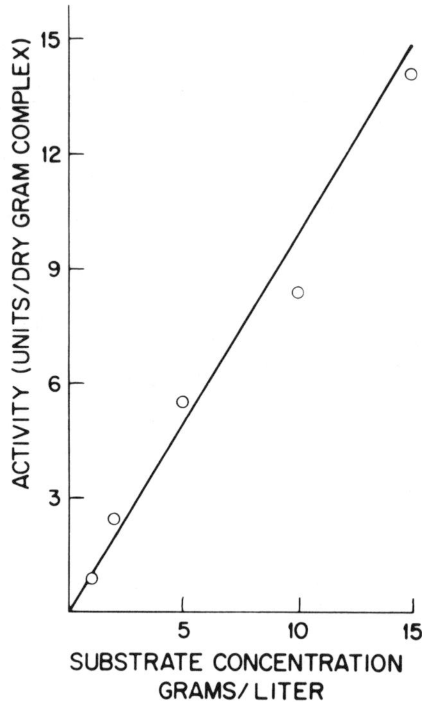

Fig. 4. Dependence of reaction rate on linear velocity.

Fig. 5, which shows the dependence of observed reaction rate on film thickness.

The spirally-wound multipore reactor configuration (3) was used in all reactor studies. Since oxygen transfer was a crucial variable, a special provision was incorporated into the reactor design to allow flow of pure oxygen countercurrent to the flow of substrate. The packing within the reactor facilitated oxygen absorption into the substrate; in this sense the overall system operated as a combined adsorber-reactor. Dissolved oxygen concentrations of 80 to 90% saturation were maintained throughout the course of the reactor runs. Referring to Fig. 4, the increased specific activity of the catalyst observed in the reactor compared to that in a shake flask was attributable, at least in part, to improved oxygen transfer in the reactor.

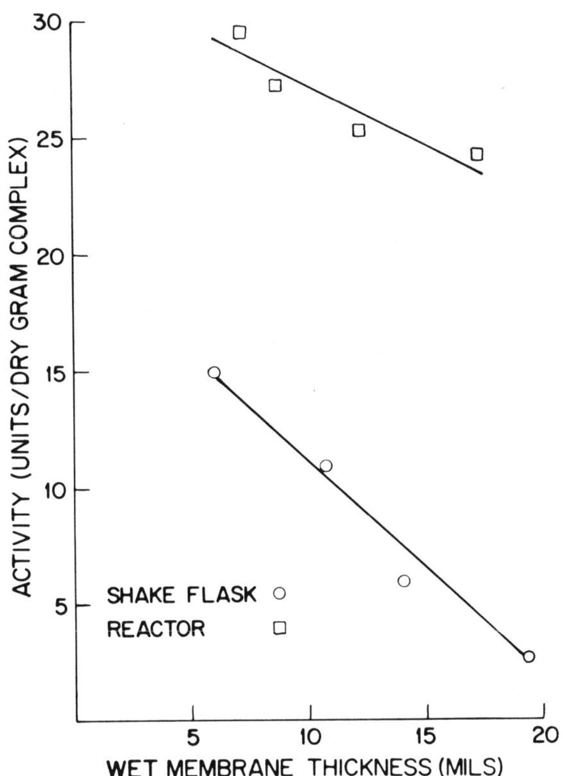

Fig. 5. Effect of membrane thickness on citric acid production rate.

Chromatographic Analysis of reaction products revealed the presence of products generated from side reactions. They included isocitric acid, oxalic acid, and trace quantities of gluconic acid. Isocitrate was perhaps the major one amounting to as much as 15 to 20% of citrate. Oxalic acid formation in citric acid fermentations is reported to be dependent both on pH and on the extent of aeration (12). By proper control of pH and dissolved oxygen levels, it might be possible to reduce the formation of oxalate.

ACKNOWLEDGMENTS

The authors are grateful to Charles Bertalan and Jeffrey Howell for their assistance in the preparation of this manuscript and to Robert Tengerdy of Colorado State University and Juan Martin of the University of Salamanca, Spain for their contributions.

REFERENCES

1. VIETH, W.R., WANG, S.S., & SAINI, R. *Biotechnol. Bioeng.* *15*:565, 1973.
2. VIETH, W.R. & VENKATASUBRAMANIAN, K. *Meth. in Enzymology 44*:243, 1976.
3. VIETH, W.R. & VENKATASUBRAMANIAN, K. *Meth. in Enzymology 44*:768, 1976.
4. VENKATASUBRAMANIAN, K., VIETH, W.R. & CONSTANTINIDES, A. in "Enzyme Engineering," vol. 3 (HE.K. Pye and H.H. Weetall, eds.), Plenum, New York, in press.
5. VIETH, W.R., VENKATASUBRAMANIAN, K. & CONSTANTINIDES, A. *Abst. 5th Intern. Ferment. Symp. Berlin* 15,26, 1976.
6. BERNATH, F.R. & VIETH, W.R. in "Immobilized Enzymes in Food and Microbial Processes," (A.C. Olson and C.L. Cooney, eds.), Plenum, New York, 1974, p. 157.
7. MORGENTHALER, J.J., PRICE, C.A., ROBINSON, J.M. & GIBBS, M. *Plant Physiol. 54*:532, 1974.
8. TREBST, A., *Meth. in Enzymology 24*:146, 1972.
9. MARIER, J.R. & BOULET, M.J. *J. Dairy Sci. 41*:1683, 1958.
10. CARLES, J., SCHNEIDER, A. & LACOUSTE, A.M. *Bull. Soc. Chim. Biol. 40*:221, 1958.
11. FERNANDES, P.M. Ph.D. Thesis, Rutgers University, 1975.
12. RHODES, A. & FLETCHER, D.L. "Principles of Industrial Microbiology," Pergamon, New York, 1966, p. 198.

STEROID CONVERSION USING IMMOBILIZED LIVING MICROORGANISMS

P.O. Larsson, S. Ohlson and K. Mosbach

Chemical Center
University of Lund
Lund, Sweden

In steroid transformations some of the advantages of immobilized microorganisms seem to be especially important, since the manufacture of steroid drugs often involves very expensive intermediates and products (1). Therefore, we have been studying two corticosteroid transformations. The first reaction is an 11-β-hydroxylation catalyzed by free or polyacrylamide immobilized *Curvularia lunata*, producing cortisol from Reichstein's S (cortexolon) (1,2). The second reaction is a 3-ketosteroid-Δ^1-dehydrogenation of cortisol to prednisolone catalyzed by free or polyacrylamide immobilized *Arthrobacter simplex* (3,4).

When the microorganisms are immobilized, certain precautions must be taken to allow for good preservation of steroid transforming activity. The microorganism should be exposed to the monomer for the shortest time possible. This can be achieved by using a rather high concentration of catalyst, for example 0.2 - 0.4% tetramethylethylenediamine and 0.1 - 0.2% ammonium persulphate. In the case of *A. simplex* the polymerization occurred within 1 minute after the mixing of monomers and microorganisms. However, a high concentration of catalyst also will lead to a high polymerization rate and consequently a rise in temperature. Efficient cooling can be provided if the entrapment of the microorganisms is carried out in sandwich-like polymerization chambers, which have a high surface to volume ratio, and finally, the polymerization medium should be buffered (pH∿7); and the concentration of microorganisms should be fairly high (∿ 10%). The reasons for the protective effect of these two factors are not fully clear. If these rules of thumb are followed, between 25 and 50% (typically 40%) of the steriod transforming activity can be retained for immobilized *A. simplex*.

The conversion of cortisol to prednisolone may be followed spectrophotometrically at 285 nm, $\Delta\varepsilon$ of 2820 M^{-1} $cm,^{-1}$ or by high pressure liquid chromatography, using 5µ LiChrosorb SI 60 and 98% chloroform plus 2% methanol. For 11-α- hydroxylation and 11-β-hydroxylation the spectra of substrate and product do not differ enough, so that HPCL analysis is preferred.

These two methods of analysis allowed the convenient determination of some basic properties of immobilized *A. simplex*. These properties did not to any appreciable extent differ from those of free bacteria. The temperature optimum, heat resistance, and optimum substrate concentration were approximately the same for *A. simplex* free and immobilized in 15% polyacrylamide, 14.25% acrylamide, and 0.75% N,N'-methylene-bis-acrylamide. However the pH dependence differed. The immobilized *A. simplex* had a flat pH profile, little influenced by the pH in the bulk medium in the range 6 - 9. The free bacteria, on the other hand, got considerably more active at higher pH, at least in short-time experiments. The reason for the flat pH profile for the immobilized microorganism may be that the polyacrylamide gel imposed rate limiting diffusional hindrance on substrates and products.

Although 40% of the Δ^1-dehydrogenase activity was retained on immobilization, the activity gradually declined when the immobilized *A. simplex* was used for dehydrogenation. Perhaps the Δ^1-dehydrogenase itself was unstable, since the isolated enzyme was very unstable (5). It is not known if the associated cofactor system plus parts of the electron transport chain remained intact or if the transport of substrate, intermediates, and products functioned properly.

We tried three different approaches to stabilize the activity. The first approach involved treatment with organic solvents, which were supposed to loosen up any permeability barriers in cell walls and membranes and consequently lead to faster transport of substrates and products. Several alcohols were tried; 0.1% butanol for 20 min gave the best result, with about 30% increase of activity. Toluene, which is commonly used for permeabilizing cells, inactivated the dehydrogenase activity.

A second approach to improved stability was to carry out the transformation in the presence of an artificial electron acceptor, such as the autooxidizable naphthoquinone menadione (6). However, this approach would make the immobilized *A. simplex* loose one of the basic advantages of immobilized microbial cells, namely the independence from externally added cofactors. Fig. 1 shows the activity as a function of the menadione concentration. 0.1 mM menadione increased the activity by a factor of about 2.5. This approach seemed promising, but unfortunately the long-term operational stability drastically decreased in the presence of menadione. This was probably due to an accelerated formation of peroxides or superoxides which would damage the dehydrogenase or some vital constituent of the cell (7).

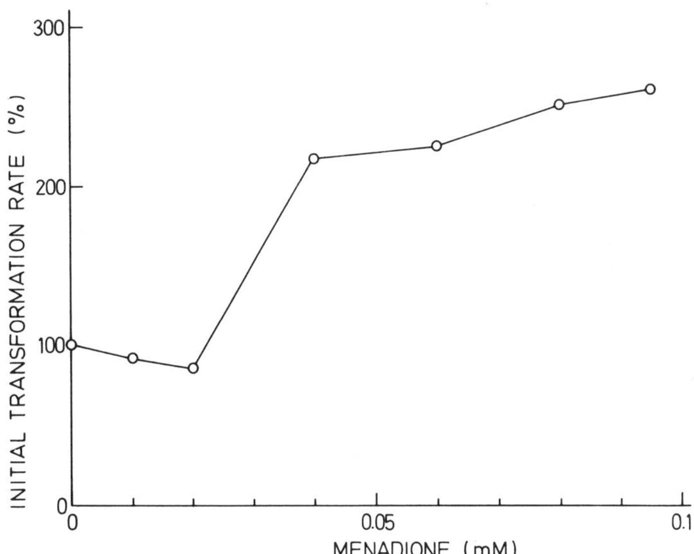

Fig. 1. Activity of immobilized *A. simplex* in the presence of menadione. The concentration of cortisol was 1 mM.

The third approach to stabilization was to treat the immobilized *A. simplex* with buffers, salts, and nutrients in the hope of reactivating inactive enzyme or of increasing enzyme synthesis. Media based on 0.1 - 1% peptone were the most satisfactory ones. Fig. 2 shows the result from an experiment comprising intermittent incubations in 1% peptone, pH 7.0, and 1 mM cortisol. In Fig. 3 are shown the results from *A. simplex* gels that were operating continuously during 4 days (10 runs). The activity increased with every batch and finally leveled off at about 30 mg of formed prednisolone per g gel (wet weight) per hr. This corresponded approximately to 720 g/kg gel/day.

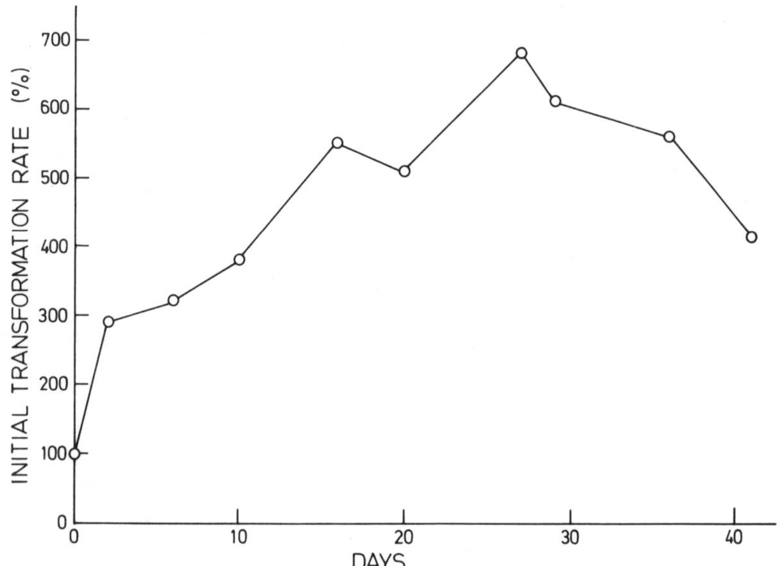

Fig. 2. Activity of immobilized *A. simplex* cells in 1% peptone. *A. simplex* gel (0.5 g wet weight) was incubated on a shaker at 21°C with 9.0 ml 1% peptone, pH 7.0, and 0.5 ml 20 mM cortisol (methanol). At 24 or 48 hr intervals the gel was filtered off, washed extensively, and assayed for Δ^1-dehydrogenase activity. After the assay the gel was washed and again incubated with fresh peptone-cortisol medium.

We investigated the cause of the increase in activity in the hope of revealing a general mechanism of activation. The lysis of cells earlier had been used as an explanation for activation (8), but this was not considered significant in the *A. simplex* system, since activation should also have been observed in buffer. Instead, activation was observed only in media which contained nutrients. This led to the hypothesis that microbial growth and/or preferential synthesis of Δ^1-dehydrogenase was responsible. Light microscope examinations were difficult to interpret in quantitative terms;

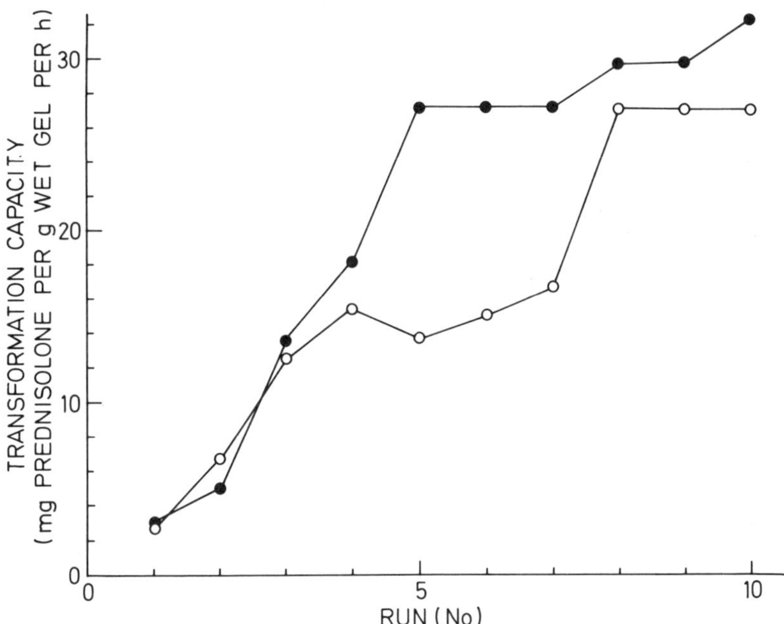

Fig. 3. Repeated batchwise transformation of cortisol. *A. simplex* gel (2 g, wet weight) was incubated in an aerated and stirred reactor with 285 ml of 0.5% peptone, pH 7.0 (●) or 285 ml of 0.1% peptone plus 0.2% glucose, pH 7.0 (o). The reaction was started by the addition of 15 ml 20 mM cortisol dissolved in methanol. After each run, i.e. when all the cortisol was converted, the gel was filtered off, washed, and again incubated with fresh peptone-cortisol medium.

but the amount of bacteria apparently had increased. After activation colonies or clusters of microorganisms were observed, in contrast to preactivation when the bacteria appeared as isolated cells. Also, the wet and dry weight of immobilized *A. simplex* preparations increased after activation; although not as much as might have been expected if microbial growth was the only cause of activation. Experiments in the presence of chloramphenicol, a known inhibitor of protein synthesis, or benzyl penicillin, a known inhibitor of cell wall synthesis and growth, showed that no activation occurred in the presence of these antibiotics. All of these experiments support the view that activation in nutrient media is a consequence of microbial growth; although a minor contribution of preferential synthesis of Δ^1-dehydrogenase in existing cells cannot be excluded.

REFERENCES

1. LARSSON, P.O. & MOSHBACH, K. *Meth. in Enzymology 44*: 183, 1976.
2. MOSBACH, K. & LARSSON, P.O. *Biotechnol. Bioeng. 12*: 19, 1970.
3. OHLSON, S., LARSSON, P.O. & MOSBACH, K. submitted for publication.
4. LARSSON, P.O., OHLSON, S. & MOSBACH, K. *Nature 263*: 796, 1976.
5. PENASSE, L. & PEYRE, M. *Steroids 12*:525, 1968.
6. LEE, B.K., BROWN, W.E., RYU, D.Y. & THOMA, R.W. *Biotechnol. Bioeng. 13*:503, 1971.
7. LAVELLE, F., MICHELSON, A.M. & DIMITRIJEVIC, L. *Biochem. Biophys. Res. Commun. 55*:350, 1973.
8. CHIBATA, I., TOSA, T., & SATO, T. *Meth. in Enzymology 44*:739, 1976.

MICROBIAL CELLS IMMOBILIZED AND LIVING ON SOLID SUPPORTS AND THEIR APPLICATION TO FERMENTATION PROCESSES

J.F. Kennedy

Department of Chemistry
University of Birmingham
Birmingham, England

Our approach has been based on the ability of water-insoluble metal hydroxides to chelate and retain peptides, proteins, etc. including enzymes. From various studies it was concluded that gelatinous titanium and zirconium hydroxide are effective matrices for enzyme immobilization. Their advantages include low cost, convenient preparation, absence of any need for pre-preparation, ability to couple enzyme at neutral pH, high retention of specific activity of the enzyme on immobilization, and the ability of modification of microenvironmental effects.

Investigations of a number of gelatinous hydrous metal oxides (frequently called hydroxides) has established (1) that hydrous titanium (IV), zirconium (IV), iron (III), vanadium (III) and tin (II) oxides at least are capable of forming with enzymes insoluble complexes which are enzymically active. From the practical viewpoint hydrous titanium (IV) and zirconium (IV) oxides proved the most satisfactory. Comparatively high retentions of enzyme specific activity may be achieved (1-3). Such hydrous oxides have also proved suitable for immobilization of amino acids and peptides (1), antibiotics with retention of antimicrobial activity (4), and polysaccharides (5). The hydrous titanium (IV) and zirconium (IV) oxides are insoluble over the normal physiological pH range.

The immobilization for hydrous metal oxides is envisaged as involving the replacement of hydroxyl groups on the surface of the metal hydroxide by suitable ligands from an enzyme or cell, resulting in the formation of partial covalent bonds. Ligands could be the side-chain hydroxyl of serine or the ε-amino of lysine. Oxygen-containing ligands are preferred to those containing nitrogen. In the case of cells the structural complexity of the cell wall ensures the availability of a great diversity of suitable ligands from both protein and carbohydrate moieties.

METHODS

Samples, each 1.3 mmol, of the metal hydroxides for use in cell immobilization were prepared from solutions of their tetrachlorides. Titanium (IV) chloride 15% w/v in 15% w/v hydrochloric acid and zirconium (IV) chloride 0.65 M in 1.0 M hydrochloric acid was neutralized to pH 7.0 by the slow addition of 2.0 M ammonium hydroxide. The samples were washed with saline solution (0.9% w/v, 3 x 5.0 ml) to remove ammonium ions and then used for cell immobilization studies.

For cell immobilization a suspension of *Escherichia coli* cells (A_{600} = 0.216) in 0.9% w/v saline was mixed with a sample of the metal hydroxide and agitated gently for 5 min at room temperature. The mixture was allowed to stand at room temperature and the suspension settled out, leaving a clear supernatant (A_{600} = 0.022), which was practically devoid of microorganisms (shown by microscopy). The immobilized cell preparation was consolidated by centrifugation at low speed and removed from the supernatant for further examination.

Saccharomyces cerevisiae and *E. coli* were immobilized in this manner. The preparations were examined for continued viability by measurement of the oxygen uptake at 25°C, in aerated 0.2 M sodium acetate buffer pH 5.0 or in 0.9% w/v saline, by use of an oxygen electrode. The rate of oxygen uptake of the immobilized cells was ∿30% that of the same number of free cells. This result showed that respiration of the cells could continue when the cells were immobilized. The reduced rate of oxygen uptake probably was caused by the diffusional resistance of the hydrous metal oxide to aerated buffer for the cells and to a decrease in the area of cell surface available for oxygen transfer. The cells were firmly attached to the surface of the hydrous metal oxide since bicarbonate, phosphate, and fluoride ions, which have been shown (1) to remove loosely bound proteinaceous and other materials from zirconium (IV) hydroxide, were singularly ineffective in the attempted release of the immobilized cells from the matrix.

To show further that the cells were firmly attached to the hydrous metal oxide, *Serratia marcescens* was incubated in nutrient medium at 25°C. This organism produced a distinctive red coloration, which enabled the immobilized cells to be distinguished readily from those of any other contaminating microorganisms. On adding cultures of *S. marcescens* to hydrous zirconium (IV) and titanium (IV) oxides, the red colored cells became associated with the insoluble matrix. The supernatant and subsequent washings with saline were almost cell free, thus demonstrating the strength of the cell-metal hydroxide interaction. When samples of these immobilized cells were added as a small inoculum to fresh culture medium, growth was observed, as detected by the large increase in numbers of red cells. At no time throughout extended studies of the growth of *S. marcescens* in various conditions of humidity, oxygen tension, temperature, medium composition, and ultraviolet irradiation-induced mutation was release of color from the cells observed. Thus, this was taken as evidence that the cells had suffered no deleterious effects on immobilization. The immobilized cells could retain their activity for weeks or months at least.

It was also found that cells could be immobilized at the lower pH of 2 to 5. Titaniom (IV) was more effective for this purpose. This enabled us to produce a small scale immobilized cell reactor for production of vinegar.

RESULTS AND DISCUSSION

The manufacture of vinegar, using *Acetobacter* species, is an industry which operates with low profit margins. There is little investment in research or in more efficient continuous processes (6). The tower fermenter is an efficient system easy to run but still of high initial cost. This could be improved by high volumetric efficiencies, i.e. high flow rates, if some means could be found to maintain an increased bacterial concentration in the fermenter. Immobilization of the cells without harm or damage, according to the above method, was predicted to achieve this desired increase in cell concentration. The cell walls of *Acetobacter* (Gram-negative) could be expected to contain mureins (peptidoglycans) as the innermost layer of the wall with proteins, lipoproteins, and lipopolysaccharides in the outer layers. Outside the cell wall there may be a capsule or slime layer, consisting of either a single polysaccharide or a polypeptide of a single amino acid. Since the acetification of ethanol takes place at about pH 3, titanium (IV) was used for the complexation. At this pH hydrous zirconium (IV) oxide is not completely precipitated whereas hydrous titanium (IV) oxide is fully precipitated. An interesting aspect is that an equilibrium will exist between immobilized and free bacteria. Thus, when the immobilized cells die, they will eventually be released from the hydrous oxide surface and be replaced by living cells.

In the 2.5 l tower fermenter used here, wort served as the ethanol source; and an innoculum of an aggregating strain of *Acetobacter* species was prepared and added to the fermenter. When the level of acetic acid in the fermenter had reached about 3% w/v, the medium delivery pump was started and the flow rate adjusted to a level that gave almost complete conversion of the ethanol available into acetic acid. Undue haste in increasing the flow rate and also serious decrease or stoppage of the air flow caused the expected fall in conversion efficiency. Adjustment of the flow and aeration rate showed that a maximum volumetric efficiency (V.E.) of 0.82 could be attained (Table 1). At this point the addition of hydrous titanium (IV) oxide began; and the volumetric efficiency gradually increased (Table 1); the aeration rate was increased only slightly. Throughout the run it was possible to increase the flow rate and V.E. while maintaining the conversion of ethanol to acetic acid between 90 and 100% for the majority of the time.

The hydrous titanium (IV) oxide had an immediately noticeable effect on the appearance of the bacteria in the fermenter; it caused a color change in the organisms from purple to brown. This did not appear to affect the performance of the fermenter. In addition over a ten day period after the commencement of hydrous titanium (IV) oxide addition, many more bacteria aggregates formed. These were spiky pellets about 2 mm in diameter. A single particle fixed by heating and stained with methylene blue revealed the presence of particles of titanium (IV) oxide embedded in the aggregate. This is not a completely true representation of the situation, since the sample had to be dried before staining, thereby causing the hydroxide particles to shrink. However, it is certain that hydrous titanium (IV) oxide was present in the aggregated particles and that the hydrous metal oxide did have a useful aggregating effect on the bacteria, as predicted. Under higher

TABLE 1

AVERAGE RATES OF PRODUCTION OF ACETIC ACID FOR MAXIMUM EFFICIENCIES FOR AGGREGATING *ACETOBACTER* STAIN

Average (over days)	Highest Efficiency (V.E.)*	Acetic Acid Produced (g/day)	Comment
21-29	0.82 (86) day 23	87 ± 7	No hydrous titanium (IV) oxide
81-88	1.64 (99) day 88	263 ± 13	hydrous titanium (IV) oxide added, average 0.75 g $TiCl_4$ daily

* Volumetric efficiency = effluent volume/day/fermenter volume; number in () is %.

magnification of the undried sample it could be seen that many bacteria were not associated with the aggregated particle (free bacteria), thus confirming the predicted equilibrium situation.

A variety of conditions for fermentation with *Acetobacter* species using immobilized cells now have been investigated in various sizes of fermenter. In the case of a non-aggregating strain of *Acetobacter* species the addition of hydrous titanium (IV) oxide resulted in an increased V.E. However, the increase was not so marked as that achieved using the aggregating strain. However, a higher V.E. was achieved using hydrous titanium (IV) oxide-cellulose chelate as the support material for cell immobilization. Hydrous titanium (IV) oxide-cellulose chelate was prepared by mixing equal weights (1.2 g) of chromatographic grade cellulose powder (Whatman CF11) and titanium (IV) chloride solution and stirring for 2 hours. The mixture was dried at 45°, ground to a powder, washed with distilled water until the washings were neutral, and then added to the fermenter as an aqueous suspension.

The effect of the chelate was not produced by a mere mixture of cellulose and hydrous titanium (IV) oxide. The added success with the chelate was therefore attributable to the altered mode of presentation of the immobilizing titanium species.

Thus it may be concluded that the addition of hydrous titanium (IV) oxide to a bacterial fermentation will cause aggregation of the bacteria and thus give a higher bacterial concentration in the fermenter. This apparently allows both a higher aeration rate and a higher dilution rate, giving dilution rates (V.E. ∿1.6) at least twice as high as those previously obtained with air aeration (V.E. ∿ 0.8) and comparable with those obtained when the fermenter was aerated with pure oxygen (V.E. 1.8).

REFERENCES

1. KENNEDY, J.F., BARKER, S.A. & HUMPHREYS, J.D. *J. Chem. Soc., Perkin I*:962 1976.
2. KENNEDY, J.F. & KAY, I.M. *J. Chem. Soc., Perkin I*:329 1976.
3. KENNEDY, J.F., BARKER, S.A. & WHITE, C.A. *Die Starke* 29:240, 1977.
4. KENNEDY, J.F. & HUMPHREYS, J.D. *Antimicrobial Agents Chemotherapy* 9:766, 1976.
5. KENNEDY, J.F., BARKER, S.A. & WHITE, C.A. *Carbohydrate Res.* 54:1, 1977.
6. FRINGS GmbH, British Patent No. 1,101,560; 1968.

MICROBIAL ELECTRODE: BOD SENSOR

S. Suzuki and I. Karube

Research Laboratory of Resources Utilization
Tokyo Institute of Technology
Yokohama, Japan

Three measurement techniques are now in use for the determination of organic pollution (1): the biochemical oxygen demand (BOD) test, the chemical oxygen demand (COD) test, and various instrumental methods most of which utilize infrared analyzers. The BOD test has remained a standard pollution monitoring tool since 1936 (2). However, the BOD test requires a 5 day incubation period at 20°C. Many papers have been published on methods for a rapid estimation of 5 day BOD (2). However, these methods are not in fact rapid or simple.

A rapid method for estimation of the 5 day BOD has been developed by using immobilized whole cells and an oxygen electrode (microbial electrode). The system for the BOD sensor is described in this paper.

MATERIALS AND METHODS

Collagen fibril suspension was prepared as described previously (3). Untreated waste waters, obtained from a Japanese government alcohol factory, a slaughterhouse, and a food factory, were diluted with 0.1 M phosphate buffer, pH 7.0, before use. A standard solution of 0.1 M phosphate buffer, pH 7.0, containing 150 mg/l glucose and 150 mg/l glutamic acid was employed as a model waste water in accord with the Japanese Industrial Standard (JIS) (4). The bacteria were isolated from soil according to the method described in JIS (4). The bacteria were cultured under aerobic conditions at 30°C for 24 hr in pH 7.0 medium, containing 10 g/l glucose, 10 g/l peptone, 10 g/l beef extract, and tap water. The immobilization of the microorganisms was accomplished using a suspension containing 1.8 g collagen fibrils and 0.6 g wet cells. The bacteria-collagen membrane was prepared by casting the suspension on a Teflon plate at 20°C. The bacteria-collagen membrane was treated with 1% glutaraldehyde solution for 1 min and dried at 4°C. The microbial electrode consisted of (a) a double membrane of which one layer was a 50 μm thick bacteria-collagen membrane and the other an oxygen permeable 27 μm thick Teflon membrane, (b) an alkaline electrolyte, (c) a platinum cathode, and (d) a lead anode. The double membranes were in direct contact with the platinum cathode and were tightly secured to the cell with rubber rings.

The schematic diagram of the microbial electrode is illustrated in Fig. 1. The microbial electrode was inserted into a 60 ml sample solution that was saturated with dissolved oxygen and stirred magnetically. For continuous use the microbial electrode was inserted into an 8 ml sensing chamber, through which 0.05 M phosphate buffer of pH 7.0 and saturated with dissolved oxygen was pumped 8 ml/min. Then waste water of BOD 44, 88, or 132 p.p.m. was transferred to the sensing chamber for 15 min at a flow rate of 1.3 ml/min.

RESULTS AND DISCUSSION

The microbial electrode was placed in 0.1 M phosphate buffer, pH 7.0, saturated with dissolved oxygen and then transferred to a solution containing organic compounds. The consumption of oxygen by the bacteria caused a decrease in the dissolved oxygen around the membrane. As a result, the electrode current decreased markedly with time until a steady state was reached. The steady state indicated that the consumption of oxygen by the bacteria and the diffusion of oxygen from the solution to the membrane were occurring at the same rate. The time required for the current to reach a steady state, i.e. the response time, depended on the thickness of the membrane, the quantity of bacteria, and the temperature. The steady state value was in all cases attained within 15 min at 30°C. When the sensor was removed from the sample solution and placed in a solution free of organic compounds, the electrode current gradually increased and returned to the initial state.

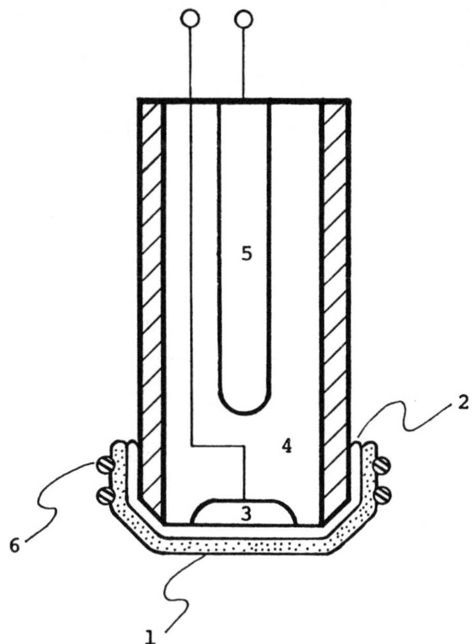

Fig. 1. Microbial electrode. 1. Bacteria-collagen membrane; 2. Teflon membrane; 3. Cathode (Pt); 4. Electrolyte (KOH solution); 5. Anode (Pb); 6. O-ring.

The optimum pH of the microbial electrode was between 6.5 and 8.0. The optimum temperature was between 30°C and 35°C.

The influence of ions was examined. At least 0.05 M of phosphate ions was needed for a rapid response of the electrode. The influence of heavy metal ions on the current output of the microbial electrode was examined using 10% of a standard solution containing 1 mM metal chloride. A slight increase in current was observed with cupric chloride; however, the microbial electrode was stable against other heavy metal ions.

A linear relationship was obtained between the concentration of the standard solution and the current in the region corresponding to a concentration of less than

10%. In this region, where the BOD is below 22, the results were estimated by this microbial electrode. The BOD estimated from the microbial electrode was reproducible within ± 7% of the relative error. The results comparing the BOD determination by the two methods are shown in Table 1.

TABLE 1

COMPARISON BETWEEN THE BOD VALUES DETERMINED BY THE CONVENTIONAL METHOD AND THOSE DETERMINED BY THE MICROBIAL ELECTRODE

Waste water*	BOD by Conventional method (p.p.m.)	BOD BY Microbial electrode (p.p.m.)	Difference (%)
Alcohol factory (1/1000)	12	11	8
Slaughterhouse (1/200)	10	9.5	5
Food factory (1/4000)	12	11	8

*Numbers in parentheses represent dilution factor.

Continuous estimation of food factory waste water BOD was carried out for 10 days. The water was transferred to the sensing chamber at 30 min intervals. The time-current relationship is shown in Figure 2. No decrease in current output was observed over a one month period. The reproducibility, using the same samples, was 3% of the relative standard deviation in the case of the diluted standard solution (BOD 132 ppm). The relative error of the BOD estimation was less than 10%. In practice the error inherent in the five day test was more than 10%.

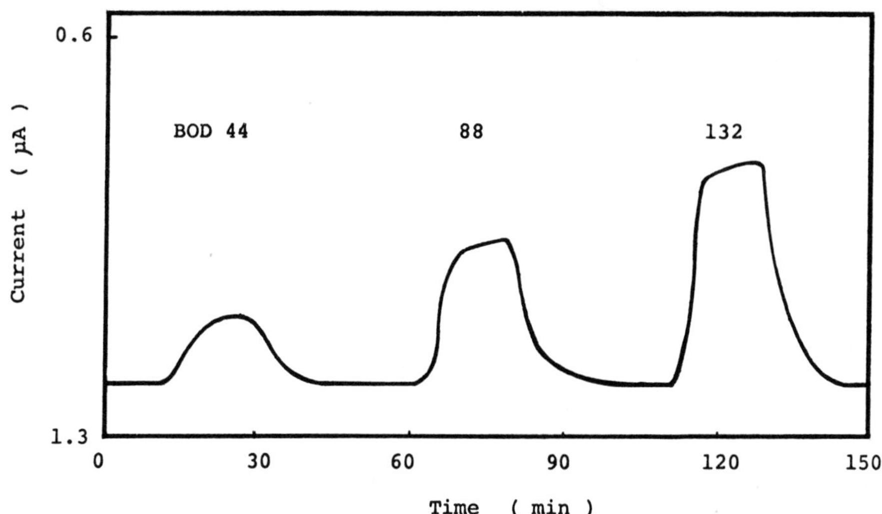

Fig. 2. Continuous estimation of BOD. Diluted food factory waste water was pumped through the electrode for 15 min at 1.3 ml/min at 30 min intervals.

Further developmental studies are being directed toward using the microbial electrode for the determination of carbohydrates, cofactors, and antibiotics in fermentation processes.

REFERENCES

1. MULLIS, M.K. & SCHROEDER, E.D. J. *Water Poll. Control. Fed.* *43*:209, 1971.
2. LEBLANK, P.J. J. *Water Poll. Control. Fed.* *46*:2202, 1974.
3. KARUBE, I., SUZUKI, S., KINOSHITA, S., & MIZUGUCHI, J. *Ind. Eng. Chem. Prod. Res. Develop.* *10*:160, 1971.
4. Japanese Industrial Standard Committee "Testing Methods for Industrial Waste Water" JIS K 0102, Tokyo, Japan, 1974, p. 33.

NEW METHOD FOR IMMOBILIZATION OF MICROBIAL CELLS AND ITS INDUSTRIAL APPLICATION

I. Chibata, T. Tosa, T. Sato, K. Yamamoto,
I. Takata and Y. Nishida

Research Laboratory of Applied Biochemistry,
Tanabe Seiyaku Co., Ltd.
Osaka, Japan

The immobilization of-microbial cells has been done mainly with the polyacrylamide gel method. We first succeeded in the industrialization of this immobilized cell system for production of L-aspartic acid (1) and L-malic acid (2). For further improvement of immobilized cell systems, we have extensively investigated new immobilization techniques.

As a result, we found that carrageenan, a polysaccharide widely used as a food additive, is a more suitable matrix for immobilization of microbial cells. The immobilization of microbial cells using carrageenan is facile and can be carried out under mild conditions.

The standard procedure was carried out as follows: 5 g (wet weight) of microbial cells were suspended in 5 ml of physiological saline at 45-50°C, and 1.7 g of kappa-carrageenan were dissolved in 34 ml of the same saline at 45-60°C. The two solutions were mixed and then cooled to 10°C for 30 min. In order to improve the gel-strength, the obtained gel was soaked in cold 0.3 M potassium chloride solution. After this treatment, the resulting stiff gel was made in to about 3 mm granules. The pore size of this gel matrix was small enough so that higher molecular weight compounds, such as enzyme-protein, did not leak out from the gel-lattice; although the substrate and the product of lower molecule weight easily passed through the gel-lattice. If the operational stability of

TABLE 1

COMPARISON OF ENZYME ACTIVITIES AND STABILITIES OF IMMOBILIZED MICROBIAL CELLS PREPARED BY CARRAGEENAN AND POLYACRYLAMIDE GEL METHODS

Microorganism	Enzyme	Enzyme activity (μmole/hr/g cells)		Half-life (day)	
		Polyacrylamide	Carrageenan	Polyacrylamide	Carrageenan
Escherichia coli	Aspartase	19,000 (29.2%)	30,400 (46.8%)	120	50
			21,400* (32.9%)		686*
Streptomyces phaeochromogenes	Glucose isomerase	3,880 (49.1%)	4,280 (54.1%)	53	53
Brevibacterium flavum	Fumarase	6,680 (34.0%)	9,920 (50.5%)	72	70

* : value after hardening with glutaraldehyde and hexamethylene-diamine

() : yield of activity after immobilization or after hardening

the immobilized cells was not satisfactory, the immobilized cells were further treated with hardening reagents, such as tannin, glutaraldehyde, or glutaraldehyde and hexamethylenediamine. As a result, more stable immobilized cells could be obtained. By this method, whole cells of *Escherichia coli* having aspartase activity, *Brevibacterium flavum* having fumarase activity, and *Streptomyces phaeochromogenes* having glucose isomerase activity were successfully immobilized. As shown in Table 1, the enzyme activities of these immobilized cells were higher than those of preparations obtained by using polyacrylamide gel. These immobilized cells were stable; and a column packed with them could be used for continuous enzyme reaction for long periods of time. The operational stabilities of immobilized *E. coli* and immobilized *S. phaeochromogenes* especially, were markedly increased by hardening with glutaraldehyde and hexamethylenediamine.

Furthermore, in this carrageenan method immobilized microbial cells of various shapes, such as cubic, disk, bead, membrane, or fiber, can be tailor-made according to the intended application.

In conclusion, this facile carrageenan method is applicable for immobilization of many kinds of microbial cells and is considered to be more advantageous for industrial purposes than the polyacrylamide gel method.

REFERENCES

1. CHIBATA, I., TOSA, T. & SATO, T. *Appl. Microbiol.* 27: 878, 1974.
2. YAMAMOTO, K., TOSA, T., YAMASHITA, K. & CHIBATA, I. *Appl. Microbiol.* 3:169, 1976.

POLYMER ENTRAPMENT OF MICROBIAL CELLS: PREPARATION AND REACTIVITY OF CATALYTIC SYSTEMS

J. Klein,* U. Hackel,* P. Schara,* P. Washausen*
F. Wagner,** and C.K.A. Martin**

Institute of Chemical Technology*
Technical University of Braunschweig
GBF m.b.H., Braunschweig-Stockheim**
Braunschweig, Federal Republic Germany

Polymer entrapment of microbial cells has proven to be an alternative to the immobilization of isolated enzymes, especially where more than one enzyme is involved in the reaction pathway. Oxidative degradation of phenol by *Candida tropicalis* may serve as an example (1,2).

Entrapment in polymeric matrices was obtained by three different procedures: (a) crosslinking of a linear styrene-maleic acid copolymer by multivalent counterions, e.g. Al^{3+}, by dropwise precipitation of concentrated polymer solution with supended cells in 0.1 M $Al_2(SO_4)_3$ (3); (b) suspension polymerisation of acrylamide (AAm) and especially methacrylamide (MAAm) in dibutyl phthalate; and (c) thin layer coating of crosslinked PAAm onto a micro- or macro- porous epoxy membrane, obtained from a water curing epoxy system in the presence of a water soluble filler. As shown in Table 1, the mechanical strength has to be optimized against the activity yield. The addition of prepolymerized PAAm gave higher mechanical stability (4).

The reactivity of the *Candida* was measured via the phenol consumption rate, which was zero order with respect to phenol. A decrease in specific activity with increasing particle diameter and increasing cell concentration could be attributed to a limited supply of oxygen towards the particle center. Based on the experimental determina-

TABLE 1

PROPERTIES OF BEADS FROM SUSPENSION POLYMERIZATION

Network composition	Mechanical Strength (relative units)	Activity Yield (%)
24% AAm, 6% Bis	4	80
24% MAAm, 4% Bis	8.5	40
16% MAAm, 4% Bis, 16% PAAm	15.4	40

TABLE 2

OPERATIONAL STABILITY OF *CANDIDA TROPICALIS*

Catalytic System	Half Life (days)
Free C. *tropicalis* Cells	20*
Immobilized Cells Types (a) and (b)	20
Immobilized Cells with Repeated Incubation	40

*Hours

tion of oxygen permeability in the polymeric matrix, a quantitative correlation between the Thiele-Modulus and the catalytic effectiveness could be established for type (a) and (b) preparations (2-4).

Polymer entrapment led to a drastic increase in operational stability, compared to free suspended cells, The catalytic stability could be further improved by reincubation of the usually resting cells with the necessary nutrients for the growth of new cell populations. This can be proven from the increasing nitrogen content (type (a) preparation) from 0.31% before to 3.45% after reincubation as well as the phosphorus content (type (b) preparation) from 0.01% before to 0.14% after reincubation.

REFERENCES

1. HACKEL, U., KLEIN, J., MEGNET, R., & WAGNER, F. *Eur. J. Appl. Microbiol.* *1*:291, 1975.
2. KLEIN, J., HACKEL, U., SCHARA, P., WASHAUSEN, P. &, WAGNER, F. *Abst. 5th Intern. Ferment. Symp. Berlin 15*:23, 1976.
3. HACKEL, U., Ph.D. Thesis, Techn. Univ. Braunschweig, 1976.
4. SCHARA, P., Ph.D. Thesis, Techn. Univ. Braunschweig, 1977.

BENZENE METABOLISM BY BACTERIAL CELLS IMMOBILIZED IN POLYACRYLAMIDE GEL

J.R. Mason, S.J. Pirt* and H.J. Somerville

Shell Research Ltd.
Sittingbourne, Kent and
Department of Microbiology*
University College London
London, U.K.

Many micro-organisms can oxygenate hydrocarbons under milder conditions than required for chemical oxygenation. These reactions provide an interesting set of model conversions in which to compare chemical with biological conversion. However, one of the greatest drawbacks with biological catalytic systems is their relative instability. The present work was carried out using a strain of *Pseudomonas putida* to evaluate whether the hydrocarbon-oxidizing system could be stabilized, and to investigate the potential for reactivation of the system after loss of activity.

Several methods of biomass estimation within gels were examined, including estimation of protein or nucleic acid content and enumeration under the electron microscope. However, the most suitable and sensitive method studied involved the assay of the haem content within gels. This was based upon the rationale that the light emitting oxidation of luminol to aminophthalate by an oxidizing radical is quantitatively proportional to the amount of haematin catalyst present (1). A linear relationship was found between the logarithm of cell dry weight and the logarithm of the peak of the light emitted as recorded with a Chance Aminco spectrophotometer.

Immobilized cells, and free cells, from cultures grown on benzene were added to preheated buffer at various temperatures; and samples were removed at intervals for the assay of activity. At 30°C free cells lost 50% of their activity in 4 hr; whereas immobilized cells lost no activity in this time. At 40°C, free cells were completely inactivated after 2 hr; and immobilized cells retained 30% of their activity after 24 hr.

Experiments were carried out to study the long-term activity of the biocatalyst under conditions of continuous biotransformation. In the presence of nitrogen and a utilizable carbon source, activity increased over a period of about a day but was eventually lost as cells multiplied to an extent where, apparently, they were lysed by physical pressures generated as a result of growth. By controlled growth and enzyme synthesis *in situ*, activity was retained over prolonged periods and, indeed, was greater than the initial activity after one month of continuous elution, with appropriate reactivation periods.

Electron microscope studies clearly indicated growth within the gel matrix. Although cells were alive, in that they had the ability to multiply, no attempts were made to demonstrate an increase in cell number because of the difficulties in separating cells from the gel. Clearly it would be desirable to demonstrate clone formation from cells grown within the gel before it could be concluded absolutely that the cells were living.

REFERENCE

1. NEUFELD, H.A., CONKLIN, G.J. & TOWNER, R.D. *Anal. Biochem.* *12*:303, 1965.

CELLULOSE BEAD ENTRAPPED WHOLE CELL GLUCOSE ISOMERASE IN FRUCTOSE SYRUP PRODUCTION

Yu-Yen Linko*, R. Viskari,** L. Pohjola** and P. Linko**

Academy of Finland* and Department of Chemistry
Helsinki University of Technology**
Espoo, Finland

Purified and whole cell glucose isomerase (GI) has been immobilized on various supports (1). Most research has involved *Bacillus* and *Streptomyces* enzymes. We have developed a simple and mild technique to entrap whole cell *Actinoplanes missouriensis* GI in cellulose (2,3). The method involves suspending dried cells in α-cellulose, dissolved in a melt of N-ethyl-pyridinium chloride (NEPC) and dimethylformamide (DMF). The GI cellulose could be regenerated in fiber or bead form by passing into water through an orifice of dia 0.8 mm, treating with glutaraldehyde for crosslinking, and vacuum drying to about 24% dry weight. This is believed to be the first time an enzyme has been immobilized within regenerated pure cellulose beads.

Isomerization of 45% glucose solution was carried out in two columns (A, 20 x 89 mm and B, 15 x 156 mm) of GI cellulose beads of bulk density about 0.67 g/ml and equal total activity. Although the linear flow rate of column A was approximately half of that of column B, the same isomerization rates were obtained, indicating no significant external diffusion effect. Similarly, a value of 0.83 M was obtained for K_{mf} and 1.02 for K_{mb} both with free and immobilized enzyme. Operational stability of the two columns was equal. A 50% conversion could be maintained for 23 days, followed by a gradual decrease that corresponded to a half-life of 42 days. Color formation during process was negligible, as indicated by the

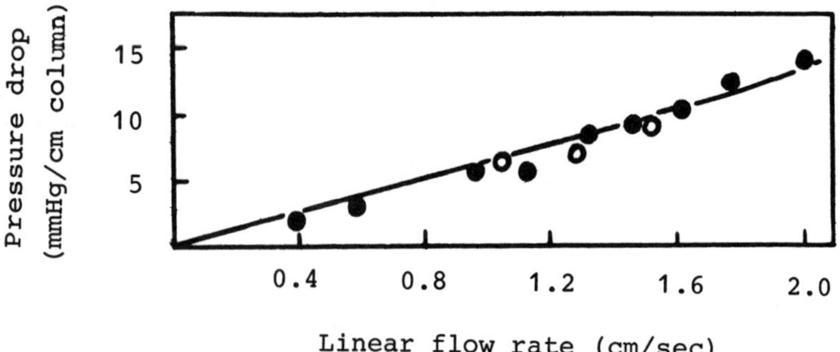

Fig. 1. Pressure drop in the column at various linear flow rates (o) increasing and (●) decreasing flow rate.

absorbance of the syrup of about 0.01 at 420 mm.

The pressure drop and linear flow rate were directly proportional; and no compression or deformation of the beads could be observed (Fig. 1). No hysteresis was noticed with decrease in linear flow rate, suggesting sufficient strength of the beads for industrial column operations. The pressure drop across the bed at a linear flow rate of 1 cm/sec was of the same order of magnitude as reported by others (4) for diisocyanate treated cellulose beads.

Scanning electron micrographs (SEM), obtained from partially dried beads, showed *Actinoplanes* cells entrapped within cellulose beads in large quantities.

Assuming recycling of NEPC (99% recovery) and DMF (98% recovery), material costs for regenerated cellulose alone were about $0.1/kg and for whole cell GI cellulose beads about $4./kg.

In conclusion, cellulose bead entrapped whole cell GI was shown by SEM to be highly loaded with cells. The kinetics, good flow properties, high operational stability, and low cost of the carrier for entrapped GI were suitable for continuous isomerization of 45% glucose solutions.

REFERENCES

1. CASEY, J.P. *Die Starke* *29*:196, 1977.
2. LINKO, Y.-Y., POHJOLA, L., VISKARI, R. & LINKO, M. *FEBS Lett.* *62*:77, 1976.
3. LINKO, Y.-Y., POHJOLA, L. & LINKO, P. *Process Biochemistry* *12*:14, 1977.
4. CHEN, L.F. & TSAO, G.T. *Biotechnol. Bioeng.* *18*:1507, 1976.

PROPERTIES OF A CELL-BOUND GLUCOSE ISOMERASE

O.J. Lantero, Jr.

Miles Laboratories, Inc.
Elkhart, Indiana, USA

The application of immobilized glucose isomerase in the production of high fructose syrup has reached commercial importance. In spite of the many immobilized forms of glucose isomerase reported in the last several years, only a few are commercially available for industrial use. The enzyme sources of all the reported immobilized glucose isomerases are microbiological. Immobilized glucose isomerases can be divided into two general types: (a) fixation of the intracellular enzyme to the structural material of the microorganism; and (b) immobilization of soluble enzyme. Recently some quantitative work on the reaction kinetics have appeared in the literature for reactor models (1-3). This paper will present some of the properties and kinetics of an immobilized intracellular glucose isomerase from *Streptomyces olivaceus*.

Glutaraldehyde treatment of *Streptomyces olivaceus* cells (4) resulted in a stable immobilized form of glucose isomerase that had a temperature optimum for activity at 70°C and a pH optimum for activity at 7.8. As with many glucose isomerases, cobalt and magnesium were required for maximum activity. Saturation of the enzyme, as determined by activity, appeared to be complete at 0.5 and 22 mM cobalt and magnesium, respectively.

Factors influencing the stability of immobilized glucose isomerase activity were investigated by exposing the enzyme at the experimental conditions at 75°C, and then measuring the residual activity at 60°C under standardized conditions in a differential batch reactor (5). Cobalt and glucose were found to stabilize the activity, whereas magnesium appeared to have no influence on stabilizing activity.

Mass transport effects were investigated. External diffusion had very little influence on the reaction rate, when the substrate linear velocity exceeded 5 cm/min. Internal diffusion became more predominate in controlling the reaction as the enzyme particle size increased.

Kinetic studies of the forward and reverse reactions were carried out in a differential batch reactor at various temperatures by varying the substrate concentration from 0.1 to 1.0 M. From the initial velocity results the kinetic parameters were estimated with the aid of the computer by the method of Wilkinson-(6). The apparent kinetic constants were subjected to Arrhenius type plots to obtain corrected apparent kinetic constants (Table 1).

An estimate of the rate constants was made from the kinetic parameters by the procedure of Sproull *et al.* (2). From an Arrhenius plot of the rate constants the activation energies were estimated as: E_1 9.68, E_{-1} 15.06, E_2 16.89, and E_{-2} 9.24 kcal/mole, respectively.

The equilibrium constants derived at temperatures of 50, 60, 65, and 70°C, starting with glucose or fructose in the presence of the immobilized enzyme, were found to be 1.00, 1.08, 1.14, and 1.21, respectively. From the van't

TABLE 1

CORRECTED APPARENT KINETIC CONSTANTS

Temp. (°C)	K_G (M)	V_G $\left(\frac{\text{mmole/min}}{\text{gm enzyme}}\right)$	K_F (M)	V_F $\left(\frac{\text{mmole/min}}{\text{gm enzyme}}\right)$
50	0.331	0.124	0.298	0.081
60	0.447	0.273	0.411	0.162
65	0.516	0.397	0.479	0.225
70	0.595	0.578	0.558	0.312

Hoff relationship a heat of reaction was estimated at 2.07 kcal/mole.

Half life studies of the immobilized glucose isomerase were conducted in an integral column reactor operated at a substrate flow rate to maintain a constant fructose conversion of 42%. At 60, 65, and 70°C the respective half lifes of 41.4, 17,3, and 6.5 days were obtained. From these results an activation energy for deactivation was estimated at 41.8 kcal/mole.

REFERENCES

1. RYU, D.Y., CHUNG, S.H., & KATOH, K. *Biotechnol. Bioeng.* *19*:159, 1977.
2. SPROULL, R.D., LIM, H.C., & SCHNEIDER, D.R. *Biotechnol. Bioeng.* *18*:633, 1976.
3. LEE, Y.Y. FRATZKE, A.R., WUN, K., & TSAO, G.T. *Biotechnol. Bioeng.* *18*:389, 1976.
4. ZIENTY, M.F. U.S. Patent No. 3,779,869; 1973.
5. FORD, J.R., LAMBERT, A.H., COHEN, W., & CHAMBERS, R.P. in "Enzyme Engineering" (L.B. Wingard, Jr., ed.), Wiley, New York, 1972, p. 267.
6. WILKINSON, G.N. *Biochem. J.* *80*:324, 1961.

Session VII
FUNDAMENTALS OF SOLID PHASE BIOCHEMISTRY

Chairmen: I. Berezin and L. Goldstein

HYDROPHOBIC INTERACTIONS IN PURIFICATION AND UTILIZATION OF ENZYMES

Wayne Melander and Csaba Horvath

Department of Engineering and Applied Science
Yale University
New Haven, Connecticut USA

The importance of the hydrophobic effect in biochemical systems has been widely recognized since Kauzmann (1) enunciated its role in determining the three-dimensional structure of proteins. The strength of hydrophobic interactions is commensurate with that of hydrogen binding and ionic effects in a typical protein (2). Numerous authors have looked for a quantitative ranking for the hydrophobic character of amino acid side chains to estimate the hydrophobicity of proteins (3-6). One theory attributed the large entropy effect to changes in the hydrogen bonding structure of water in the vicinity of solute molecules (7-10); while another found the free energy of hydrocarbons in water to be proportional to the number of solvent molecules surrounding the solute and also to the solute surface area as determined from molecular models (11-14).

All theories for the hydrophobic effect agree that it arises from the properties of water which are related to the interaction between the solvent and the solute molecules. The phenomenon, however, need not be restricted to water. In the solvophobic effect (15,16), the process of bringing a molecule into solution is divided into two hypothetical steps; a cavity of shape and size appropriate for the solute is created in the solvent followed by interaction of the solute with the solvent. The energy required for the first process is calculated from the cavity area and microscopic surface tension. This treatment has been adapted to interpret retention data obtained in reversed-phase chromatographic systems with charged and uncharged analytes (17,18), the effect of salts on protein behavior in the salting-out process, and in hydrophobic affinity chromatography (17) and data on the formation of protein aggregates (20).

The theory of hydrophobic interactions has been extensively described elsewhere (15-20). In simplified form the free energy of association ΔG_a^o is described as follows:

$$\Delta G_a^o = \alpha_o + \alpha_1 A \sigma m + \alpha_2 A \qquad \text{(Eq. 1)}$$

where α_o, α_1 and α_2 are the appropriate system parameters, A is the molecular surface area of the other species investigated, m is the salt molality, and σ is the surface tension increment. The electrostatic interaction term is expected to remain constant when the ionic strength and dielectric constant of the medium is fixed and the electrostatic properties of the associating species are invarient, as in the case of homologues. Eq. 1, which is linear in A and σm, is used to interpret the effect of molecular surface area on enzyme inhibition by small molecules of hydrophobic character and the effect of salt in enzyme immobilization on hydrophobic surfaces.

APPLICATIONS

The solvophobic theory has been successfully used to shed light on phenomena involved in the salting-out of proteins and on the effect of salt on retention in hydrophobic affinity chromatography (19). The linear dependence of the logarithm of the protein solubility on the salt concentration in the salting-out regime and the salting out coefficient both are linearly dependent upon the molal surface tension increment, σ. Indeed this behavior was observed in all cases examined. In hydrophobic-affinity chromatography the logarithm of the binding constant was found to be linearly dependent on the salt molality, m, at sufficiently high salt concentration as predicted by the theory.

Further support for the validity of this approach to the interaction of biological molecules can be gained by analyzing data on the inhibition of enzymes by small molecules having hydrophobic nature. When an enzyme is inhibited by low molecular weight hydrocarbons or simple alcohols, the energy of interaction, which is proportional to the logarithm of the inhibition constant, is expected to depend according to Eq. 1 upon the change in the molecular surface area which occurs when the complex forms. If the inhibitor fits into a large pocket or groove on the enzyme surface, this linear relationship is expected also to be valid for the surface area of the inhibitor. Indeed, Canady and Royer (21) found this to be true for the inhibition of chymotrypsin, α-chymotrypsin, and alcohol dehydrogenase by aromatics, and their data validate Eq. 1.

The competitive inhibition of pepsin by aliphatic alcohols has been investigated by Tang (22), who tabulated the inhibition constants for seven alcohols. We have calculated the non-polar surface areas of the inhibitors as described previously (17) and have plotted the logarithm of the inhibition constants against the hydrophobic surface area. In agreement with Eq. 1 a straight line is obtained as shown in Fig. 1. Numerous similar data published in the literature have been analyzed in the framework of the solvophobic theory and the results show that Eq. 1 is widely applicable in treating hydrophobic binding of small molecules to proteins (unpublished work). Since immobilization of enzymes by hydrophobic interactions on non-polar supports has been found to have promising practical applications (23), we have investigated certain aspects of this approach and attempted to interpret the results within the framework of the solvophobic theory.

The support used in this study was octadecyl-silica having mean particle diameter of 10 micron. Such materials, which contain a molecular fur of octadecyl chains covalently bound to the surface of silica gel, are widely used in high performance liquid chromatography (24). The enzyme under investigation was adenosine deaminase (E.C. 3.5.4.4) from Sigma.

The enzyme was adsorbed onto the hydrophobic support in the presence of different salts at constant ionic strength of the bathing solution. 1.5 ml of adenosine deaminase solution containing 20 units of enzyme in 0.1 M phosphate buffer, pH 6.8, were used to contact 0.1 g of Partisil ODS 2 (Whatman, Clifton, N.J.) octadecyl-silica having a carbon content of 15% (w/w). Prior to contacting the particles with the enzyme solution they were wetted with a 1 to 1.5 mixture (v/v) of the above buffer and methanol. In various experiments KCl, K_2SO_4, KSCN, and KI were added to the enzyme solution so that the ionic strength of these salts was 0.5 in all experiments except the controls which were carried out without added salt.

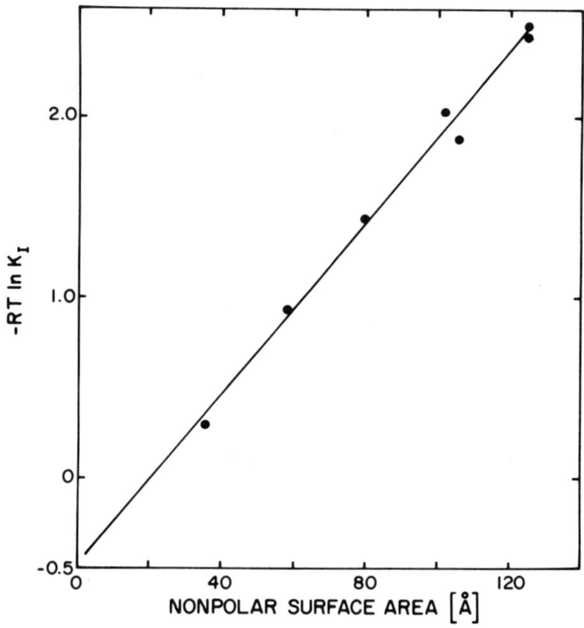

Fig. 1. Inhibition of pepsin by aliphatic alcohols (22). The free energy of association for the inhibitor enzyme, ΔG_a^o, is related to the inhibition constant, K_I, by ΔG_a^o equals - RT $\ln K_I$. The nonpolar surface area, A, of the alcohols was calculated from their hydrocarbonaceous moiety (17). The plot illustrates Eq. 1.

The slurry of the octadecyl-silica particles in the enzyme solution was stirred for 3 min, allowed to rest two hr, and centrifuged. The enzyme activity of the supernatant was measured spectrophotometrically by using 10^{-4}M adenosine solution in the same buffer. The rate of absorbance change at 285 nm was determined and converted to the concentration scale using the difference absorbancy coefficient 0.30 (25). Appropriate controls showed no inhibition of the enzyme by the added salts.

The results obtained without and with the addition of salts is shown in Fig. 2. The data indirectly demonstrate that the logarithm of the equilibrium constant for the adsorption of the enzyme is proportional to the surface

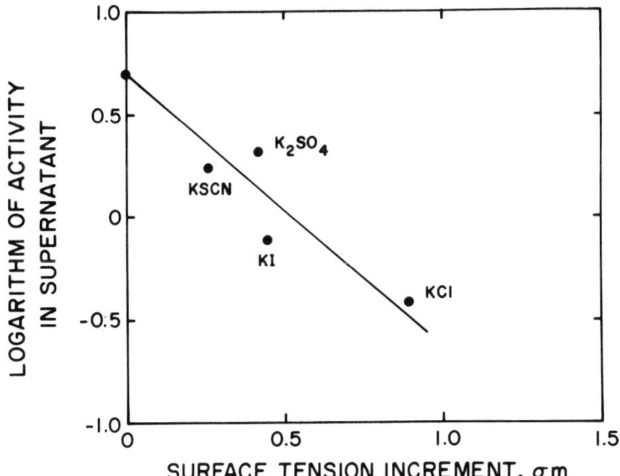

Fig. 2. Linear dependence of the equilibrium constant on the surface tension increment of various salt solutions for the hydrophobic adsorption of adenosine deaminase on octadecyl-silica at constant ionic strength. Under the conditions of the experiment, the equilibrium constant was inversely proportional to the enzyme activity in the supernatant.

tension increment, σm, of the salt solution as suggested by Eq. 1. The results of these experiments also show that both the concentration and the nature of the salt are important in determining the extent of protein binding to a hydrophobic surface.

The adenosine deaminase immobilized by the above methods has also been used in a chromatographic enzyme reactor. 50 mg of the immobilized enzyme were placed at the top of a reversed phase column packed with 10 μm Partisil ODS 2. The column dimensions were 25 x 0.46 cm. In a liquid chromatograph, schematically depicted in Fig. 3, the column was used with 0.1 M phosphate buffer, pH 6.8, with an eluent flow rate of 2.4 ml/min, and at 25°C. Various quantities of adenosine were injected, using a 20 μl injection valve. The reaction took place at the inlet of the column; and the unreacted substrate and inosine, the product of the reaction, were separated on

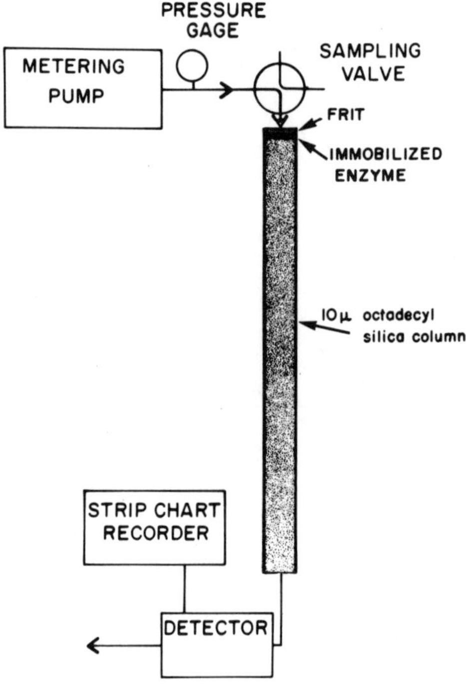

Fig. 3. Chromatographic enzyme reactor system.

the basis of hydrophobic interactions. Chromatograms were obtained; and the widths and areas of the adenosine and inosine peaks were measured. In calculating the input concentration of adenosine and the inosine concentration at the outlet of the reactor it was assumed that no dilution occurred in the catalytically active segment of the column. The data were analyzed by using the integrated rate equation

$$\ln\left(\frac{A_o - I}{A_o}\right) = \frac{I}{K_m} - V_m t \qquad \text{(Eq. 2)}$$

where A_o, I, K_m, V_m and t were the initial adenosine concentration, the concentration of inosine produced,

Michaelis constant, saturation rate of reaction, and the residence time in the catalytic section of the column. Fig. 4 presents a plot of data according to this analysis. The straight lines do appear to confirm the validity of Eq. 2. The K_m obtained was 40 μM, which compared well with that for free enzyme of 35 μM (26). Thus, diffusional limitations seemed unlikely to affect the rate of reaction.

The immobilized adenosine deaminase was quite stable with a half-life of about 300 hr. Under our operating conditions about 8 x 10^4 bed volumes of feed were passed through the column at 30°C before the conversion fell below 60% of its initial value. These results clearly demonstrated that for adenosine deaminase adsorption a suitable hydrophobic support offers a viable immobilization method for technical applications.

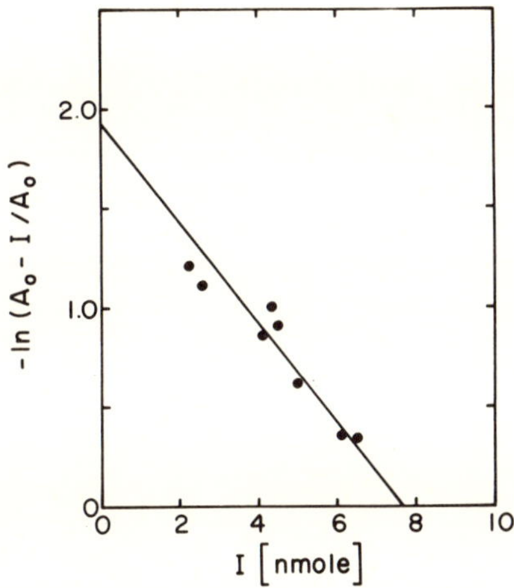

Fig. 4. Data obtained with the adenosine deaminase reactor according to the integrated form of the Michaelis-Menten equation given by Eq. 2. The slope of the line is - $1/K_m$.

CONCLUSIONS

The solvophobic theory, as adapted to molecules which carry charges or are complex dipoles, gives a framework within which a variety of experimental data can be interpreted. These include the salting-out of proteins, salt effects in hydrophobic affinity chromatography, adsorption on hydrophobic surfaces, and inhibition of enzyme activity by nonpolar or amphiphilic molecules and the aggregation of proteins. As a consequence of this treatment, it is possible to place the Hofmeister series on a rational quantitative basis. The order of the molal surface tension increments represent this scale which has been shown to be colinear with empirical lyotropic scales established for various salts (19). Since salting-out and hydrophobic affinity chromatography are widely used for isolation and purification of proteins, the understanding of the effect of salt in these processes is of technological significance. As enzymes are increasingly used in industrial processes, the salt effect on the inhibition of enzymes by certain compounds should also be considered when operating conditions for an enzyme reactor are established.

The results presented here suggest that adsorbtion onto appropriate non-polar supports offers a very simple method for immobilization of certain enzymes; and sufficient stability of the product can be obtained without covalent attachment of the enzyme to the support.

The development of chromatographic enzyme reactors (CER) could open new avenues for the use of immobilized enzymes. Such reactors can be designed so that the enzymic reaction and the separation of the substrate and product take place in series, thus, novel effects can be obtained, as have been shown with other types of chromatographic reactors (27).

ACKNOWLEDGMENTS

The authors thank Avi Nahum for assistance in measuring the stability of enzyme reactors. This work was supported by grants GM 20,993 and GM 22735 from the National Institutes of Health.

REFERENCES

1. KAUZMANN, W. *Adv. Protein Chem. 14*:1, 1959.
2. TANFORD, C. *J. Am. Chem. Soc. 84*:4240, 1962.
3. TANFORD, C. "The Hydrophobic Effect: Formation of Micelles and Biological Membranes," Wiley, New York, 1973.
4. WETLAUFER, D.B., MALIK, S.K., STOLLER, L. & COFFIN, R.L. *J. Am. Chem. Soc. 86*:508, 1964.

5. BULL, H.B. & BREESE, K. *Arch. Biochem. Biophys. 161*: 665, 1974.
6. BIGELOW, C.C. *J. Theoret. Biol. 16*:187, 1967.
7. NEMETHY, G. & SCHERAGA, H.A. *J. Chem. Phys. 36*: 3382, 1962.
8. NEMETHY, G. & SCHERAGA, H.A. *J. Chem. Phys. 36*:3401, 1962.
9. NEMETHY, G. & SCHERAGA, H.A. *J. Phys. Chem. 66*:1773, 1962.
10. FRANK, H.S. & EVANS, M.W. *J. Chem. Phys. 13*:507, 1945.
11. EYRING, H. & JHON, M.S. "Significant Liquid Structures," Wiley, New York, 1969.
12. HERMANN, R.B. *J. Phys. Chem. 75*:363, 1971.
13. AMIDON, G.L., YALKOWSKY, S.H., ANIK, S.T. & VALVANI, S.C. *J. Phys. Chem. 79*:2239, 1975.
14. HARRIS, M.J., HIGUCHI, T. & RYTTING, J.H. *J. Phys. Chem. 77*: 2694, 1973.
15. SINANOGLU, O. in "Molecular Associations in Biology," (B. Pullman, ed.), Academic Press, New York, 1968.
16. HALICIOGLU, T. & SINANOGLU, O. *Ann. N.Y. Acad. Sci. 158*:308, 1969.
17. HORVATH, C., MELANDER, W. & MOLNAR, I. *J. Chromatogr. 125*:129, 1976.
18. HORVATH, C., MELANDER, W. & MOLNAR, I. *Anal. Chem. 49*:142, 1977.
19. MELANDER, W. & HORVATH, C. *Arch. Biochem. Biophys. 183*:200, 1977.
20. HORVATH, C. & MELANDER, W. *J. Solid State Biochem.*, in press.
21. ROYER, G. & CANADY, W.J. *Arch. Biochem. Biophys. 124*:530, 1968.
22. TANG, J. *J. Biol. Chem. 240*:3810, 1965.
23. BUTLER, L. *Arch. Biochem. Biophys. 171*:645, 1975.
24. HORVATH, C. & MELANDER, W. *J. Chromatogr. Sci. 15*: 393, 1977.
25. SMILEY, K.L., JR., BERRY, A.J. & SUELTER, C.H. *J. Biol. Chem. 242*:3655, 1967.
26. ZIELKE, C.L. & SUELTER, C.H. in "The Enzymes," vol. 4 (P.D. Boyer ed.), 3rd Ed., Academic Press, New York, 1971, p. 47.
27. MAGEE, E.M. *Ind. Eng. Chem. Fundamentals 2*:33, 1963.

RELATIVE IMPORTANCE OF DIFFUSION LAYER RESISTANCE AND MICROENVIRONMENTAL EFFECTS ON THE EFFECTIVENESS OF IMMOBILIZED ENZYME REACTORS

V. Kasche,* A. Kapune,* and H. Schwegler**

Biology* and Physics** Departments
University of Bremen
Bremen, Federal Republic Germany

One goal of the physical-chemical description of immobilized enzyme reactors is to obtain enzyme independent relations where the efficiency and time dependence is expressed in terms of the properties of the environment (1). Previously the macro- and microenvironmental effects have been treated independently of each other instead of simultaneously. Therefore, it is necessary to describe the more realistic situation where the micro and macro effects are interrelated. The differential equations describing these systems do not have analytical solutions, so that numerical procedures must be applied. The collocation method has many advantages, as no restrictive assumptions are used and the solutions converge to the real solution (2). The aim of the present work was to investigate the relative importance of the macro- and microenvironmental factors for the steady-state effectiveness and time dependence (substrate conversion) of reactors with immobilized enzymes that catalyze the conversion of a single substrate by Michaelis-Menten kinetics. The theoretical results, obtained with the use of the collocation method, were compared with experimental data.

CALCULATION OF STEADY-STATE EFFECTIVENESS FACTORS

The diffusion layers surrounding different enzyme containing particles are assumed not to influence each other. The layer thickness and the mass transport through the layer are functions of the velocity u of the particles relative to the solvent (3). This velocity varies between the free fall velocity u_o and a maximum velocity u_{max}, depending on the solvent velocity (4). The rate of mass transfer through the diffusion layer at steady state equals the rate of substrate conversion within the enzyme containing particle. This gives the mass-conservation relation (in dimensionless form) as:

$$\left.\frac{dC_s}{dz}\right|_{\text{at } z=1} = Sh\ (1 - C_{s,i}) \qquad \text{(Eq. 1)}$$

where z is the ratio of the radial distance r to the particle radius R; C_s is the concentration of substrate relative to the concentration at an r of infinity. The subscript i refers to the boundary between the macro- and microenvironment; and Sh is the Sherwood Number given by (3):

$$(Sh)^2 = 4 + 1.21\ \frac{(u\ 2\ R)^{2/3}}{D_s} \qquad \text{(Eq. 2)}$$

The substrate concentration at z = 1 is determined by Eq. 1 and the coupling reaction-diffusion within the enzyme containing particles (5). The steady-state effectiveness factor, η, is the ratio of the rates of reaction for equal amounts of immobilized and free enzyme under identical experimental conditions. The collocation method was used to calculate η for different γ, Sh, and Φ-values as described in (2). The η-values varied marginally when the number of collocation points was increased from two to three. Some results of these calculations are shown in Fig. 1. Over a wide range of operating conditions both the macro- and micro- environmental effects influence the effectiveness factor. Above a Sh of $\sim$50, η is influenced only marginally by changing the properties that modify the macroenvironment. As the Sherwood number can be approximated by 2R/(diffusion layer thickness), it follows that the diffusion layer thickness is negligible compared with the dimensions of the enzyme containing particles.

CALCULATIONS OF OPERATIONAL EFFECTIVENESS FACTORS

For most enzyme reactors normal operating conditions are not those given in Fig. 1. A property of more practi-

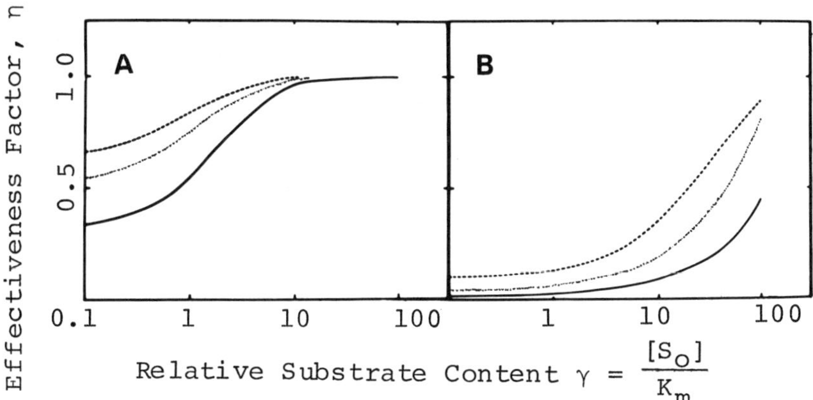

Fig. 1. The steady-state effectiveness factor η calculated by the 3-point collocation method. Microenvironmental parameter Thiele modulus Φ and macroenvironmental parameter Sherwood Number Sh. A: $\Phi^2 = 10$; B: $\Phi^2 = 1000$; Sh: 100(---), 10 (·····), 2(——)

cal interest is the time required to obtain a certain degree of substrate conversion as a function of Φ^2 (Φ^2 is directly proportional to the enzyme content) and Sh. These times may be calculated based on the data in Fig. 1, when the enzyme reactor is operated under nearly steady-state conditions for decreasing γ-values. Then the rate of substrate conversion v, or $(d\gamma/dt)$, equals $(V_{max}/K_M)\ (\eta\gamma)/(1+\gamma)$. The time required to obtain a certain degree of substrate conversion T is

$$T = (K_M/V_{max}) \int_{\gamma_2}^{\gamma_1} ((1 + \gamma)/\eta\gamma)\ d\gamma \qquad \text{(Eq. 3)}$$

The value of this integral can be obtained by numerical integration, using the data in Fig. 1. An operational effectiveness factor, η_o, can be defined as the ratio of the times required to obtain a defined degree of substrate conversion with the same amount of free, T_f, and immobilized, T_m, enzyme under otherwise identical conditions. It is a suitable measure for the efficiency of an operating reactor.

EXPERIMENTAL

α-Chymotrypsin (Worthington) and trypsin (Merck 3.5 U/mg) were bound to Sepharose as described previously (5,6). The purity of the enzymes was checked with affinity chromatography (6). The trypsin contained 60% β-form and 10% α-form; α-chymotrypsin contained <10% impurities. The substrates N-acetyl-l-tyrosine-ethyl-ester, N-benzoyl-arginine-ethyl-ester were from Merck. All other chemicals were analytical.

Enzyme activities were determined in buffer of 0.05 ionic strength, due to the buffering components Tris-HCl and 0.2 M Nacl, pH 8.0, at 25°C using a pH-stat as described previously (5). The active enzyme content was measured by fluorometric titration of the active groups (7). The average particle radius R was determined using a microscope. At the stirring speeds used in the activity determinations the activity was practically independent of stirring speed (5). The corresponding Sh were determined from a calculated curve of η as a function of Sh at low substrate contents ($\alpha \geq 1$) A value of 16 ± 2 was obtained. The Sh under these conditions increased with the sedimentation velocity (and density) of the particles.

RESULTS AND DISCUSSION

Experimental effectiveness factors are compared with theoretical values in Fig. 2. The Thiele modulus for the experimental system was calculated using k_{cat} and k_M values for the free enzyme (8). The diffusion coefficient for the substrate in the gel particle was calculated as described in (9), using 50 Å for the average pore diameter. The experimental data fit the theoretical data within the range of experimental error. This demonstrates the predictive capacity of the collocation method in computing η-values for enzyme reactors at steady-state conditions. Although this was demonstrated for a spherical symmetry, the procedure can be applied also for enzyme supports of other shapes.

Experimental operational effectiveness factors η_o for immobilized trypsin are compared with calculated η_o values in Fig. 3. Even here the collocation method is shown to be suitable in predicting reactor performance. An ideal enzyme reactor should have η- and η_o-values as close to 1 as possible. To obtain such reactors the

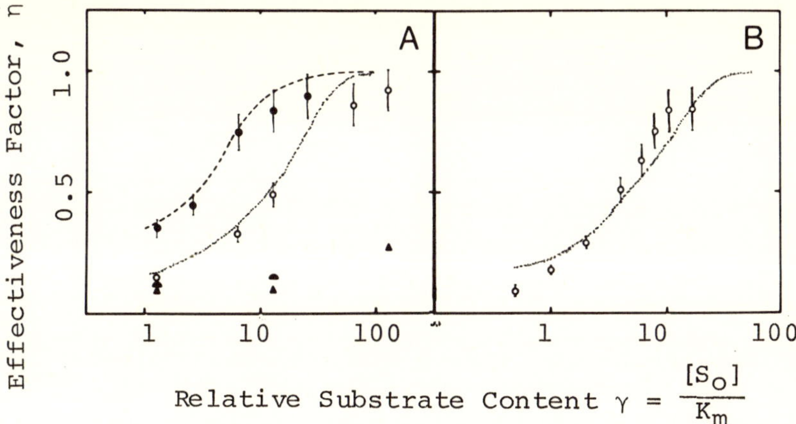

Fig. 2. Comparison of experimental and theoretical steady state effectiveness factors. A: ● Trypsin immobilized in Sepharose Cl 2B (Φ^2 = 75, R = 47μ)-- theoretical; O Trypsin immobilized in Sepharose 4B (Φ^2 = 300, R = 32μ) ... theoretical; B: O Chymotrypsin immobilized in Sepharose 4B (Φ^2 = 150, R = 60μ... theoretical.

micro- and macroenvironmental properties that give the largest contributions to the reduction of η and η_o must be changed. Based on the findings reported here, this can be done as follows: when the microenvironment does not perturb the intrinsic properties of the enzyme, the steep substrate concentration gradient in the outer shell of the gel particle causes the loss of efficiency; a large proportion of the enzyme in the interior of the particle is practically not used. Thus, changing the *microenvironment* so that the enzyme is located in the outer shell of the particle would increase the efficiency considerably. Reducing the particle size gives an increase in η (5). The increase is, however, partly compensated by *macroenvironmental* changes as Sh (and η) decreases with R (Eq. 2) (unpublished observation). Thus, particle size reduction should not be as effective in increasing η as the change in enzyme distribution. Increasing the relative velicity between particle and solvent, so that Sh is about 50, should give maximum efficiencies in stirred reactors. Larger Sh have a marginal influence on η and η_o. This is, however, difficult to achieve using enzyme supports of low density (Sepharose), as has been shown

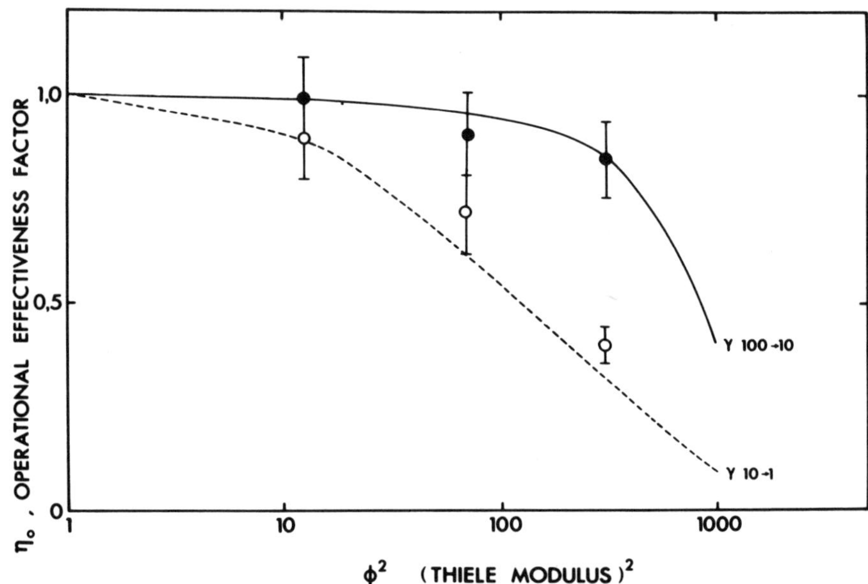

Fig. 3. Operational effectiveness factors as function of enzyme content (Φ^2) for trypsin immobilized in Sepharose. ● experimental for substrate conversion from γ of 100 to 10, —— theoretical; O experimental for substrate conversion from γ of 10 to 1,--- theoretical.

here. In column or fluidized bed reactors the diffusion layers of different particles are superimposed, and cannot easily be reduced to negligible dimensions by increasing the flow rate. In this case spatial separation of the enzyme containing particles by inert particles or by immobilization of the enzymes containing shells on a coarse net work in the column should yield higher η and η_o. The latter change also should give less flow resistance. Values for η and η_o for these reactor types should be possible to calculate by the collocation method.

ACKNOWLEDGMENT

This work has been supported by the Bundesministerium fur Forschung und Technologie.

REFERENCES

1. GOLDSTEIN, L. *Meth. in Enzymology 44*:397, 1976.
2. RAMACHANDRAN, P.A. *Biotechnol. Bioeng. 17*:211, 1975.
3. SATTERFIELD, C.N. "Mass Transfer in Heterogenous Catalysts," M.I.T. Press, Cambridge, Mass., 1970.
4. LEVICH, V.G. "Physicochemical Hydrodynamics," Prentice Hall, Englewood Cliffs, N.J., 1962.
5. KASCHE, V., LUNDQVIST, H., BERGMAN, R. & AXEN, R. *Biochem. Biophys. Res. Commun. 45*:615, 1971.
6. KASCHE, V., AMNEUS, H., GABEL, D. & NASLUND, L. *Biochim. Biophys. Acta 490*:1 1977.
7. GABEL, D. & AXEN, R. *Meth. in Enzymology 44*:383, 1976.
8. AXEN, R. MYRIN, P.A. & JANSON, J.C. *Biopolymers 2*:401, 1970.
9. RENKIN, E. *J. Gen. Physiol. 38*:225, 1954.

STUDIES OF HEMOGLOBIN AND ALLOSTERIC ENZYMES UNDER ARTIFICIAL CONFORMATIONAL CONSTRAINTS

D. Guillochon, C. Bourdillon and D. Thomas

Laboratoire de Technologie Enzymatique
University de Technologie
Compiegne, France

The concept of a conformational change of regulatory enzyme is not well-defined; and the study of the effect of artificial conformational constraints on enzyme behavior is of interest. In accord with Koshland (1), it seems reasonable to attempt to "freeze" some of the conformational states, possibly by tight chemical immobilization of the regulatory enzymes inside an artificial membrane. The hemoglobin modification of conformation is generally discussed as a general model for allosteric enzymes, and it is of interest to compare the relative effect of immobilization on an allosteric enzyme and on hemoglobin.

Goldbeter (2) predicted that some allosteric enzymes could induce some "dissipative structures" at a supra cellular level when coupled with diffusion. The effect was experimentally demonstrated by Hess *et al.* (3). Colosina *et al.* (4) have described the study of ethylisocyamide equilibrium of matrix-bound hemoglobin. Benko *et al.* (5) have studied by proton magnetic relaxation matrix bound hemoglobin and methaemoproteins bound to monodisperse polystyrene latex particles (6).

The present paper deals with the comparison between immobilized pyruvate kinase and immobilized hemoglobin. In both cases the membranes produced by a cocrosslinking method are homogeneous in structure and exhibit mechanical properties similar to those of cellophane.

MATERIAL AND METHODS

Pyruvate kinase (PK) membranes were produced by a method already described (7). A solution of 0.02 M phosphate buffer pH 6.8, containing 30 mg/ml albumin, 2 mg/ml glutaraldehyde, and 1 mg/ml PK was spread perfectly flat on a glass plate. The crosslinking proceeded at room temperature for 2 hours. The plate and film were dipped into distilled water and the membrane easily separated from the plate. The film was rinsed until the rinse water no longer adsorbed at 280 nm. No activity and no proteins were observed in the rinse water.

Human hemoglobin solutions were prepared as described previously (8). For producing hemoglobin membranes a solution of 0.02 M phosphate buffer pH 6.8, containing 60 mg/ml albumin, 4.5 mg/ml glutaraldehyde, and 43 mg/ml human hemoglobin was spread on a glass slide and dried. The coating was obtained and rinsed with 10 mg/ml glycine in 0.02 M phosphate buffer pH 6.8. No hemoglobin was found in the rinse solution by spectrophotometric measurement between 450 and 650 nm.

Pyruvate kinase activity was measured by coupling with lactate dehydrogenase (LDH). The measurements were done in a 3 ml quartz cuvette with 0.2 M Tris-maleate buffer pH 8, containing 1 mg/ml ADP, 0.05 mg/ml $MgCl_2$, 0.233 mg/ml NADH, and 30 I.U./ml LDH. Enzyme membranes (20 cm^2) were tested in a 30 ml batch reactor. The rate of NADH consumption was recorded spectrophotometrically at 340 nm with a continuous flow cuvette.

Hemoglobin binding of oxygen was determined spectrophotometrically. Since the spontaneous oxidation of immobilized hemoglobin is very quick, oxyhemoglobin, deoxyhemoglobin and methemoglobin were present together in the membrane; and spectrophotometrical measurements were performed at three wavelengths: 540, 560 and 575 nm. The glass side bearing the hemoglobin coating was introduced into a continuous flow cuvette, perpendicularly to the spectrophotometer beam. Spectra of HbO_2, Hb, and Hb^+ were obtained with the same coating in each experiment in order to evaluate the oxygen binding ($\overline{Y}$). The Hb spectrum was obtained in the presence of excess sodium dithionite 0.05 M); the HbO_2 spectrum was obtained after quick oxygenation following removal of the dithionite; and the Hb^+ spectrum was obtained after treatment of the membrane with 0.05 M potassium ferricyanide. The pO_2 was measured with a Clark oxygen electrode in a continuous flow of 0.2 M phosphate buffer pH 6.8. The electrode was connected to a regulation device (pO_2-stat) that injected 0.005 M dithionite. The oxygen binding of native hemoglobin was measured under similar conditions; but the oxidation was neglected and the pO_2 was regulated with simple nitrogen bubbling. All the experimental devices were thermostated at 23°C.

EXPERIMENTAL RESULTS

The active membranes were studied by scanning and transmission electron microscopy (9). The membranes were homogeneous in thickness; and the surfaces were regular. A dense structure was observed with no pores or holes until the proteins could be distinguished. The enzymes were tightly immobilized within the solid-phase structure. After immobilization the enzymes were quantitatively stable for weeks.

It was demonstrated by gel chromatography that the enzyme molecules were involved in the proteic macro polymers. The MW was 1 million at the beginning of the crosslinking process before the insolubilization step (10). The proteins were chemically linked and not entrapped in the structure. The composition of a dry membrane (W/V) was found to be 5% glutaraldehyde and 95% protein by nitrogen titration (Carlo Erba C.H.N. analyzer). About twenty bonds linked each protein molecule to the insoluble membrane.

The activity of the native and immobilized pyruvate kinase was studied as a function of substrate concentration (Fig. 1). The results obtained with an active membrane exhibiting low diffusional limitations (Thiele modulus smaller than 1) (7) were similar to the behavior of the free enzyme. With higher diffusional limitations obtained with a membrane three times thicker, a quite important difference of behavior was observed. The apparent enzyme substrate affinity was lower; the phenomenon was due to the diffusional limitations; and was discussed previously for Michaelian enzymes (7).

By using Hill plots (Fig. 2) the same Hill number was obtained for native and membrane enzyme with low diffusional limitations. With higher diffusional limitations a biphasic curve was obtained. A modification of the slope was observed for the activity corresponding to one half of V_{max}. The same Hill number was obtained for con-

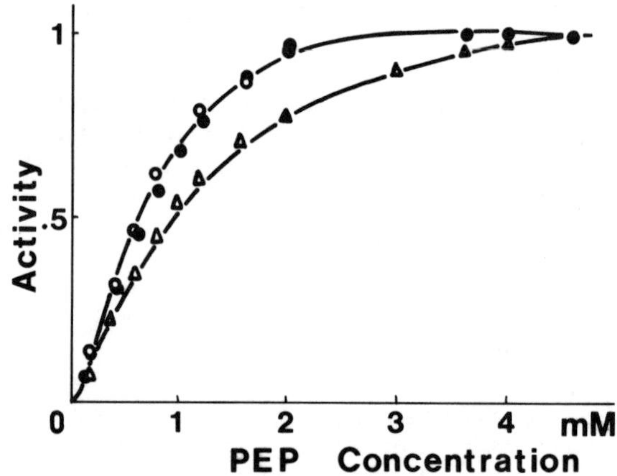

Fig. 1. Pyruvate kinase activity in arbitrary units as a function of phosphoenol pyruvate (PEP) concentration. Native enzyme in solution (●) and immobilized within membranes exhibiting low (0) and high (triangle) diffusional limitations; 6 mM Mg^{2+} and no K^+; 20 cm^2 membrane area and 0.12 I.U. maximum enzyme activity; respectively. The membrane for the triangles was 3 times thicker than for the circles.

centrations higher than the apparent K_M; and a smaller one was observed for lower concentrations. The binding of O_2 by native and immobilized haemoglobin was studied as a function of the pO_2 level (Fig. 3). A deep modification of the behavior was observed. Hill plots are given in Fig. 4. After immobilization the apparent affinity of heamoglobin for oxygen is dramatically increased and the cooperativity is reduced. The Hill numbers of native and immobilized haemoglobin were found equal to 2.5 and 1.7, respectively. When immobilized in the absence of oxygen, the haemoglobin gave a Hill number equal to 1.6; but the apparent affinity was smaller than in the previous case.

DISCUSSION

It is shown that diffusional limitations give a biphasic aspect to the Hill plots of an allosteric enzyme.

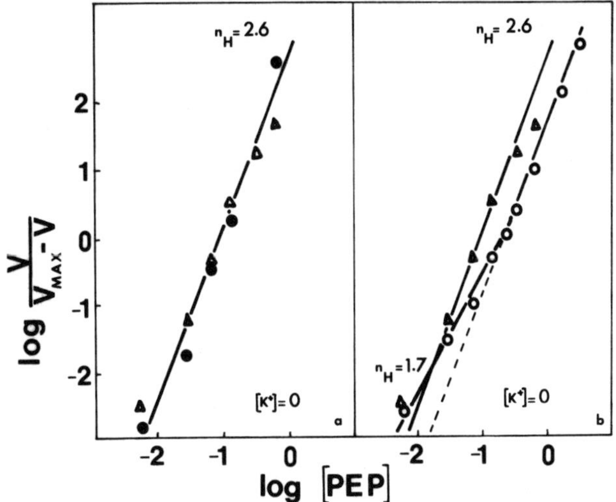

Fig. 2. Hill-plot of native (●) and immobilized pyruvate kinase within membranes with low (triangle) and high (O) diffusional limitations. Low diffusional limitations (Fig. 2a); higher diffusional limitations (Fig. 2b).

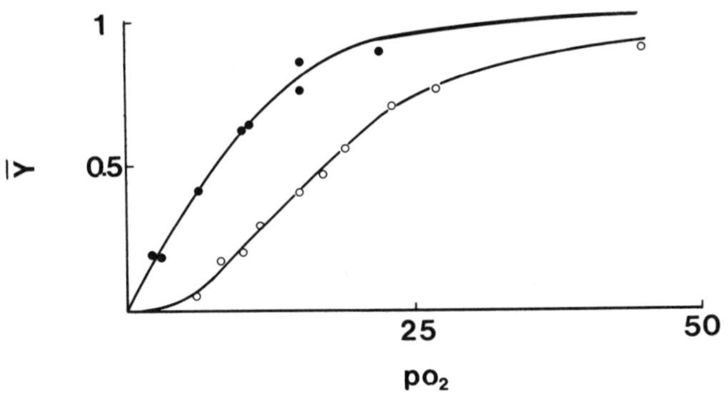

Fig. 3. Native (O) and immobilized hemoglobin (●) binding as a function of pO_2 in the bulk solution. The hemoglobin was immobilized in the presence of oxygen.

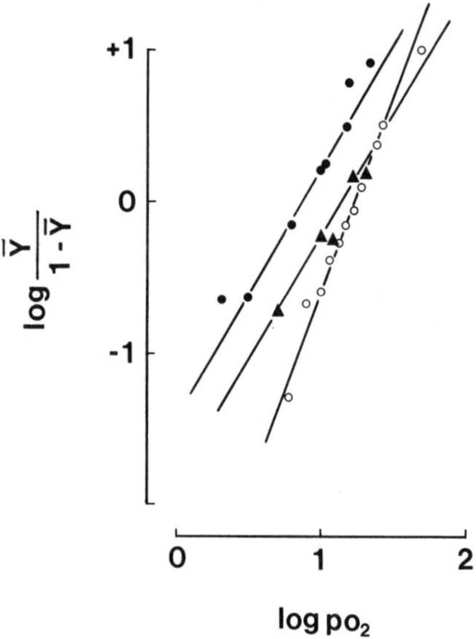

Fig. 4. Hill plots dealing with the binding of native hemoglobin (O) and of hemoglobin immobilized in presence (●) and absence (triangle) of oxygen.

The problem of conformational changes of proteins, especially allosteric enzymes, has attracted considerable attention (1) and the introduction of artificial conformational constraints by immobilization seems to be a new tool of investigation.

From the results obtained with pyruvate kinase the immobilization under low diffusional limitations does not give evidence for a modification of the cooperativity. With hemoglobin there is no doubt that oxygen binding induces changes in quaternary structure of the protein, i.e., a rearrangement of the subunits with respect to each other, as shown by X-ray crystallography (11). It is demonstrated here that immobilized human hemoglobin in the presence of oxygen under the conditions described does not exhibit the same cooperativity. The protein conformation exhibiting the highest affinity seems frozen by immobilization. From the results presented in this paper, pyruvate kinase does not exhibit so deep a modification of conformation. Under these conditions the use of hemoglobin as a model for allosteric enzymes is highly questionable.

REFERENCES

1. KOSHLAND, D.E. in "The Enzymes," vol. 2, 3 Edit., (P.D. Boyer, ed.) Academic Press, New York, 1970, p. 342.
2. GOLDBETER, A. *Proc. Nat. Acad. Sci. 70*:3255, 1973.
3. HESS, B., BOITEUX, A., BUSSE, H. & G. GERISH in "Membranes, Dissipative Structures and Evolution," (G. Nicolis and R. Lefever, eds.) Wiley, New York, 1975, p. 137.
4. COLOSINO, A., STELANINI, S., BRUNORI, M. & ANTONINI *Biochim. Biophys. Acta 328*:74, 1973.
5. BENKO, B., VUK-PAVLOVIC, S. & K., POMMERENING *Stud. Biophys 55*:3, 1976.
6. BENKO, B., VUK-PALOVIC, S., DEZELIC, G. & S., MARICIC J. *Colloid. Interface Sci. 52*, 1975.
7. THOMAS, D., BOURDILLON, C., BROUN, G. & J.P., KERNEVEZ *Biochemistry 13*:2995, 1974.
8. ANTONINI, E. & M. BRUNORI "Hemoglobin and Myoglobin in Their Reactions with Ligands," vol. 21, North-Holland, Amsterdam, 1971, p. 3.
9. BOURDILLON, C., BARBOTIN, J.N. & D., THOMAS *FEBS Lett. 68*:27, 1976.
10. REMY, M.H., WETZER, J. & D. THOMAS in "Analysis and Regulation of Immobilized Enzyme Systems," (D. Thomas and J.P. Kernevez, ed.), North-Holland, Amsterdam, 1975, p. 283.
11. MUIRHEAD, H. & M. PERUTZ *Nature 199*:633, 1963.

DETERMINATION OF MICHAELIS MENTEN PARAMETERS FOR IMMOBILIZED ENZYMES: DETECTION AND MEASUREMENT OF DIFFUSION

Paul F. Greenfield and Robert L. Laurence

Department of Chemical Engineering
University of Massachusetts
Amherst, Massachusetts, USA

One of the simplest relationships to describe the kinetic behavior of enzymes in solution is the Michaelis Menten equation. However, it has been criticized by statisticians for a number of reasons (1,2). If there is any uncertainty in the v or S readings, the Michaelis Menten relation should be written

$$\bar{v} = \frac{V_{max}\,\bar{S}}{K_m + \bar{S}} + \bar{e} \qquad \text{(Eq. 1)}$$

where $\bar{v}$ and $\bar{S}$ are the vectors of the dependent and independent observations, respectively and $\bar{e}$ is a vector of the associated errors. Errors in both $\bar{S}$ and $\bar{v}$ may be represented in this fashion, although a least squares analysis theoretically assumes that error occurs only in the dependent variable. It can be seen from Eq. 1 that the normal Michaelis Menten linear transformations ignore the fact that the additive error in any of the linear models is not the same as $\bar{e}$ (3). Another improvement would be to use a least squares analysis with some form of weighting; although the error in predicting the kinetic parameters may be only slightly better than that obtained with unweighted data.

The effect of diffusional resistance on Michaelis Menten kinetics and on the determination of the kinetic parameters is not always appreciated. Thus the purpose of this paper is to point out the effect of unrecognized diffusional resistance on the kinetic parameters of an enzyme immobilized on a spherical particle. The basic concepts of film and pore diffusional resistances have been well reviewed. However, we should point out that one of the problems of film diffusion lies in estimating the film transfer parameter, k_m. Pore diffusion, in turn, is generally accounted for by the use of an effectiveness factor, η, which is the ratio of the actual rate of reaction to that obtained in the absence of pore diffusion. The effectiveness factor can be related to the Thiele modulus, Θ_m, which is dependent on the geometry of the immobilized enzyme-support arrangement. With either form of diffusional resistance, K_m and V_{max} become apparent values.

METHODS

The effect of dominant film diffusion on the rate of change of concentration of substrate or product was determined by calculating values of the velocity and substrate concentration at the particle surface for a range of bulk substrate concentrations. Values of apparent K_m and apparent V_{max} were obtained by linear least squares analysis of the data after the normal linear transformations, such as Lineweaver-Burke. Estimates of the parameters were made for high concentration data, low concentration data, and for the combination of low and high concentrations.

For the effect of dominant pore diffusion on the linearized plots, an expression was obtained of the reaction velocity as a function of substrate concentration for a range of values of the Thiele modulus. The results were plotted and analyzed in an equivalent fashion to the film diffusion data. For the situation where both film and pore diffusion were significant, the expressions were solved implicitly by the Newton-Raphson method to obtain the diffusion limited reaction velocity. The results were plotted as described above.

The effect of film and pore diffusion on the integrated form of the Michaelis Menten equation also was determined.

RESULTS AND DISCUSSION

In the case of film diffusional resistance the apparent K_m was increased by 2 to 3% at high substrate concentrations and 6 to 13% at low concentrations. The K_m in the absence of diffusion was taken as 1.0 mM; the high and low concentration ranges were 2.5 to 50 mM and 0.05 to 1.0 mM, respectively. The above results were for a Damkohler number, N_D, of 0.2. This number is a measure of the severity of film diffusional resistance and is defined as

$$N_D = \frac{(\text{apparent } V_{max})}{k_m \, a_m \, (\text{apparent } K_m)} \qquad \text{(Eq. 2)}$$

where a_m is the surface area of support per unit volume. In the absence of film diffusional resistance, N_D is zero; it increases with greater film resistance. For increased film resistance, N_D of 1.0, the apparent K_m was increased by 25 to 40% at high substrate concentrations and 217 to 245% at low concentrations. The apparent V_{max} was little effected at an N_D of 0.2, being 1.0 mM/sec. However, at an N_D of 1.0, the apparent V_{max} was increased by 1 to 2% at high substrate concentrations and 60 to 72% at low concentrations. Although the Lineweaver-Burke plots gave the smallest deviations from linearity, they consistently gave the worst estimates of the apparent K_m and V_{max}.

For pore diffusional resistance the same concentration ranges and diffusion-free K_m and V_{max} values were used. At a Thiele Modulus of 0.5, the apparent K_m was increased by 3 to 6% at high concentrations and 10 to 11% at low concentrations. The apparent V_{max} was little effected at either concentration range. At greater film diffusional resistance, Θ_m of 2.0, the apparent K_m was increased 49 to 64% at high and 145 to 149% at low concentrations. The apparent V_{max} was increased 1 to 3% and 18 to 20% for the same concentration ranges. Again, the Lineweaver-Burke plots were the most misleading. A more thorough description of the effects of pore diffusion on Lineweaver-Burke plots is given elsewhere (4,5).

In some cases the integrated Michaelis Menten-equation showed curvature due to diffusional resistances more readily than for the non-integrated linearized forms.

ACKNOWLEDGMENT

The research was supported by Grant No. GI 34976, RANN, National Science Foundation.

REFERENCES

1. EISENTHAL, R. & CORNISH-BOWDEN, A. *Biochem J. 139*: 715, 1974.
2. WALTER, C. *J. Biol. Chem. 249*:699, 1974.
3. WILKINSON, G.N. *Biochem. J. 80*:324, 1961.
4. HAMILTON, B.K., GARDNER, C.R., & COLTON, C.K. *A.I.Ch. E.J. 20*:502, 1974.
5. GONDO, S., ISAYAMA, S., & KUSONOKI, K. *Biotechnol. Bioeng. 17*:423, 1975.

STUDIES WITH IMMOBILIZED TOXINS

P.V. Sundaram

Abteilung Chemie
Max Planck Institut fur Experimentelle Medizin
Gottingen, Federal Republic of Germany

Pathogenic bacteria cause illness as a consequence of specific toxins that they secrete into the host cell. Recent studies have shown that a group of bacterial and plant toxins resemble each other in their general mode of action, despite the difference in the pathology of the diseases they produce.

Diphtheria toxin is a zymogen of MW 63,000 (1). The inherent toxicity of the molecule is released only after an enzyme cleavage of the toxin into two fragments followed by reduction of a disulphide bridge (2). Fragment A of MW 24,000 is the toxic part; and Fragment B of MW 39,000 is believed to bind to the cell membrane and mediate the entry of Fragment A into the cell to inhibit protein biosynthesis. Cholera toxin, produced by *Vibrio cholerae*, is a zymogen of 82,000 MW which is believed to split into Fragment A (MW 27,000) and Fragment B (MW 55,000) of four subunits of proposed structure AB_4 (3). Fragment A needs Fragment B to enter the cell and trigger the increase in cyclic AMP levels, through the activation of the adenylate cyclase system.

Our investigation was started in order to understand the details of the molecular mechanism of action of these toxins. For this reason the toxins were immobilized; and their properties on cell free systems and whole cells were studied.

METHODS

Diphtheria toxin isolated from *Corynebacterium diphtherae* (Behring Werke GmbH) was used. Elongation factor II (Ef II) was isolated from rat liver according to the method described for human tonsils (4). ^{14}C-NAD at 274 ci/mole was from Amersham. Toxin activity was assayed by incubating 2 to 5 μM ^{14}C-NAD, 1.52 μM EF II, and 1 to 2 μg whole toxin in 0.05 M pH 7.4 Tris buffer and 0.25 M sucrose containing 10 μM 2-mercaptoethanol at 20°C for 2 to 12 min (5). Immobilized toxin was assayed by replacing whole toxin in the above mixture with small strips of nylon-DT or Nylon-PEI-DT.

Nylon cloth, as 2 x 2 cm strips, in 2 ml of dimethyl sulphate was heated in a boiling water bath for about 40 sec; the reaction was stopped by immersing the flask in melting ice. The reacted nylon strips were then washed with ice cold methanol followed by ice cold water. Toxin was coupled directly to the O-alkylated nylon in pH 7.4 Tris buffer, containing 10 μM mercaptoethanol, and also to nylon-polyethyleneimine (Nylon-PEI) copolymer by cross linking with glutaraldehyde at pH 7 (6). PEI was first coupled to nylon in pH 7.8 phosphate buffer as a 1% v/v solution made from the commercial PEI solution (7).

RESULTS AND DISCUSSION

Diphtheria toxin coupled directly to nylon and to nylon-PEI copolymer retained its catalytic activity in a cell free system, as shown by its capacity to ADP-ribosylate EF II. Kinetic plots of ADP-ribosylation of EF II catalyzed by nylon-DT are shown in Figs. 1 and 2. The variation of EF II concentration produced a biphasic plot with a typical sluggish initial part up to 0.52 μM EF II concentration followed by a burst at higher concentrations. Consequently double reciprocal plots of these data gave several values for $K_{m(app)}$, when data above or below 0.52 μM was extrapolated back to obtain these values (Table 1). Values for K_m and $K_{m(app.)}$ for NAD also are mentioned in Table 1. The kinetics of the reaction catalyzed by Nylon-PEI-DT showed a more pronounced initial lag. This could be due to a diffusional effect that was accentuated by the electrostatic effects of the positively charged nylon-PEI derivative.

The immobilized diphtheria toxin was used repeatedly to catalyse the transfer of the ADPR moiety to EF II.

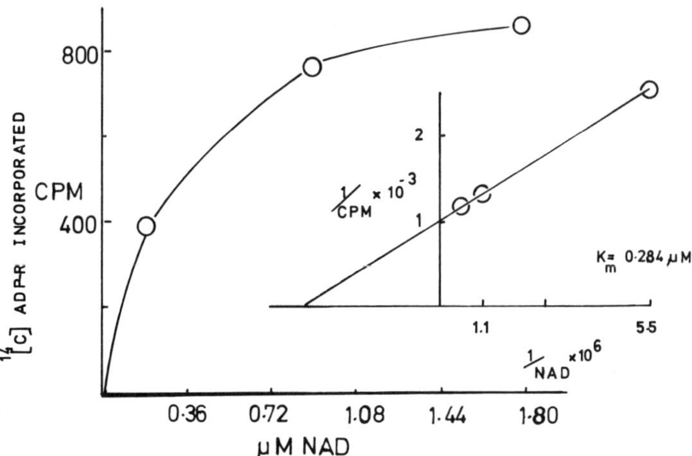

Fig. 1. Kinetics of ADP-ribosylation as a function of NAD concentration with immobilized diphtheria toxin.

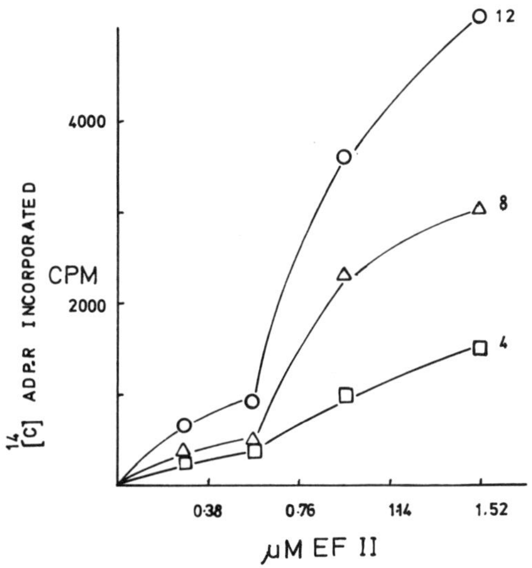

Fig. 2. Kinetics of ADP-ribosylation as a function of EF II concentration with immobilized diphtheria toxin. (square) 4 min, (triangle) 8 min, (circle) 12 min end point assays.

TABLE 1

MICHAELIS CONSTANTS OF FREE AND IMMOBILIZED DIPHTHERIA TOXIN IN ADP-RIBOSYLATION OF EF II*

Agent	Free Toxin (μM)	Nylon Toxin** (μM)	Nylon-PEI-Toxin** (μM)
NAD	0.15	0.284	--
EF II	0.71	--	--
Low S	--	0.43	0.56
High S	--	1.15	2.0

* Counting efficiency for ^{14}C was 60% in all the experiments.

** Apparent K_m

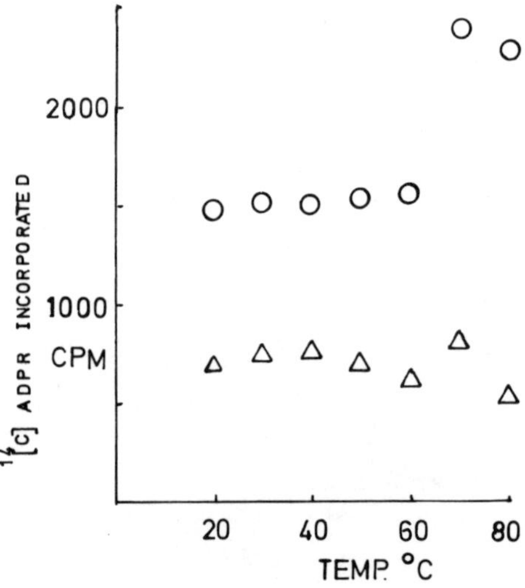

Fig. 3. Effect of incubation at various temperatures prior to assay of free and immobilized toxin. (square) free diphtheria toxin; (triangle) nylon -DT.

After several times it lost about 25% of its activity. Exposure at pH 7.4 to various temperatures for 5 min showed that nylon-DT did not exhibit an activation at 70° to 80°C similar to that of the free toxin (Fig. 3). This property, originally reported by Honjo *et al.* (8), was probably diminished upon immobilization due to a restricted freedom of the protein molecule.

The main finding in this study was that immobilized diphtheria toxin retained its catalytic activity *in vitro*, as shown by its ability to transfer the ADP-ribosyl moiety to EF II. Coupling the toxin in the presence or absence of 2-mercaptoethanol did not appear to make a vast difference in the reactivity of the immobilized toxin. Early experiments with the cholera toxin immobilized on nylon by similar methods failed to elicit activity.

ACKNOWLEDGMENTS

The work on diphtheria toxin was carried out in collaboration with D. Bermek of Istanbul University. That on cholera toxin in collaboration with S. van Heyningen of the University of Edinburgh Medical School is in progress.

REFERENCES

1. COLLIER, R.J. & KANDAL, J. *J. Biol. Chem.* *246*:1496, 1971.
2. GILL, D.M. & PAPPENHEIMER, A.M. *J. Biol. Chem.* *246*: 1492, 1971.
3. COLLIER, R.J. *Bact. Rev.* *39*:54, 1975.
4. BERMEK, E. & MATTHAEI, M. *Biochemistry* *10*:4906, 1971.
5. BERMEK, E. *FEBS Lett.* *23*:95, 1972.
6. SUNDARAM, P.V. *Nucleic Acids Res.* *1*:1587, 1974.
7. SUNDARAM, P.V. in "Biomedical Applications of Immobilized Enzymes and Proteins" vol. 2 (T.M.S. Chang, ed), Plenum, New York, 1977, p. 317.
8. HONJO, T., NISHIZUKA, Y., KATO, I. & HAYAISHI, O. *J. Biol. Chem.* *246*:4251, 1971.

IMMOBILIZED PROTEIN MODIFICATION REAGENTS

William H. Scouten

Department of Chemistry,
Bucknell University
Lewisburg, Pennsylvania, USA

Solid phase biochemistry is a rapidly growing area including immobilized enzymes, affinity chromatography, protein synthesis and protein sequencing. The research here describes a new potential application of solid phase reactions to biochemistry, namely the use of immobilized protein modification reagents. These reagents offer the same increased ease of handling and improved yields of most other solid phase applications in biochemistry. In addition, they may, in many cases, be used as topology probes, since the reaction of immobilized reagents with proteins is often sterically restricted to the surface of the protein. A small number of such immobilized protein reagents have appeared in the literature. Rose Bengal (1) and methylene blue (2,3) have been employed as immobilized photocatalysts, while immobilized α-keto halides (4) have been employed in the selective cleavage of protein methionyl residues.

We have coupled dehydrolipoamide to amino alkyl controlled pore glass beads (CPG) via an amide linkage and have employed the reagent as a reducing agent for protein thiols (5). Papain, for example, is activated by this reagent; although much of the activated papain remains bound to the glass matrix, presumably via hydrophobic interaction between the hydrophobic active site and the alkyl portion of lipoic acid.

Dihydrolipoamide glass has numerous additional uses. It chelates Hg^{++}, Pt^{++} and Au^{+}, while not chelating other

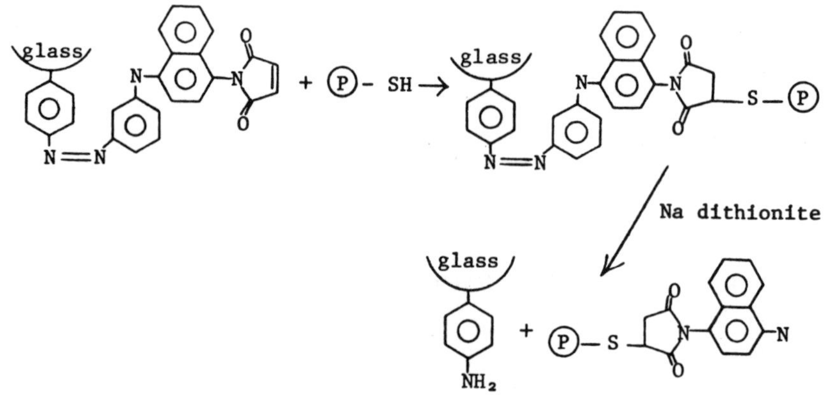

Fig. 1. Anilinonaphthalene maleimide fluorescent sulfhydryl reagent.

heavy metals. It also may be employed to keep solutions, fermentations, etc., anaerobic while not producing potentially toxic soluble products.

A second immobilized protein reagent we have prepared is a glass coupled anilinonaphthalene maleimide (ANM) which serves as a fluorescent sulfhydryl reagent. To prepare the reagent, ANM (Molecular Probes, Roseville, NM), is added to diazotized aryl amine CPG prepared by the method of Lewis and Scouten (2). Zirconia coated CPG may be used to obtain improved stability at alkaline pH. The diazotized ANM-glass is non-fluorescent, but when reacted with a protein with a surface thiol, it yields a fluorescent, immobilized, protein. The protein-ANM complex is removed from the glass by reduction with sodium dithionite or similar reducing agents, as shown in Fig. 1.

REFERENCES

1. KAYE, N.M.C. & WEITZMAN, P.D.C. *FEBS Lett.* *62*:334, 1976.
2. LEWIS, C. & SCOUTEN, W.H. *Biochim. Biophys. Acta* *444*: 326, 1976.
3. LEWIS, C. & SCOUTEN, W.H. *J. Chem. Ed.* *53*:395, 1976.
4. SHECHTER, Y., RUBINSTEIN, M., & PATCHORNICK, A. *Biochemistry* *16*:1424, 1977.
5. SCOUTEN, W.H. & FIRESTONE, G.L. *Biochim. Biophys. Acta* *453*:277, 1976.

ON THE MECHANISM OF PROTEIN CROSS-LINKING WITH GLUTARALDEHYDE

Sven Branner-Jorgensen

NOVO Industri A/S
Bagsvaerd, Denmark

Four theories have been developed to explain the cross-linking reaction of glutaraldehyde with proteins: (a) simple Schiff-base formation; (b) reaction of glutaraldehyde pplymers to form Michael-type additions to the protein (1); (c) glutaraldehyde polymers forming conjugation stabilized Schiff-bases with the protein (2); and (d) pyridinium ion theory (3). We have tested these different mechanistic proposals with the following dialdehydes: succinic, glutaric, 2,4-dimethylglutaric (2,4-DMGA), and adipic.

The gelforming ability of the aldehydes in about 10% crude protein solution with 0.15 M dialdehyde at pH 7 took several hr for succinic, a few min for glutaric and 2,4-DMGA, and several days for adipic. The development of UV adsorption after 24 hours reaction in about 0.1% crude protein solution with 5 mM dialdehyde at pH 7 in a 1 cm cell showed essentially a zero extinction coefficient for succinic, 17.5 at 268 nm for glutaric, 22.6 at 273 nm for 2,4-DMGA, and 181 at 243 nm for adipic.

The theories based on α,β-unsaturated glutaraldehyde polymers as the reactive reagent were not consistent with the fact that 2,4-DMGA reacted almost as readily as glutaraldehyde. 2,4-DMGA simply could not form unsaturated polymers. The same type of restriction ruled out the pyridinium ion theory. Also, the development of the chromophore at 273 nm with 2,4-DMGA was against this theory. The main conclusion was that only the simple Schiff-base formation was valid as the initial step in the cross-linking reaction. It is noteworthy that none of the theories could account for the exceptional reactivity of glutaraldehydes relative to succinic and adipic aldehyde.

I suggest that the many corss-links formed, when a lysine-rich protein is reacted with glutaraldehyde, stabilizes the imine bonds in the complex against hydrolysis. If one imine bond is hydrolyzed, the high concentrations of the reactants (aldehyde and amine) in the microenvironment will force the reforming of the imine bond. This contrasts to what can be expected in free solution. The reasons that glutaraldehydes are the most reactive reagents can at present only be explained by (a) solvation effects, (b) a suitable chain length to facilitate inter-molecular cross-links, and (c) possibly stabilization due to a cyclic hydrate such as:

Protein-HN O NH-Protein

REFERENCES

1. RICHARDS, F.M. & KNOWLES, J.R. *J. Mol. Biol.* 37:231, 1968.
2. MONSAN, P., PUZO, G. & MAZARGUIL, H. *Biochemie* 57: 1281, 1975.
3. HARDY, P.M., NICHOLLS, A.C. & RYDON, H.N. *J. Chem. Soc. Perkin I 1976*:958.

SYNTHESIS OF WATER SOLUBLE POLYMERS WITH COVALENTLY BOUND GENERAL LIGANDS

A.F. Buckmann, M. Morr and M.-R. Kula

Gesellschaft fur Biotechnologische Forschung mbH

Braunschweig-Stockheim, Federal Republic Germany

The partition of biological macromolecules, cells, and cell organelles in aqueous polymer two-phase systems is a well known phenomenon (1). One of our research activities is concentrated on the design of liquid-liquid separation techniques based on this phenomenon for the isolation and purification of enzymes (2). The partition of the constituents of an enzyme extract in an aqueous two-phase system, consisting of polyethyleneglycol (PEG) and dextran or PEG and potassium phosphate, can be attributed mainly to nonspecific physico-chemical interactions. In principle the specificity might be enhanced by confining to one of the phases polymers with appropriate covalently bound ligands (affinity partition). Based on PEG, methods to synthesize water soluble polymers with covalently bound general ligands (NAD-, NADH-, AMP-, ATP- and pyridoxal-5-phosphate analogues) have been explored. The PEG-NAD(H) derivatives have been tested in affinity partition experiments with formate dehydrogenase and formaldehyde dehydrogenase. Their stability at different pH values and reutilization was investigated.

PEG (MW 6700) was modified as shown in Fig. 1. The N(I)-[amino-ethyl] derivatives of NAD^+, AMP, and ATP have been synthesized by reaction with ethyleneimine and purified by ion exchange with 65, 70, and 35% yields, respectively; they were then coupled to carboxyl-PEG by the carbodiimide method. After gel filtration of the reaction mixture, preparations of PEG-N(I)-[AE]-NAD^+, PEG-N(I)-[AE]-AMP, and PEG-N(I)-[AE]-ATP could be obtained

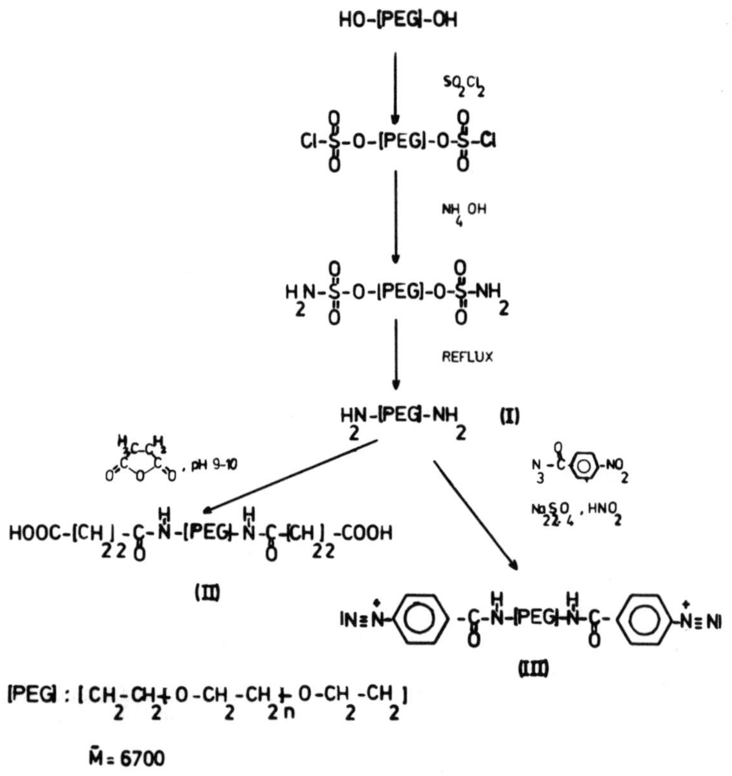

Fig. 1. Modification of PEG

with 0.4-0.5 mmol nucleotide/mmol PEG with 48, 37, and 20% as overall yields, respectively. This left 50-60% of the PEG-COOH molecules unmodified. PEG-N-(I)-[AE]-NADH, PEG-N^6-[AE]-AMP, PEG-N^6-[AE]-ATP, PEG-N^6-[AE]-NADH, and PEG-N^6-[AE]-NAD^+ could be synthesized by reduction with $Na_2S_2O_4$, rearrangement, and enzymatic oxidation respectively with negligible losses.

Pyridoxal-5-phosphate (PLP) has been bound to PEG derivative III at the 6 position (0.2 mmol PLP/mmol PEG). PEG-N(I)-[AE]-NAD(H) and PEG-N^6-[AE]-NAD(H) functioned as coenzymes in an assay with yeast alcohol dehydrogenase (more than 90% reducible or oxidizable). Without removing the unmodified carboxyl-PEG the PEG-N^6-[AE]-NAD(H) preparations were tested in preliminary affinity partition experiments with formaldehyde dehydrogenase (FADH) and formate dehydrogenase (FDH).

The enzymes were separated from the PEG-N^6-[AE]-NAD(H) derivatives by an adsorption procedure with 100% recovery of the enzyme activity, while the PEG ligands could be reused. The stability of the PEG-N^6-[AE]-NAD(H) derivatives at pH 7.0, 8.0 and 9.0 during 6 days incubation at room temperature at a concentration of 1 mM was checked by enzymatic assay. Remarkable stability of the PEG-N^6-[AE]-NAD^+ was observed at basic conditions.

REFERENCES

1. ALBERTSSON, P.A. "Partition of Cell Particles and Macromolecules," Wiley, New York, 1971.
2. M.-R. KULA *et al.*, this volume.

SOLUBLE NAD^+-DERIVATIVES OF HIGH MOLECULAR WEIGHT IN ENZYMIC RECYCLING SYSTEMS

B. Dolabdjian, G. Grenner, P. Kirch, and H.-L. Schmidt

Institut fur Chemie Weihenstephan
Technical University Munchen
Freising-Weihenstephan, Federal Republic Germany

The quantitative turnover of many dehydrogenase reactions can be afforded in most applications only if coenzyme recycling is available. Although the immobilization of coupled coenzyme depending enzyme systems is possible (1); the reactors thus obtained are expensive and not very effective. Therefore, the concept of this work was to use membrane enclosed soluble enzymes together with soluble but non-dialysable coenzyme derivatives.

Such derivatives have been synthesized by the following procedures: (a) direct coupling of NAD^+ to epoxy-modified dextran, (b) binding of N^6-(2-aminoethyl)-NAD^+ (2) to dextran by the BrCN-method, and (c) attaching N^6-(2-aminoethyl)-NAD^+ to unmodified and to HIO_4-oxidized dextran by means of adipic acid dihydrazide (3). Some of the derivatives obtained were effective hydrogen transmitters in batch experiments (4); but in reactors the turnover rate was not quite suitable for practical use. The most effective derivative until now was that prepared by the first method mentioned above. Pilot experiments in model recycling systems also were performed with unmodified NAD^+. Analysis was done using a dehydrogenase electrode, consisting of a Clark oxygen electrode in contact with a solution of a dehydrogenase and NADH-oxidase enclosed in a membrane. The concentration of the substrate of the dehydrogenase outside the membrane could be determined from the decrease of the cathode current.

So far ethanol, lactate, malate, glucose, and glutamate have been determined in the range of 1 to 15 μM solutions with response times of 10 min. The life times of the electrodes were at least one week. The response time did not change when dextran - NAD^+ was used; however the sensitivity decreased, probably due to the reduced turnover rate.

NAD^+ - recycling systems for the micropreparative syntheses of ^{15}N-labelled L-amino acids from oxo-acids and $^{15}NH_3$ with ethanol as the coupled hydrogen donor were tested. An Amicon ultrafiltration apparatus was the reaction chamber (5). L-glutamate, L-alanine, L-serine, and L-aspartate were synthesized in the mmole scale; however the use of unmodified NAD^+ demanded very dense filters that resulted in low flow rates. By use of the dextran-NAD^+ the flow rate could be raised. Yet turnover decreased due to the lower biological activity of the hydrogen transmitter.

The coupled system isocitrate dehydrogenase/glutamate dehydrogenase was investigated as a model for the irreversible synthesis of L-glutamate without an additional hydrogen donor. This system did not work with $NADP^+$-derivatives. Again, in a coupled system for amino acid synthesis with glutamate dehydrogenase and a transaminase, the alcohol dehydrogenase reaction proved to be the most convenient hydrogen donor system. Excess ethanol gave a favorable equilibrium and did not interfere with the isolation of the products.

REFERENCES

1. MATTIASSON, B. & MOSBACH, K. *Biochim. Biophys. Acta* *235*:253, 1971.
2. GRENNER, G., SCHMIDT, H.-L., & VOLKL, W. *Hoppe-Seyler's Z. Physiol. Chem.* *357*:887, 1976.
3. JUNOWICZ, E. & CHARM, S.E. *Biochim. Biophys. Acta* *428*:157, 1976.
4. SCHMIDT, H.-L. & GRENNER, G. *Eur. J. Biochem.* *67*:295, 1976.
5. DAVIES, P. & MOSBACH, K. *Biochim. Biophys. Acta* *370*:329, 1974.

THE USE OF ENZYMES FOR DESIGN OF LIGHT SENSITIVE SILVERLESS MATERIALS: PHOTOENZOGRAPHY

N.F. Kazanskaya

Department of Chemistry
Moscow State University
Moscow, USSR

A process for producing a silverless photo image using enzymes is proposed. The photo layer is made of a photosensitive inactive and stable 12 hr half-life derivative of a proteolytic enzyme and a film-forming carrier (1-4).

The photosensitive derivatives and the wavelength of the activating light are: papain dimer (240-260 nm), cis-cinnamoyl-α-chymotrypsin,-trypsin,-subtilisin (300-320 nm), n-NO_2-cis-cinnamoyl-α-chymotrypsin (320-350 nm), n-dimethylamino-cis-cinnamoyl-chymotrypsin (370-420 nm), and others. The short impulse of light (<1 sec) isomerises the cis-cinnamoyl group in the active center of the acylenzyme. The trans isomer is unstable ($t_{1/2}$ = 1 sec) at the pH optimum of the enzyme.

The trans-acylenzyme hydrolysis and the active enzyme begins a catalytic reaction. The effective guantum yield of such a process depends on the guantum yield of the first act of the light (∿0.1). The rate constant of the catalytic reaction is 100 to 1000 sec^{-1}; and the time to carry out the proteolytic reaction usually is 10 min. The quantum yield equals 10^4 to 10^5. The enzyme concentration in the layer may be substantially increased due to an autocatalytic reaction of the enzyme when a zymogen and the light sensitive enzyme derivative are mixed. The guantum yield in this case is 10^8 to 10^{10}.

If the carrier is gelatine, the enzyme cleaves the carrier during the development of the image. The peptides are washed out with water; and a relief image is obtained. The resolution is about 20 points per mm.

If the carrier is an uncleavable polymer or a microcapsule, the enzyme hydrolyses the substrate introduced into the photo layer or present in a developing buffer solution. The good substrates in this case are the esters of indoxyl. Indoxyl oxidates form fast in the presence of air during the hydrolysis of the ester; and crystals of indigo precipitate at the point of light action. A half-tone image will appear in such a case. The resolution is equal to the size of the microcapsule.

REFERENCES

1. MARTINEN, K., VARFOLOMEYEV, S.D., BEREZIN, I.V. *Eur. J. Biochem.* *19*:242, 1971.
2. BEREZIN, I.V., AISINA, R.B., VARFOTCHAHEEV, S.D., & KAZANSKAYA, N.F. *Doklady Akad. Nauk SSSR* *219*:1255, 1974.
3. AISINA, R.B., BRONNIKOV, I.E., KAZANSKAYA, N.F., & BEREZIN, I.V. *Bioorganitcheskaya Chimia* *1*:402, 1975.
4. AISINA, R.B., KAZANSKAYA, N.F., LOSHKEVITCH, T.V., & BEREZIN, I.V. *Biochimia* *42*:119, 1977.

CONTINUOUS MEASUREMENTS ON IMMOBILIZED CELLS BY A MASS FILTER

J.C. Weaver, F.M. Reames, L. DeAlleaume, C.R. Perley, and C.L. Cooney

Massachusetts Institute of Technology
Cambridge, Massachusetts, USA

The possibility of a continuous assay based on direct measurement of volatile reactants of an enzyme-catalyzed reaction has been described (1,2). Extension of this analytical approach to immobilized cells should offer the same potential for speed and sensitivity. Also, it should allow a variety of species, such as carbon dioxide, HCHO, methanol, or acetic acid, to be monitored directly. Or, by downstream continuous enzymatic conversion, using a suitabale immobilized enzyme, nonvolatile species, such as glucose, pyruvate, lactate, and acetylcholine, could be used. The rationale for pursuing such measurements includes the possibility of studying the effects of many stimuli on immobilized cells. It also includes applications directed at environmental measurements, including the possibility of continuous, rapidly responding bioassays for environmental monitoring.

To date we have used primarily *Saccharomyces cerevisiae* (Baker's yeast) as a model system, with the cells immobilized by entrapment. We describe briefly two examples of the types of measurements that are readily performed. In both cases pulses of glucose as substrate are injected upstream from the reaction module, and either carbon dioxide or ethanol production is monitored downstream. The cells can be injected or removed from the reaction module by a syringe. As expected, for a given large magnitude of glucose pulse, both the carbon dioxide and ethanol product pulse size are proportional to the total cell number. Upon calibration, therefore, a rapid determination of cell number can be made readily using a sample volume of ∿ 0.1 ml. In the present configuration a calibration curve (ethanol count rate vs cell number) is linear over the range 10^7 to about 10^9 cells.

A second example is the ready determination of a response curve to a given substrate going to a specific product, as glucose to either carbon dioxide or ethanol. Unlike the microbe thermistor (3) response curve utilizing *S. cerevisiae* entrapped in polyacrylamide, where $k_{1/2}$ ∿2 mM glucose, our response has a half-rate concentration of 17 mM. This indicates that our combination of dializer membrane and 0.01 cm module depth results in a greater diffusional limitation than does the polyacrylamide gel through which flow occurs in the microbe thermistor.

ACKNOWLEDGMENT

This work was supported partially by grant GM22633 from the National Institutes of Health.

REFERENCES

1. WEAVER, J.C., MASON, M.K., JARRELL, J.A., & PETERSON, J.W. *Biochim. Biophys. Acta. 438*:296, 1976.
2. WEAVER, J.C. in "Biomedical Applications of Immobilized Enzymes and Proteins," vol. 2 (T.M.S. Chang, ed.), Plenum, New York, 1977, p. 207.
3. MATTIASSON, B., LARSON, P.-D., & MOSBACH, K. *Nature 268*:519, 1977.

ENZYME ACTIVITY AT A GAS SOLID INTERPHASE: OXIDATION OF METHANOL TO FORMALDEHYDE*

Miguel Cedeno and Mario Waissbluth

Centro de Investigacion y Estudios Avanzados
Instituto Politecnico Nacional
Mexico, D.F., Mexico

*Presented by J. Limon-Lason

Methanol-consuming yeasts have an FAD-dependent methanol oxidase, which catalyzes the reaction between methanol and oxygen to form formaldehyde and hydrogen peroxide. The catalase that is present in the same peroxisomes decomposes the hydrogen peroxide; it also acts as a formaldehyde producer by oxidation of the methanol (1,2). A subsequent enzymatic step incorporates formaldehyde into different metabolic pathways through an $NADH_2$ - dependent formaldehyde dehydrogenase, which requires a mechanism for NAD regeneration. Therefore, non-viable cells will not decompose formaldehyde since the NAD cannot be regenerated (1).

As the substrates and products of the methanol oxidation reactions can exist in the gas phase, we have explored the possibility of having enzyme catalysis at a gas-solid interphase, in the absence of liquid or vaporized water, using non-viable, lyophylized yeast cells grown in methanol under conditions such that they had a high content of methanol oxidase.

METHODS

A strain of methanol-consuming yeast isolated from grapes (3) was used. It was grown continuously in a 14-l fermentor at pH 4.5. To enhance its methanol oxidase content (4), a low dilution rate was used (0.04 Hs^{-1}). Periodically, the outflow collected in ice was centrifuged, washed in 0.05 M tris buffer, pH 8.5, containing 1 mM EDTA, resuspended in deionized water, and lyophylized. The stock was stored in vacuum sealed ampoules at -20°. The enzyme activity could be maintained intact in that way for several months.

The enzyme activity in the liquid phase was monitored by assaying the formaldehyde production at pH 8.5 in the presence of 1% methanol (v/v) in baffled erlenmeyers. For the gas-phase assays, 6.2 g of cells were packed in a plug-flow reactor, which was placed in a thermostated bath. A continuous mixture of pre-dried air containing methanol vapor was passed through the reactor. The methanol dosage was controlled with a syringe pump. The gas outflow was passed through a 70 ml absorber, where the formaldehyde assays were made using the Hantzch reagent (5). Further qualitative determinations of formaldehyde were made with resorcinol (6), Tollens reagent (7), and Schiff reagent (5). It was found that the rate of formaldehyde production always fell to zero as soon as the methanol feed was turned off.

Blank runs were made (a) in the absence of CH_3OH, (b) in the absence of cells, and (c) with the methanol oxidase activity of the cells pre-destroyed by a 30 min incubation with 1 mM $HgCl_2$ at 60°C prior to washing and lyophylization. All blank runs gave undetectable levels of formaldehyde in the absorber. The same batch of cells was used for more than 1000 hr without any detectable loss in activity. It should also be noted that the optimum temperature was above 80°C, which was 40°C higher than the optimum temperature of the enzyme in the liquid phase.

RESULTS AND DISCUSSION

The initial reaction velocity increased both with the concentration of methanol in the air and with temperature. The methanol oxidase did catalyze the conversion of methanol to formaldehyde in the gas phase and in the absence of water. The system showed exceptional stability and good resistance to high temperatures.

REFERENCES

1. KATO, N., TANI, Y., & OGATA, K. *Agr. Biol. Chem.* *38*: 675, 1974.
2. SABAIO, F., ATSUO, T., SUSUMA, K., SHIGEKI, Y., YUTAKA, T., & MASAKO, O. *J. Bacteriol.* *123*:317, 1975.
3. ZUNIGA, V. *M.S. Thesis*, CIEA del IPN; Mexico, 1977.
4. SAHM, H., B. DIX. L. EGGELING & R. ROGGENKAMP. *Abst. 5th Intern. Ferment. Symp.*, Berlin, 1976.
5. WALKER, F. "Formaldehyde," 3rd Edit., Reinhold, New York, 1966, p. 469.
6. VOGEL, A.I. "A Textbook of Practical Organic Chemistry Including Qualitative Organic Analysis," 3rd Edit., Longmans, New York, 1962, p. 325.
7. HELMKAMP, H. "Sinopsis de Quimica Organica," 2nd Edit., McGraw-Hill, New York, 1968, p. 68.

SOME REACTIVE CARRIERS AND IMMOBILIZED ENZYMES

G. Manecke, R. Pohl, J. Schluensen, and H.G. Vogt

Institut fuer Organische Chemie
Freie Universitaet Berlin
Berlin, Federal Republic of Germany

Reactive polymeric carriers with neutral resp. polyionic matrices were synthesized for the immobilization of enzymes by covalent binding. Effects influencing the immobilization and the properties of the immobilized enzymes were investigated.

The copolymerization of acrylic acid with 3- or 4-isothiocyanato styrene and divinyl benzene (DVB) as crosslinking agent in the presence of inert solvents gave carrier of Fig. 1-1 (1) with a macroreticular structure.

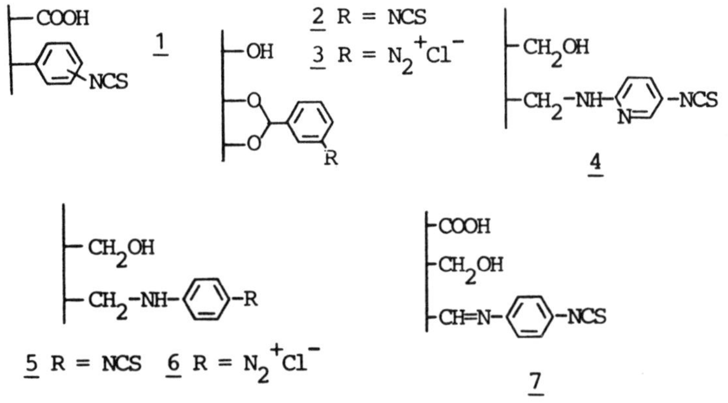

Fig. 1. Various carriers to which enzymes can be coupled by covalent linkage.

TABLE 1

INFLUENCE OF THE COMPOSITION OF CARRIER 1-1 ON THE AMOUNT OF BOUND PAPAIN AND ITS ACTIVITY

Molar Ratio NCS:COOH	Reactive Groups (mmole/g carrier)	Bound Papain* (mg/g carrier)	Retained Activity** (%)	K_M(app) (mole/dm^3)
1:3	2.3	139	36	0.014
1:6	1.5	347	23	0.014
1:9	1.1	294	44	0.014
		native papain		0.040

*Coupled at pH5.
**BAEE as substrate, pH 6.

The influence of the composition of carrier 1-1 on the amount of bound enzyme is shown in Table 1 for papain.

Reactive poly(vinyl alcohol) (PVA) carriers (2) shown in Fig. 1-2 and 1-3 were synthesized using (a) gels of PVA crosslinked with terephthalaldehyde, (b) tubes consisting of vinylacetate-ethylene copolymers coated with PVA, or (c) PVA containing fiber material (SWP- synthetic pulp, Hoechst, Mitsui Zellerbach K.K., Japan). The content of the reactive groups of carrier 1-3 (gel, process (a)) influenced the amount of bound enzyme. In the case of papain an optimum binding ability was achieved at a content of 1.4 mmole reactive groups per g carrier. The immobilization of enzymes on the inner surface of tubes (process (b)) was performed with trypsin. Using DL-BAPA as substrate the tryptic activity was 0.15 U/m tube material. Some results for the immobilization of enzymes on synthetic pulp are given in Table 2.

TABLE 2

IMMOBILIZATION OF ENZYMES ON SYNTHETIC PULP (SWP R830)

Carrier	Immobilized Enzyme Type	Immobilized Enzyme Amount (mg/g pulp)	Retained activity (%)	Substrate
1-2	Trypsin	12	3-4	BAEE, pH 8
1-3	Trypsin	46	7	BAEE, pH 8
1-3	Glucose Oxidase	11	0.4-0.8	Glucose, pH 7

Reactive poly(allyl alcohol) carriers shown in Fig. 1-4, 1-5, 1-6, and 1-7, were synthesized starting from polyacrolein, which was reacted with amines. Allyl alcohol groups were formed by analogous polymer reduction; in the case of carrier 1-7 Cannizzaro conditions were used; and reactive isothiocyanato- resp. diazonium groups were introduced. The amount of bound enzyme strongly depended on the coupling pH. In the case of trypsin carrier 1-5 (NCS-groups) showed an optimum binding ability at pH 8; whereas carrier 1-6 (N_2^+-groups) showed a minimum binding ability at this pH due to the reactivity resp. pH-stability of the reactive component.

The influence of the matrix on the shift of the pH-optimum, an effect caused by the microenvironment and by product accumulation, was investigated for papain immobilized on carriers 1-4, 1-5, and 1-7 using BAEE as substrate. In all cases the pH-optimum was shifted in the alkaline direction. The polyanionic matrix of carrier 1-7 caused the biggest shift.

REFERENCES

1. MANECKE, G. & SCHLUENSEN, J. *Meth. in Enzymology* 44: 107, 1976.
2. MANECKE, G. & VOGT, H.G., *Makromol. Chem.* 177:725, 1976.

IMMOBILIZATION OF PROTEINS ON POLYHYDRAZIDES

Meir Wilchek and Talia Miron

Department of Biophysics
The Weizmann Institute of Science
Rehovot, Israel

Immobilized proteins have been prepared by covalent coupling of proteins to various insoluble carriers. The supports used were polyamino acids, cellulose, polyacrylamide derivatives, beaded glass, maleic anhydride copolymers (EMA), dialdehyde starch, beaded agarose, crosslinked dextrans and nylons (1). Different reactions were used to couple the protein to the carriers, including the azide, carbodiimide, diazonium salt, acyl chloride, cyanogen bromide, isothiocyanate, anhydride and other methods (1).

No attempt was made to compare the efficiency of these polymers as carriers by coupling the proteins to different matrices using uniform coupling methods. In the following report we describe the conversion of the above-mentioned carriers to hydrazides and the subsequent coupling of proteins after activation with glutaraldehyde (2). The final coupling step is shown in the general scheme below, whereas the preparation of the various immobilized hydrazides is described in the text.

$$\text{C}\!\!\wr\!\!-\overset{\overset{\text{O}}{\|}}{\text{C}}-\text{NHNH}_2 + \underset{\text{excess}}{\text{CHO}(\text{CH}_2)_3\text{-CHO}} \rightarrow \text{C}\!\!\wr\!\!-\overset{\overset{\text{O}}{\|}}{\text{C}}-\text{NHN}{=}\text{CH}(\text{CH}_2)_3\text{-CHO} + \text{NH}_2\text{-Pr}$$

$$\xrightarrow{\text{pH } 6-9} \text{C}\{-\overset{\text{O}}{\overset{\|}{\text{C}}}-\text{NHN=CH}(\text{CH}_2)_3-\text{CH=N-Protein}$$

The hydrazides were prepared by coupling hydrazine, adipic dihydrazide, or linear polyacrylhydrazide to the respective carriers. Adipic dihydrazide was coupled in order to check the efficiency of a carrier containing a spacer interposed between the solid matrix and the proteins. Polysaccharides were activated with cyanogen bromide, and either hydrazine or adipic dihydrazide was added. In the case of carboxymethyl cellulose, carboxyl groups were first methylated followed by reaction with the hydrazine derivatives. EMA and dialdehyde starch were reacted directly with hydrazine derivatives. Beaded glass was first converted to an amine, activated with cyanogen bromide, followed by interaction with hydrazine derivatives. Nylon was partially hydrazinolyzed or reacted first with triethyloxonium tetrafluoroborate, followed by the hydrazine derivatives (3). Enzymes, antibodies, lectins and small ligands were then coupled to the glutaraldehyde activated polyhydrazides. High yields of coupling, ranging between 35 mg/g of trypsin for Sephadex adipic hydrazide to 380 mg/g on EMA adipic hydrazides, were obtained. High enzymatic activity was also observed depending on the amount of enzyme bound. The bound proteins exhibited increased stability to heat, 6 M urea, and storage, compared to that of the respective soluble enzyme. A comparative study on the efficiency of the different hydrazide carriers is currently being performed.

REFERENCES

1. ZABORSKY, O. "Immobilized Enzymes," CRC Press, Cleveland, 1973.
2. MIRON, T., CARTER, W.G., & WILCHEK, M. *J. Solid Phase Biochem.* *1*:225, 1976.
3. MORRIS, D.L., CAMPBELL, J., & HORNBY, W.E. *Biochem. J.* *147*:593, 1975.

STUDIES OF INSOLUBLE ORGANIC/INORGANIC COMPOSITES OF GLUCOSE ISOMERASE

A. Rosevear,* C.A. Kent,* A.R. Thomson,* and C. Bucke**

Biochemistry Group*
AERE Harwell, Oxon and
Tate and Lyle Ltd.**
Reading, Berks, U.K.

The full potential of many immobilized enzymes has not been realized because of the poor physical characteristics of the enzyme particles. To overcome this we have used a porous pre-fabricated support matrix as a rigid skeleton within which the much less robust enzyme gel can be deposited. The physical strength of the composite is provided by a macroporous inorganic bead, so that even very weak enzyme gels can be formed into easily handled particles.

The success of the composite depends on enmeshing the enzyme gel within the aerogel support. This is achieved by soaking the enzyme solution so it enters the internal pores of the beads. The enzyme gel is then precipitated with a solution of a polyphenol in acetone; and the gel is stabilized with a miscible cross-linking agent, such as glutaraldehyde. Spent reagent and unbound enzyme can be washed from the composite in a fluidized bed.

The technique has been applied to a variety of enzymes and is particularly successful with partially purified industrial enzymes. In this paper, we describe a joint project to immobilize soluble glucose isomerase for the continuous isomerization of glucose syrups.

The soluble enzyme was extracted from lyophilized cells of *Actinoplanes missouriensis*, either by lysis and precipitation with acetone or by disruption of the cells using a Dynomill. A concentrated solution of the extract was soaked with preformed macroporous beads of celite or titania (0.4 - 0.8 mm). The optimal bound activity was achieved by precipitating the enzyme gel with cold 2% tannic acid in acetone. Glutaraldehyde (0.7%), which was included in the acetone, slowly diffused into the gel to cross-link the enzyme.

The resulting enzyme composite had the same physical robustness as the prefabricated support and could be used in packed beds with minimal pressure drop. The operational stability of the enzyme composite was related both to temperature and to the concentration of cobalt in the substrate solution. The half-life of the system at 60° using 47.3% w/v glucose and 3 mM magnesium as substrate rose from 2 days in the absence of cobalt or 20 days with 0.3 mM cobalt to 46 days with 1.5 mM cobalt.

Using industrial glucose syrup (40% w/w) with 3 mM magnesium and 1 mM cobalt at 65° and pH 7.5, it was possible to produce 42 DI isomerose at an initial flow rate of > 2 effective volumes/hr. The half life of this system was about 18 days if buffered substrate was used; but it fell to 10 days without buffer. The pressure drop across a 42 x 1.75 cm reactor (100 ml) was about 1 lb/square inch during normal operation.

The excellent physical properties of this enzyme composite give it significant advantages over existing systems; although it will be necessary to reduce the need for cobalt in the substrate if the method is to compete successfully with systems using crude homogenates.

Session VIII
APPLICATION OF AFFINITY METHODS AND NEW PURIFICATION PROCEDURES

Chairmen: E. Fischer and M. Wilchek

PRESSURE-DRIVEN AFFINITY SORPTION OF TRYPSIN INHIBITORS

Harry P. Gregor and Paul W. Rauf

Department of Chemical Engineering and Applied Chemistry
Columbia University
New York, NY, USA

The enzyme-coupled ultrafiltration membrane system described by Gregor and Rauf (1) is characterized by a very high loading of bound enzyme, with the specific activity of the native enzyme largely retained. The system has unique advantages. Since the feed streams are pumped rapidly under pressure, the contact times are extremely short, such that diffusion control of the process is virtually eliminated. The nature of the membrane system lends itself to the use of high speed, repetitive processing.

In pressure-driven affinity sorption, the ligand is bound chemically to the inner pore surfaces of high flux ultrafiltration membranes of appropriate pore diameters. First, the feed stream is passed through to saturate the ligands; then a wash displaces excess feed. The desorption solution is pumped through; the decoupling agent is washed out; and the next cycle is begun. The ultrafiltration membrane matrix must be capable of high flux at relatively low (<50 lb/in^2) pressures, must be chemically stable, and also must be capable of fabrication on the kind of support papers and fabrics customarily used in ultrafiltration and reverse osmosis modules. Its flux range should be 10 to 1000 μm/sec/atm. With relatively low molecular weight enzymes, such as trypsin, loadings of about 20% (on a dry weight basis) are possible; with higher molecular weight proteins loadings to 50% are possible. In general, the pore diameters which must be employed are at least 3 times the diameter of the ligand molecules. Usually, the loading of most ligands is in the range of 1 to 5 g/ft^2.

The thickness of ultrafiltration membranes ranges from 5 up to 500 μm, with the concentration of enzymatic ligands in the membrane about 40 mg/ml of pore volume. These pressure-driven systems are extremely fast, with residence times of 0.01 to 1 sec. The rate of ultrafiltration or solution passage is dictated primarily by the stability of the ligands and the substances being separated, because shear denaturation can readily occur at high flow rates in pores of molecular dimensions. Non-fouling sulfonic acid membranes (2) may have a particular advantage if particulate matter is present.

EXPERIMENTAL

25 μm thick cellulosic membrane was prepared (3) that had a flux of 60 μm/sec/atm and a water content of 80%. The film was activated with cyanogen bromide (40 mg/100 ml) at 50 lb/in^2 and then coupled with trypsin (Bovine Pancreas, Worthington, 40 mg/100 ml, pH 7.8 phosphate buffer) at 50 lb/in^2 to achieve a loading of 2.4 mg protein/11.3 cm^2 disc area, of about 22 mg dry weight. The enzymatic activity, with respect to the substrate TAME (Worthington), was 65% of that of the same amount of native enzyme assayed in free solution. Fig. 1 shows a typical result wherein a solution of Kunitz inhibitor (MW 21,500, 1 mg inhibits 1.58 mg trypsin, Worthington) was processed by the affinity membrane. The feed solution contained 15.2 mg of inhibitor in 50 ml of pH 7.8 phosphate buffer. The series of sorption, wash, desorption and wash steps were carried out at a pressure adjusted to give a flux of 5 ml/min. During the sorption cycle 1.20 mg of inhibitor was bound, calculated from the volume passed and feed and effluent absorbance. This corresponded to 79% saturation, based upon the rated binding capacity of the inhibitor. Then, 50 ml of water was passed through the cell, followed by decoupling with 6 M urea at pH 2. A sharp elution in 15 ml resulted, with 1.17 mg of inhibitor desorbed. Thus, the weight of Kunitz inhibitor sorbed and desorbed was about half the weight of coupled trypsin protein. There was a rapid rinsing of excess feed from the membrane and a rapid decoupling process, with minimal wash-out times and solution volumes; this is characteristic of ultrafiltration membranes with a tortuosity approaching unity (2).

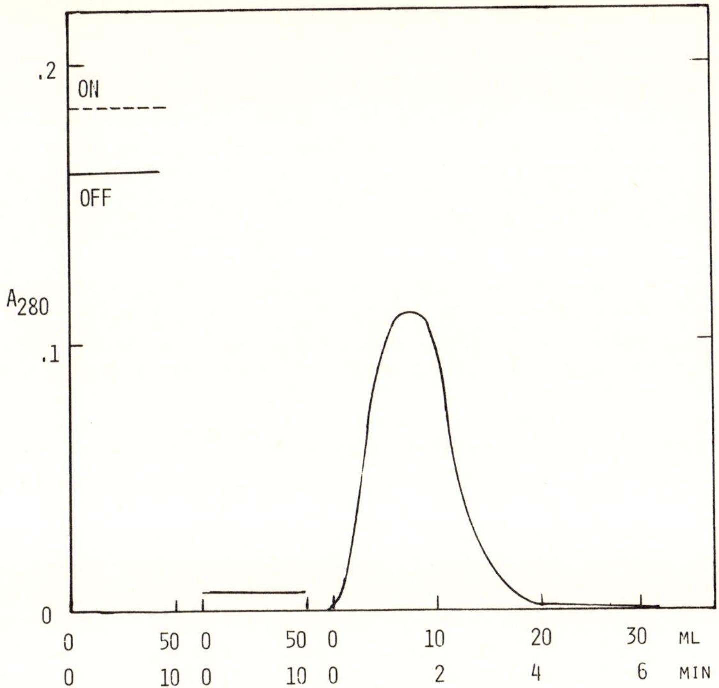

Fig. 1. Sorption-desorption with absorbance measured at 280 nm. Feed solution indicated by ON and effluent by OFF, for 50 ml of solution. A wash solution of 50 ml was then followed by the desorption solution, whose volume is shown. Times correspond to volumes passed.

With a more porous matrix membrane, flux of 300 μm/sec/atm 27 μm thick and 80% water content, 1.5 mg of trypsin was bound and exhibited 81% of the activity of native material in solution with substrate TAME. Sorption-desorption experiments performed with 1 mg of Kunitz inhibitor dissolved in 50 ml buffer, also containing 9 mg of 20,000 MW dextran as an impurity, gave a sorption step which removed 0.85 mg of inhibitor; this corresponded to 90% saturation capacity. With this high flux membrane the displacement of feed was accomplished readily; and elution with 6 M urea, pH 2 proceeded rapidly to completion in 6 min with 30 ml displacing 0.71 mg of inhibitor.

The errors of these assays suggested that substantially all of the inhibitor was desorbed.

With 6.5 mg of Kassel bovine basic pancreatic inhibitor (MW 6513, 1 mg inhibits 5.5 mg trypsin, Worthington) dissolved in 50 ml of buffer, 0.5 mg of inhibitor was sorbed. This corresponded to 183% of its rated inhibitory capacity.

Finally, a series of repetitive experiments were performed using a Millipore cell (Fig 2) consisting of a small reservoir to which coupling, decoupling and washing solutions could be introduced and 5 lb/in^2 nitrogen pressure could be applied, and with a tube extending just above the membrane to keep the volume above the membrane at a minimum so mixing of successive solutions was minimized. For a matrix membrane of high flux (300 μm/sec/atm), 40 μm thick, and 78% water content, 1.4 mg of trypsin was coupled, and showed 51% activity to TAME. The sorption solution contained 6 mg of human α-1-antitrypsin (MW 54,200 to 58,000, made from partially purified pooled human plasma, about 20% pure, 0.65 mg inhibits 0.10 mg trypsin, Worthington) in 50 ml of buffer. Over some 6 complete cycles of 8 min each, the absorbance of the decoupling eluate showed an average deviation of ± 5%; and sorption washes showed negligible amounts of protein present. In an average experiment, 2.8 mg of inhibitor was desorbed; this corresponded to about 31% of rated

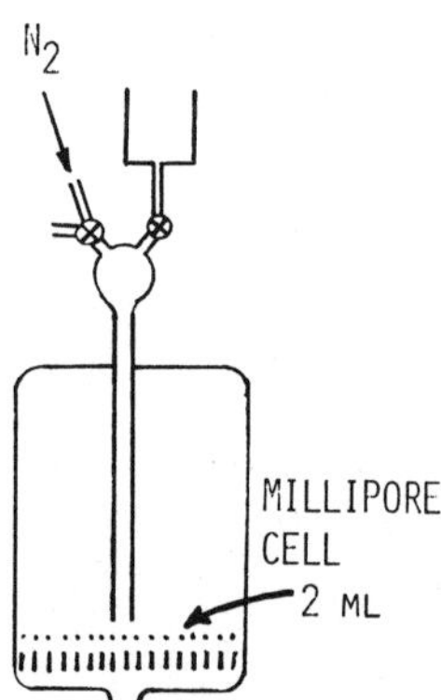

Fig. 2. Standard Millipore cell modified for sequential sorption-desorption experiments. The volume of liquid above the membrane was held to approximately 2 ml.

saturation. The eluted products were shown to be better than 90% pure by subsequent SDS gel electrophoresis analyses.

Some affinity sorption experiments also were performed on the purification of goat antialbumin, with a specific antibody concentration as received of 6.47 mg/ml (Atlantic Antibodies, Westbrook, Me.). This preparation had about 10% of its protein as the antibody. A cellulosic membrane having a flux of 500 μm/sec/atm, 50 μm thick, and 80% water content was activated in the standard manner by cyanogen bromide. Then, pH 7.5 phosphate buffer, containing 6 mg of human serum albumin in 20 ml, was passed through at 70 lb/in^2 for coupling. Following washing, a sorption solution containing 1 ml of the antibody solution, diluted with pH 7.5 phosphate buffer, was passed through. The feed was displaced with water; and then a 2 M sodium chloride decoupling solution was passed, all at 20 lb/in^2. The elution curve is shown in Fig. 3. The antibody concentration was estimated to be >85% of the total protein present.

DISCUSSION

Pressure-driven affinity sorption systems show promise for these two widely different areas of applicability, first for the production of large amounts of biochemicals and second for high speed analysis. For the former use, if one postulates a ligand stable for 6 months (most membrane coupled, enzymatic ligands show much more), with 20% ligand loading in a 30 min cycle with a membrane 100 μm thick, then 1 ft^2 of ultrafiltration membrane would contain about 1 g of enzyme ligand. Then, exclusive of the cost of an enzymatic ligand or the decoupling chemicals, the equipment investment cost for this kind of device should be approximately $2/kg of product made. Conventional spiral-would ultrafiltration equipment is extremely simple and lends itself readily for use on a large scale for this application. High speed, automated analyses similarly are possible with this affinity sorption system, and cycle times of but a few minutes are possible if rapid coupling-decoupling reactions are available.

We no longer employ cellulose as the matrix polymer because of its susceptibility to cellulases and the difficulties in obtaining reproducibility for analytical purposes. Wholly synthetic matrices are superior and

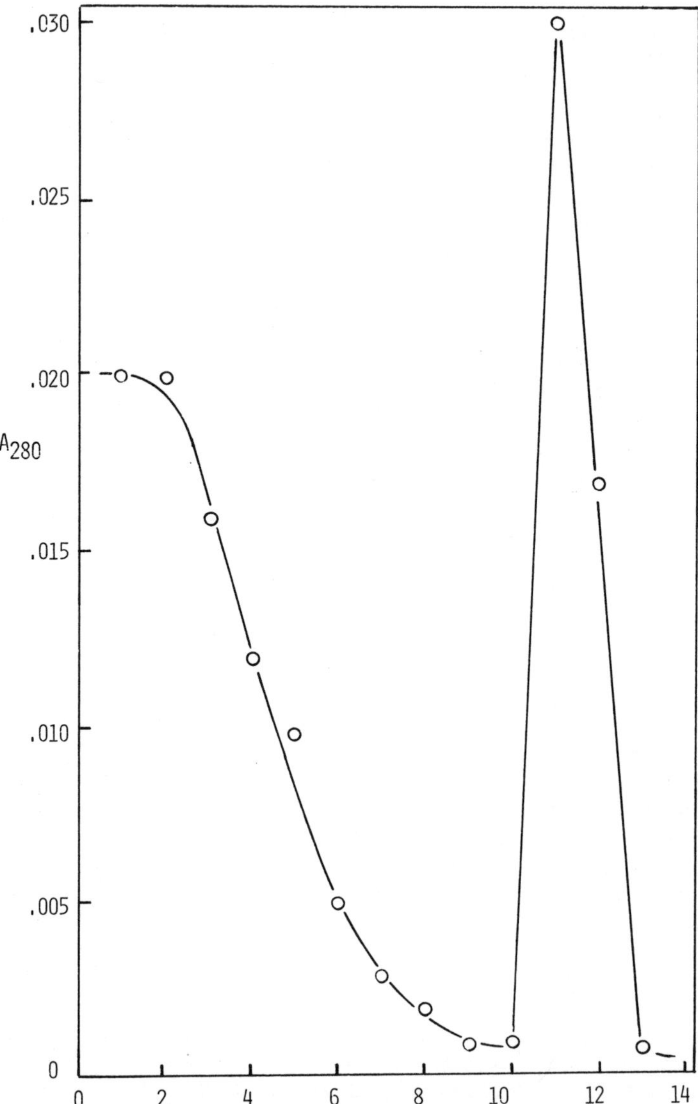

Fig. 3. Absorbance of effluent of wash followed by desorption solution. Numbers on abcissa denote collection tubes, each 20 ml.

allow for the appropriate control of all important parameters.

ACKNOWLEDGMENTS

This study was supported, in part, by a gift from Photovolt Corporation.

REFERENCES

1. GREGOR, H.P. & RAUF, P.W. *Biotechnol. Bioeng.* *17*:445, 1975.
2. GRYTE, C.C. & GREGOR, H.P. *J. Polymer Sci. Physics Ed.* *14*:1839, 1855, 1976.
3. RAUF, P.W., Ph.D. Thesis, Columbia University, 1974.

MEASUREMENT OF ATP AND LIGAND-ATP CONJUGATES BY ENZYMIC CYCLING WITH CO-IMMOBILIZED HEXOKINASE AND PYRUVATE KINASE

K.K. Yeung, R.J. Carrico, J.E. Christner, and R.C. Boguslaski

Ames Research & Development Laboratories
Miles Laboratories, Inc.
Elkhart, Indiana, USA

Various methods have been devised for monitoring specific protein-binding assays with enzymic reactions (1-4). In one method the ligand is coupled covalently to NAD; and this conjugate is measured at low levels by means of enzymic cycling reactions (2). The limit of sensitivity of specific protein-binding assays is determined in part by the sensitivity of the enzymic cycling reactions. In principle, the sensitivity achievable with this method should be dependent on cycling rates and reaction times; however, it is often limited by the level of cofactor contaminating the enzymes which are used at extraordinarily high levels.

Several reports have shown that immobilized enzymes catalyze sequential reactions at rates similar to or greater than the rates obtained with soluble enzymes (5-7). Therefore, we investigated the feasibility of cycling low levels of ATP and N^6[2-(2,4-dinitrophenyl) aminoethyl]adenosine 5'-triphosphate (N^6-DNP-ATP) with co-immobilized enzymes in small columns (3). By co-immobilizing the enzymes, contaminating ATP/ADP could be washed away with buffer and the enzymes could be reused.

Enzymic cycling was conducted by the pyruvate kinase catalyzed reaction of ADP and phosphoenolpyruvate to form ATP plus Pyruvate. The ATP then was reacted with glucose and hexokinase to form ADP plus glucose-6-phosphate. When these reactions had proceeded for a definite period, they were stopped by elution of products and residual substrates from the column. Then glucose-6-phosphate (G6P) was determined by fluorometric measurement of NADPH formed by reaction of G6P, NADP, and G6P dehydrogenase.

The kinetics of cycling of ATP/ADP and N^6-DNP-ATP/ADP were studied with the soluble and co-immobilized enzymes. Then, specific protein-binding reactions involving N^6-DNP-ATP, antibody to the 2,4-dinitrophenyl residue and N-DNP-β-alanine were monitored with the co-immobilized enzymes.

EXPERIMENTAL

Enzyme assays were conducted at 25°C in 0.1 M Tris-HCl, pH 7.8. In all cases one unit of enzyme activity was that which generated 1 μmole of product/min.

Sepharose 4B was activated with CNBr (8). The activated Sepharose was collected on a scintered glass funnel under vacuum; and approximately 1 ml (0.75g) was transferred into 2 ml 0.1 M sodium phosphate buffer, pH 7.5, containing known amounts of hexokinase and pyruvate kinase. The mixture was stirred overnight at 5°C; then the supernatant was decanted for measurements of soluble hexokinase and pyruvate kinase activities. The levels of bound enzymes were calculated by the differences between the amounts added to the activated Sepharose and the amounts in the supernant.

The co-immobilized enzymes were poured into 1 cm diameter polyethylene columns to provide about 0.4 ml beds. A cycling reagent was prepared containing 0.1 M Tris-HCl, pH 7.8, 2 mM glucose, 1 mM phosphoenolpyruvate, 50 mM KCl, 2 mM $MgCl_2$, 7.7 mM sodium azide, and designated levels of ATP (or ligand-ATP conjugates). A 100 μl aliquot of this reagent was applied to the top of a column of the co-immobilized enzymes described above. The enzymic cycling was allowed to proceed for 30 min at 25°C. Then reaction products and residual substrates were eluted with 2.0 ml of 0.1 M Tris-HCl buffer, pH 7.8, containing 7.7 mM NaN_3. The effluent was collected; 100 μl of 1 mM NADP was added; and the fluorescence was measured. Then 2 μl (0.2 unit) glucose-6-phosphate dehydrogenase was added; after 15 min incubation at room temperature the increase in fluorescence due to formation of NADPH was determined.

Control reactions, without added cofactors, were carried through the procedure in order to establish a correction for background fluorescence and cycling due to contiminating ATP and ADP.

Enzymic cycling with soluble hexokinase and pyruvate kinase was conducted at 25°C in 0.1 ml reaction mixtures with the same composition as the cycling reagent described above, except that the enzymes were included.

RESULTS

Kinetic parameters for the reactions of ATP/ADP and N^6-DNP-ATP/ADP with hexokinase and pyruvate kinase are presented in Table 1. The Michealis constants and the reaction velocities for the ligand-cofactor conjugates are lower than those for the corresponding natural cofactors.

TABLE 1

COENZYMIC ACTIVITIES OF LIGAND-ATP/ADP CONJUGATES

Enzyme	Cofactor	Km (M)	Relative Velocity
Hexokinase	ATP	2×10^{-4}	50
	N^6-DNP-ATP	4×10^{-5}	3.6
Pyruvate kinase	ADP	3×10^{-4}	500
	N^6-DNP-ADP	3×10^{-5}	2.3

The ratio of enzymes which gave the maximum velocities were conducted (9) with ATP/ADP and the soluble enzymes. The optimal hexokinase:pyruvate kinase ratio was 1:0.8 (Fig. 1A). The ratio determined with N^6-DNP-ATP was 1:9 (Fig. 1A). The cycling velocity measured with 10 nM ATP/ADP was essentially independent of enzyme concentrations at levels above 9.0 units hexokinase and 7.8 units pyruvate kinase/0.1 ml.

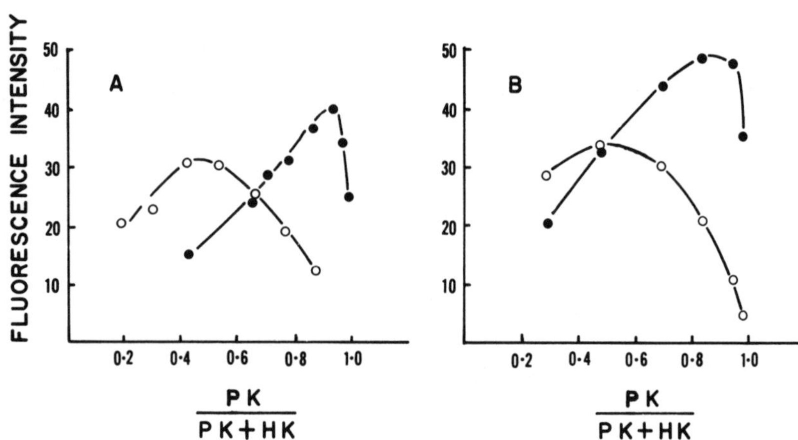

Fig. 1. Determination of optimal pyruvate kinase:hexokinase ratios for cycling of ATP and N^6-DNP-ATP. (A) 10 nM ATP (O) and 83 nM N^6-DNP-ATP (●) were each cycled with various proportions of soluble hexokinase and pyruvate kinase for one hour. (B) Hexokinase and pyruvate kinase were co-immobilized on Sepharose and used to cycle 2.5 nM ATP (O) and 85 nM N^6-DNP-ATP (●) for 0.5 hour. PK designates pyruvate kinase activity and HK is hexokinase activity.

Optimal ratios were also determined with co-immobilized enzymes. For these measurements, hexokinase and pyruvate kinase were co-immobilized in various proportions at levels such that the sum of the two activities was about 790 units/ml packed Sepharose. The optimal ratios determined with ATP and N^6-DNP-ATP were virtually the same as those found for the soluble enzymes (Fig. 1B). Furthermore, the cycling velocity measured with the co-immobilized enzymes and ATP/ADP was constant at enzyme levels above 206 units hexokinase and 251 units pyruvate kinase/ml Sepharose.

Cycling velocities determined wth the co-immobilized enzymes were proportional to the concentration of ATP/ADP over the range studied, 0.3 to 3 nM. Cycling velocities measured similarly with N^6-DNP-ATP/ADP were also proportional to the concentration of this cofactor in the range 10 to 120 nM.

Various levels of antibody to DNP were incubated for 15 min at 25°C with 100 nM N^6-DNP-ATP or 3 nM ATP in 0.3 ml of cycling reagent. Then 100 µl aliquots of these reaction mixtures were applied to columns of the co-immobilized enzymes; and cycling velocities were determined. The rates measured with 100 nM N^6-DNP-ATP decreased as the antibody concentration increased (Fig. 2); it was completely abolished at the highest level of antibody employed. Cycling of ATP was not affected by the antibody, indicating that the DNP ligand was necessary for inactivation of N^6-DNP-ATP.

Competitive binding reactions were carried out with antibody to DNP, N-DNP-β-alanine, and 100 nM N^6-DNP-ATP. The antibody was employed at a level which gave 75% inhibition of the cycling reaction. As the concentration of N-DNP-β-alanine increased, the cycling rate increased (Fig. 3). N-DNP-β-alanine was measured quantitatively at levels as low as 10 nM.

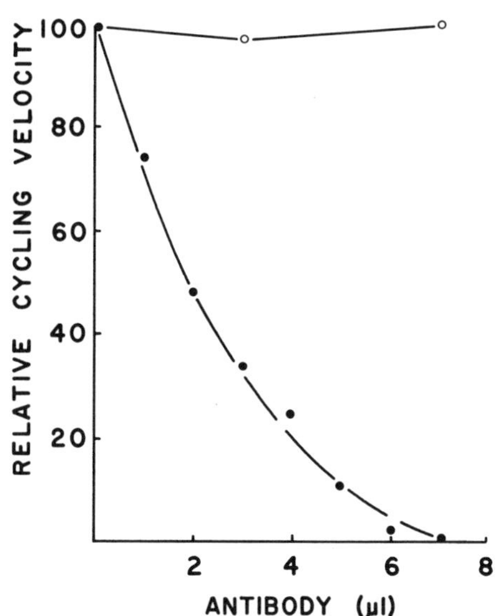

Fig. 2. Cycling velocities measured with N^6-DNP-ATP in the presence of antibody to DNP. N^6-DNP-ATP (●) and ATP (O). The cycling velocities are presented relative to the values measured in the absence of antibody. Each point is the average of two measurements.

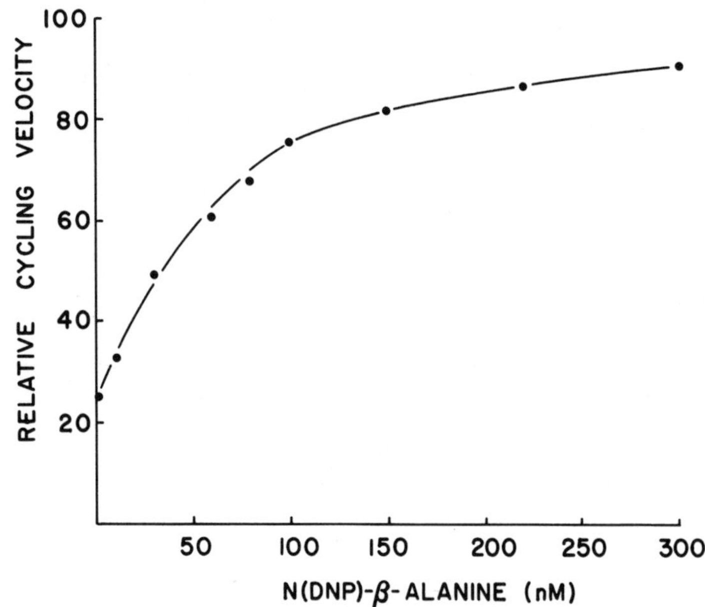

Fig. 3. Cycling velocities as a function of N-DNP-β-alanine levels in competitive binding reactions. Various levels of N-DNP-β-alanine were incubated at 25°C with 100 nM N^6-DNP-ATP and antibody to DNP. After 15 min 100 μl aliquots were assayed for N^6-DNP-ATP activity with co-immobilized enzymes. Each point represents the average of measurements from duplicate columns. The cycling velocities are presented relative to the rates measured in the absence of antibody.

DISCUSSION

It soon became evident that supports which bound high levels of active enzymes provided systems with the highest cycling rates. CNBr activated Sepharose was chosen for the present work because it bound sufficient enzymes to provide cycling rates which were not limited by enzyme levels. The level of contaminating ATP/ADP in the co-immobilized enzymic cycling system was about 1% of that associated with the soluble enzymes.

The maximum cycling rate obtained with the soluble enzymes and ATP/ADP was 5300/hour. This is faster than the rate, 3800 cycles/hour, measured by Breckenridge (10) with a similar system at 25°C. By comparison, the cycling rate obtained with the co-immobilized system was 5900/0.5 hour, which was significantly faster than the rates for the soluble enzymes. This observation was consistent with earlier reports (8-10) that the initial rates of sequential enzymatic reactions were higher with co-immobilized enzymes than with soluble enzymes. This improved efficiency can be used advantageously in enzymic cycling assays where the cycled substrates are typically present at concentrations much lower than the Michaelis constants.

Competitive binding assays monitored with N^6-DNP-ATP and co-immobilized enzymes determined DNP-β-alanine quantitatively at levels as low as 10 nM. This limit of sensitivity was dependent on the cycling rate of N^6-DNP-ATP/ADP, which was 110/0.5 hour. If the cycling rate was proportional to the relative velocity/Km (11), then the data in Table 1 indicate that ATP/ADP should cycle 60 times faster than N^6-DNP-ATP/ADP. The result of this calculation agreed reasonably well with the rate observed with the co-immobilized enzymes and ATP/ADP, 5900 cycles/0.5 hour, which was 54 times greater than the rate for N^6-DNP-ATP/ADP.

Ligand-ATP conjugates are considerably less active than ATP with kinases, while ligand-NAD conjugates retain high activities with dehydrogenases relative to NAD(2). Probably, cycling rates which could be achieved with ligand-NAD conjugates would be higher than those obtained with N^6-DNP-ATP/ADP (11). Unfortunately, ligand-NAD conjugates are more difficult to synthesize than ligand-ATP conjugates.

REFERENCES

1. RUBENSTEIN, K.E., SCHNEIDER, R.S., & ULLMAN, E.F. *Biochem. Biophys. Res. Commun.* *47*:846, 1972.
2. CARRICO, R.J., CHRISTNER, J.E., BOGUSLASKI, R.C., & YEUNG, K.-K. *Anal. Biochem.* *72*:271, 1976.
3. CARRICO, R.J., YEUNG, K.-K., SCHROEDER, H.S., BOGUSLASKI, R.C., BUCKLER, R.T., & CHRISTNER, J.E. *Anal. Biochem.* *76*:95, 1976.
4. ENGVAL, E. & PERLMANN, P. *Immunochem.* *8*:871, 1971.
5. MOSBACH, K. & MATTIASSON, B. *Acta Chem. Scand.* *24*:2093, 1970.

6. SRERE, P.A., MATTIASSON, B., & MOSBACH, K. *Proc. Nat. Acad. Sci. U.S.A.* *70*:2534, 1973.
7. BOUIN, J.C., ATALLAH, M.T., & HULTIN, H.O. *Biochim. Biophys. Acta* *438*:23, 1976.
8. MARCH, S.C., PARIKH, I., & CUATRECASAS, P. *Anal. Biochem.* *60*:149, 1974.
9. CHA, S. & CHA, C.M. *Mol. Pharm.* *1*:178, 1965.
10. BRECKENRIDGE, B. MCL. *Proc. Nat. Acad. Sci. U.S.A.* *52*:1580, 1964.
11. KATO, T., BERGER, S.J., CARTER, J.A., & LOWRY, O.H. *Anal. Biochem.* *53*:86, 1973.

AFFINITY THERAPY

Meir Wilchek

Department of Biophysics
The Weizmann Institute of Science
Rehovot, Israel

The chemicals or drugs used for cancer chemotherapy can reach and destroy cancer cells via the circulatory system. However, the chemical agents that are effective in destroying cancerous cells are usually toxic to normal cells also. Consequently, the slow progress in cancer chemotherapy can be attributed to this effect. This problem may be overcome by designing drugs which will be preferentially concentrated in the target tissue.

Two different approaches in this direction are based on the extraordinary specificity of immunological reactions. One approach concerns the production of antibodies to specific marker antigens on cancer cells. Such antibodies, with the help of complement, should kill the cancerous cells leaving the normal cells untouched. The second approach utilizes antibodies as carriers for cytotoxic drugs in order to destroy tumor cells chemically. The first approach is termed immunotherapy; while the second approach we prefer to call affinity therapy, since potential carriers can be other molecules possessing affinity to certain cells. In order for affinity therapy to succeed, both the carrier and the drug must retain their activity after they are chemically linked. Particularly the activity and specificity of the carrier must remain intact, since the conjugated drug may regain its activity following its release into the target cell.

This paper deals mainly with daunomycin and adriamycin coupled to antibodies against BSA or antibodies specific to several experimental murine tumor cells (Fig. 1) (1). Three different methods of covalent binding tof the drug to antibody were checked. In all cases the amino group of the sugar moiety of the drug was used.

Water-soluble carbodiimide was employed to bind the drug through its amino group to the carboxyl groups of the antibody (2). About 4 residues of drug per antibody were bound. Unfortunately, this method gave conjugates which were almost devoid of drug activity. Dialdehydes, in particular glutaraldehyde, also were used to crosslink the amino groups of the protein with that of the drug. This method gave higher ratios of drug to antibody, about 7-10 moles per mole; but most of the drug and the protein was aggregated due to protein cross linking. In the third method periodate cleavage of the C_3-C_4 bond of the amino sugar residues of the drug was also tested. This cleavage produced aldehyde groups capable of reacting with amino groups on the protein. Subsequent reduction of the resultant Schiff base with sodium borohydride gave a stable conjugate. An average of 2-5 moles of drug per antibody was obtained in different runs. The bond was stable over several weeks of storage at neutral pH; and the drug was not dissociated from the antibody. If the $NaBH_4$ treatment was omitted, the binding was unstable, indicating Schiff base formation and not a cyclic structure. No binding occurred when the drug was mixed with the protein without periodate oxidation.

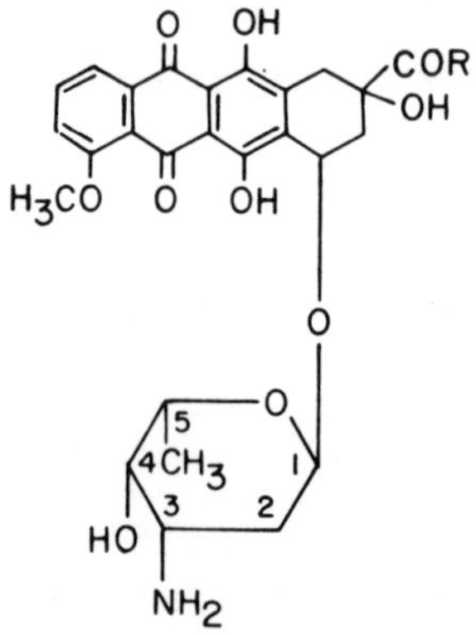

R = CH_2OH, adriamycin

R = CH_3, daunomycin

Fig. 1. Structure of drug.

In order for this approach to succeed, the drug and antibody should keep their potency when bound together. The drug activity of the conjugated daunomycin antibody was tested *in vitro* on tumor and normal cells, by measuring the inhibition of cellular RNA synthesis expressed by [^{3}H]-uridine incorporation into the cells (3). In these studies it was found that the conjugates retained substantial amounts of both drug and antibody activity. The drug activity of these conjugates was less than that of free daunomycin at low concentrations; but conjugated and free drug produced the same inhibition of RNA synthesis at higher concentrations. In addition, the free drug was more active upon short incubation time, whereas prolonged incubation produced equivalent maximal effects with both the free and bound drugs. In a drug-antibody conjugate comprising 2 moles of drug per mole antibody, 64% antibody activity was retained. At 6 moles per mole only 25% activity was left.

To determine whether daunomycin covalently bound to antitumor antibodies exhibited selective cytotoxicity against the tumor cells, daunomycin was conjugated to immunoglobulins containing antibodies against two different mouse lymphoid tumors. The drug was shown to preferentially affect cells against which the antibodies were prepared, as measured by inhibition of uridine incoropration. An example of one such experiment is presented in Table 1.

The periodate oxidation method yielded conjugates which contained about 2-5 moles of drug per mole of antibody. This amount seemed to be too low for *in vivo* studies, since large quantities of antibodies would be required. We therefore developed alternative methods of binding daunomycin to antibodies, which would result in conjugates comprising higher degrees of substitution with concomitant retention of antibody and drug activity.

In order to increase the amount of drug coupled to the antibody, dextran was used as a bridge between the antibody and daunomycin. The dextran was oxidized with sodium periodate to give the corresponding polyaldehyde, and daunomycin was coupled to part of the aldehydes via its amino sugar. This complex was further conjugated to the lysines of the antibody. This conjugate could be used directly or could be stabilized by further reduction with sodium borohydride. Separation between the free drug, daunomycin-dextran, and the daunomycin-dextran antibody was achieved by gel filtration on Sephadex G-100. Dex-

TABLE 1

SPECIFIC CYTOTOXICITY OF DAUNOMYCIN LINKED TO ANTI-PC-5 IMMUNOGLOBULIN

Incubated With	Inhibition of [^{3}H]-Uridine Incorporation (%)		
	PC5	Rat lymphoma	YAC
Daunomycin-anti-PC5	60[(a)]	63[(a)]	20[(b)]
Daunomycin-anti-BSA	7[(b)]	14[(b)]	-
Daunomycin-anti-B-leukemia	16[(b)]	14[(b)]	62[(a)]
Free daunomycin	32[(c)]	53	67

(a),(b),(c): All (a) values different from all (b) values by $p < 0.001$; student t-test; (a) values different from (c) values by $p < 0.001$.

trans of different molecular weights ranging from 10,000 up to 500,000 were used. The drug-dextran-antibody preparations were tested both for their drug and antibody activity. The drug activity was found to decrease to about 20-50% of the free drug; but prolonged incubation time resulted in the same maximal activity as that of the free drug. The antibody activity was also about half that of the original antibody. By this method, however, about 20 moles of drug per mole of antibody were bound.

Another approach was to couple the periodate oxidized drug to linear polyacrylhydrazide (PAH). The polyacrylhydrazide daunomycin was linked to the antibody by glutaraldehyde. This was performed in several steps: glutaraldehyde was linked to the drug-PAH complex, the excess reagent was removed by gel filtration on Sephadex G-25, and the activated complex was then reacted with the antibody. Analysis of the drug and antibody activities showed that both functions were almost entirely lost.

The use of t-RNA as a bridge between the drug and antibody was based on the idea that since the drug could bind to the t-RNA noncovalently to form a complex, the drug would be slowly released at the target cells. Commercial *E. coli* t-RNA was oxidized by $NaIO_4$. The antibodies were coupled to the aldehyde group of the periodate oxidized t-RNA. The conjugate was further stabilized by reduction with sodium borohydride. The drug was then allowed to react with the t-RNA antibody conjugate. The resultant complex was separated from free drug by gel filtration. The complex retained most of its drug activity and about 50% of its antibody activity.

Other methods of coupling, presently under study, include coupling the drug to other polymers followed by covalent attachment to the antibody or binding DNA to the antibody.

The studies described here emphasize that antibodies are capable of locating and binding to the tumor cells against which they were prepared, and of delivering conjugated cytotoxic agents. The main difficulty is to obtain suitable antibody preparations which will be specific exclusively for tumor cells and which will not interact with the normal cell population. Other compounds with affinity to specific cells or organs that may be used as carriers for affinity therapy may include plasma proteins (4), lectins (5), nucleic acids (6), hormones (7), transport systems of nutrients such as amino acids, peptides, sugars and vitamines (8), polysaccharides, and cells or liposomes (9).

ACKNOWLEDGMENTS

This work is being investigated in collaboration with E. Hurwitz, R. Maron, A. Bernstein, R. Arnon and M. Sela and supported by Contract No. CB-64049 from the National Cancer Institute, N.I.H., USA.

REFERENCES

1. HURWITZ, E., LEVY, R., MARON, R., WILCHEK, M., ARNON, R., & SELA, M. *Cancer Res.* *35*:1175, 1975.
2. WILCHEK, M., FRENSDORFF, A., & SELA, M. *Biochemistry* *6*:247, 1967.
3. ROSENBERG, S.A., LEVY, R., SCHECHTER, B., FICKER, S., & TERRY, W. *Transplantation* *13*:541, 1972.

4. WADE, R., WHISSON, M.E., & SZEKERKE, M. *Nature* *215*: 1303, 1961.
5. KITAO, T. & HATTORI, K. *Nature* *265*:81, 1977.
6. TRONET, A., DEPREZ-DE CAMPENEORE, D., & DE DUVE, C. *Nature New Biol.* *239*:110, 1972.
7. VARGA, J.M., ASATO, N., LANDE, S., & LERNER, A.B. *Nature* *267*:56, 1977.
8. AMES, B.N., AMES, G.F., YOUNG, J.D., TSUCHIYA, D., & LECOCQ, J. *Proc. Nat. Acad. Sci. U.S.A.* *70*:456, 1973.
9. GREGORIADIS, G. *Nature* *265*:407, 1977.

PRINCIPLES OF MULTI-ENZYME PURIFICATION BY AFFINITY CHROMATOGRAPHY

C-Y. Lee, A. Leigh-Brown, C. Langley, B. Pegoraro, J. Lopez-Barea and D. Charles

Laboratory of Environmental Mutagenesis
National Institute of Environmental Health Sciences
Research Triangle Park, North Carolina, USA

Following our previous reports on the use of 8-substituted adenine nucleotide derivatives as general ligands for affinity chromatography (1-3), several important new principles have been developed regarding multi-enzyme purifications from crude extracts by general ligand affinity chromatography.

First of all, we found that 8-(6-aminohexyl)-amino-ATP-Sepharose columns exhibited multi-functional affinity properties (2,4). They could be employed to purify as many as ten different NAD^+-dependent dehydrogenases and kinases (2-4) from many tissue homogenates. Dehydrogenases (lactate dehydrogenase, malate dehydrogenase and α-glycerol-phosphate dehydrogenase) could be separated from kinases (aldolase, phosphoglycerate kinase and pyruvate kinases) in mouse kidney homogenates by biospecific elution with 0.1 mM NADH and 5mM ATP, respectively (2). Because of the ion exchange properties of this affinity gel, phosphoglucose isomerase also could be adsorbed and eluted biospecifically with 10 mM glucose-6-phosphate.

When two enzymes in the tissue homogenate exhibited distinct different affinity to the column, the strongly bound enzyme could be separated from the weakly bound enzyme by a saturation-readsorption technique. The leaked low affinity enzyme from the first affinity column could be readsorbed onto a second affinity column. Based on this principle, α-glycerol-phosphate dehydrogenase was purified and effectively separated from chicken muscle extract lactate dehydrogenase by an 8-(6-aminohexyl)-amino-AMP-Sepharose column. Glutathione reductase was purified and separated from glucose-6-phosphate dehydrogenase and malic enzyme in kidney homogenate on an 8-(6-aminohexyl)-amino-2',5'-ADP-Sepharose column.

One could also purify low affinity enzymes from the crude homogenate by an affinity filtration method, which combined the principles of gel filtration and affinity chromatography. The low affinity enzymes in the homogenate could be separated effectively from the strongly bound enzymes and other non-adsorbed proteins in the affinity column by a gel filtration of the homogenate through the affinity column because of the retardation of the low affinity enzymes through weak enzyme-ligand interactions. Three enzymes from the homogenate of *Drosophila*, α-glycerol-phosphate dehydrogenase, malate dehydrogenase, and alcohol dehydrogenase, were successfully purified from an 8-(6-aminohexyl)-amino-AMP-Sepharose column based on this methodology. The purity of the purified enzyme can be greater than 50% by this single-step affinity filtration procedure.

With the introduction of these new principles for general ligand affinity chromatography, the following enzymes from mouse and *Drosophila* were purified to homogeneity by single or two-step affinity column procedures: lactate dehydrogenase-A and lactate dehydrogenase-X (4); phosphoglycerate kinase-A and phosphoglycerate kinase-B (4); cytoplasmic and mitochondrial isocitrate dehydrogenase; malic enzyme; glucose-6-phosphate dehydrogenase, glutathione reductase, and phosphoglucose isomerase from mouse tissues; alcohol dehydrogenase; malate dehydrogenase and α-glycerol-phosphate dehydrogenase from *Drosophila*.

REFERENCES

1. LEE, C-y., LAPPI, D.A., WERMUTH, B., EVERSE, J. & KAPLAN, N.O. *Arch. Biochem. Biophys. 163*:561, 1974.
2. LEE, C-Y. & JOHANSSON, L-J. *Anal. Biochem. 77*:90, 1977.
3. LEE, C-Y., LAZARUS, L.H., KABAKOFF, D.S., LAVER, M.B., RUSSELL, P.J. & KAPLAN, N.O. *Arch Biochem. Biophys. 178*:8, 1977.
4. PEGORARO, B. & LEE, C-Y. *Biochem. Biophys. Acta*, in press.

AFFINITY CHROMATOGRAPHY OF PROTEOLYTIC ENZYMES

V.M. Stepanov

Institute of Genetics and Selection of
Industrial Microorganisms
Moscow, USSR

Two types of specific ligands attached to Sepharose or macroporous silica were used for affinity chromatography of proteinases: binding site and catalytic site directed. Several binding site-directed ligands were used. Peptides, such as 6-aminohexanoyl(6-Ahx)-D-PheOCH$_3$ or 6-Ahx-L-Phe-D-Phe-OCH$_3$ and their analogs, were used for purification of swine, bovine, or horse pepsins, calf chymosin, and aspergillopepsin A (carboxylic proteinase from *Asp. awamori*) (1,2). Pronounced electrostatic effects were detected for the sorbents containing these ligands (K_i was about 1-3 mM) (3). Peptide antibiotics, such as Gramicidin S (cyclopeptide containing hydrophobic amino acids), might be regarded as a substrate analog. Due to the specific conformation and the presence of D-amino acid, it was stable to proteolysis and acted as a weak inhibitor for various proteinases (K_i of the order 1-5 mM). Other bacterial polypeptides might also be used as the ligands. Various pepsins (swine, bovine, horse) and aspergillopepsins A and F (from *Asp. foetidus*) were purified on these sorbents (4,5), as well as subtilisin, metalloproteinase (6), and intracellular serine proteinase from *Bac. subtilis* (7). Gramicidin S-Sepharose also was used to purify carboxylic proteinase from the carnivorous plant *Nepenthes*. Also, mono-2,4-dinitrophenyl-hexamethylenediamine and its analogs might serve as the ligands for hydrophobic chromatography of carboxylic proteinases (8).

Catalytic site-directed ligands were designed to interact specifically with the functional groups of the catalytic site of enzymes. The formation of a covalent bond with the active site is not an obligatory trait of this procedure. The sorbent of this type, for serine proteinases, was prepared by carbodiimide mediated coupling of omega-amino-methyl-phenylboronic acid with CH-Sepharose (9). The resulting CHPB-Sepharose was used for the purification of trypsin, chymotrypsin, and a new type of subtilisin. The procedure enabled subtilisin BPN' to be isolated directly from the culture filtrate of *Bac. subtilis*. The involvement of the active site of the process was proven by abolition of the sorption after enzyme inactivation with phenylmethyl-sulfonylfluorides.

REFERENCES

1. STEPANOV, V.M., LAVRENOVA, G.I. & SLAVINSKAYA, M.M. *Biokhimiya 39*:384, 1974.
2. STEPANOV, V.M., LAVRENOVA, G.I., ADLY, K., GONCHAR, M.V., BALANDINA, G.N., SLAVINSKAYA, M.M., & STRONGIN, A.YA. *Biokhimiya 41*:294, 1976.
3. CHERNAYA, M.M., ADLY, K., LAVRENOVA, G.I. & STEPANOV, V.M. *Biokhimiya 41*:732, 1976.
4. STEPANOV, V.M., LAVRENOVA, G.I., RUDENSKAYA, G.N., GONCHAR, M.V., LOBAREVA, L.S., KOTLOVA, E.K., STRONGIN, A.YA., BARATOVA, L.A. & BELYANOVA, L.P. *Biokhimiya*, *41*:1285, 1976.
5. STEPANOV, V.M., LOBAREVA, L.S., RUDENSKAYA, G.N., BOROVIKOVA, V.P., KOVALEVA, G.G., & LAVRENOVA, G.I. *Bioorganicheskaya Khimiya 3*:831, 1977.
6. VAGANOVA, T.I., LASTOVETSKAYA, L.V., STRONGIN, A.YA., LUBLINSKAYA, L.A., & STEPANOV, V.M. *Biokhimiya 41*:2229, 1976.
7. STEPANOV, V.M., STRONGIN, A.YA., ISOTOVA, L.S., ABRAMOV, Z.T., ERMAKOVA, L.M., LUBLINSKAYA, L.A., BARATOVA, L.A., & BELYANOVA, L.P. *Biochem. Biophys. Res. Commun. 77*:298, 1977.
8. STEPANOV, V.M., LAVRENOVA, G.I., BOROVIKOVA, V.P., & BALANDINA, G.N. *J. Chromatog. 104*:373, 1975.
9. AKPAROV, V.KH. & STEPANOV, V.M. *J. Chromatog.*, in press.

GUANOSINE NUCLEOTIDE ANALOGUES AS GENERAL LIGANDS IN AFFINITY CHROMATOGRAPHY

Peter E. Brodelius and Nathan O. Kaplan

Department of Chemistry
University of California at San Diego
La Jolla, California, USA

The development in recent years of biospecific adsorbents of the general ligand type used in affinity chromatography has proven most valuable. The most widely used ligands for enzyme purification have been analogues of adenosine mono- and dinucleotides (1-5). We now report on the synthesis of a new guanosine nucleotide analogue, 8-(6-aminohexyl)-amino-GMP and its utilization for the purification of IMP dehydrogenase from *Aerobacter aerogenes*.

8-BrGMP was synthesized with a good yield according to the method of Ikehara *et al.* (6). An increased yield of the product was obtained by a modified purification method, which omitted the charcoal treatment by running the crude reaction mixture over an ion-exchange column after adjustment of pH. Elution from the column gave pure 8-BrGMP.

The synthesis of 8-(6-aminohexyl)-amino-GMP from 8-BrGMP was also performed in principal as outlined before (6). However, this substitution reaction was carried out in an aqueous solution instead of methanol in order to increase the solubility of 8-BrCMP. Even though the reaction time was 20 hr at 115°C, no appreciable hydrolysis of the phosphate group could be observed. 8-(2-aminoethyl)-amino-GMP also was synthesized in the same way using ethylene diamine.

8-(2-aminoethyl)-amino-GMP and 8-(6-aminohexyl)-amino-GMP were immobilized on BrCN-activated Sepharose.

These preparations, which contained 110-120 μmoles of nucleotide per g of dry polymer (determined by phosphate analysis), were used for affinity chromatography of IMPDH from *Aerobacter aerogenes*. Model studies with partially purified IMPDH showed that the enzyme was bound to a column containing the immobilized hexyl-analogue of GMP. Subsequent elution of the enzyme activity from the column could be obtained by several nucleotides such as IMP, GMP, NAD^+, or NADH. On the other hand, when the enzyme was applied to a column containing the immobilized ethyl-analogue of GMP, the enzyme activity was quantitatively recovered in the void volume. From this chromatographic behavior we concluded that the enzyme bound biospecifically to the immobilized GMP-analogue with the longer spacer arm and not nonspecifically to the support itself or to ionic groups which might have been introduced on the carrier during the immobilization procedure.

IMPDH was successfully purified from crude preparations by affinity chromatography. In order to investigate various elution systems, samples containing approximately 10 U of IMPDH activity, were applied to a column of GMP-Sepharose. As could be seen from SDS-gel electrophoresis relatively pure IMPDH (at least 95% pure) was obtained by elution with either IMP or NADH. It is interesting to note that the small impurities, which were present in these preparations, were not the same for the IMP and NADH eluted enzyme. Thus, in principle, it would be possible to increase the purity of the enzyme with double affinity chromatography, eluting once with NADH and once with IMP. In fact, after such double affinity chromatography, a higher specific activity for IMPDH was obtained as follows: crude extract 0.083 U/mg; after ammonium sulfate treatment 0.18 U/mg; after elution with pulse of IMP 2.12 U/mg; after elution with gradient of IMP 2.78 U/mg; and after double affinity chromatography 3.12 U/mg. It should be pointed out that a procedure like this may have a general applicability for increasing the purity of enzymes purified by general ligand affinity chromatography.

Finally we would like to point out that the immobilized GMP-analogue may be useful as an affinity adsorbent for other guanosine nucleotide binding proteins.

ACKNOWLEDGMENT

This work was supported by grants from the American Cancer Society (BC-60) and USPHS National Cancer Institute (5 R01 CA11683-08).

REFERENCES

1. OHLSSON, R., BRODELIUS, P., & MOSBACH, K. *FEBS Lett.* *25*:234, 1972.
2. BRODELIUS, P., LARSSON, P.-O, & MOSBACH, K. *Eur. J. Biochem.* *47*:81, 1973.
3. BARRY, S., BRODELIUS, P., & MOSBACH, K. *FEBS Lett.* *70*:261, 1976.
4. LEE, C.-Y., LAPPI, D.A., WERMUTH, B., EVERSE, J., & KAPLAN, N.O. *Arch. Biochem. Biophys.* *163*:561, 1974.
5. LEE, C.-Y. & KAPLAN, N.O. *Arch. Biochem. Biophys.* *168*:665, 1975.
6. IKEHARA, M., TAZAWA, I., & FUKUI, T. *Chem. Pharm. Bull.* *17*:1019, 1969.

PHYSICAL-CHEMICAL FACTORS INFLUENCING THE RESOLUTION IN AFFINITY CHROMATOGRAPHY

V. Kasche,* D. Gabel,** and H. Amneus***

Biology* and Chemistry** Departments
University of Bremen, Bremen, Federal Republic Germany and
Department of Physical Biology***
Uppsala University, Uppsala, Sweden

When affinity chromatography is used for analytical or preparative purposes, such as for the separation of modified enzyme molecules from native enzymes, a biospecific adsorbent with a high resolution is required. The resolution is determined by the following physical-chemical factors: (a) distribution coefficient in the sorption chromatographic procedure, (b) specificity of the binding between adsorbent and enzyme and its modulation by desorbing ligands, and (c) kinetics of the adsorption and desorption. The influence of (a) and (b) on the resolution has been reported previously (1-3). The kinetics of adsorption and desorption are reported here. The number of adsorbent molecules, determined by stationary methods, is generally higher than the corresponding number determined by dynamic methods (loading a chromatographic column). The static capacity may be up to three times higher than the dynamic (operational) capacity. This is caused by adsorbent molecules being bound at locations in the matrix that have large steric hindrance for the diffusion of the enzyme. This does not perturb the association constants determined by equilibrium measurements (2). The kinetics of the adsorption and desorption process are markedly influenced by steric hindrance of diffusion (Fig. 1). Adsorbent sites with high diffusional resistance (low rate constant k) are hardly used as binding sites at flow rates of 0.01 cm/sec. The small fraction that is used gives rise to tailing, due to the decrease in the desorption

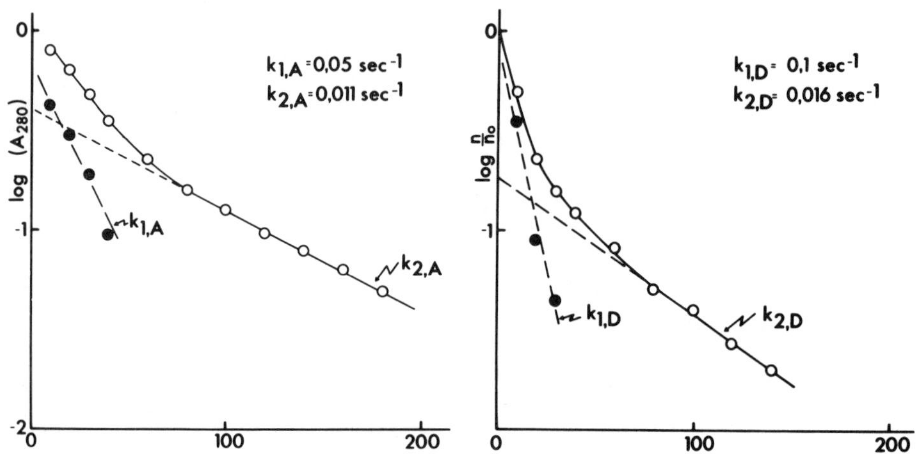

Fig. 1. Kinetics of adsorption and desorption at 25°C. A: To 50 mg α-chymotrpysin (CT) in 100 ml pH 8.0 buffer 10 ml soybean trypsin inhibitor (STI) bound to Sepharose (stationary capacity 50 mg CT) was added at time t=0. D: Solution pH adjusted to ∿3.0 by addition of HAc at time t = 0. The number of bound enzyme molecules, n, was calculated from adsorbance changes. Horizontal axis in sec.

rate. Optimum resolution is obtained when operational capacities are used to determine the load of enzyme in the affinity chromatographic procedure (>10% of capacity).

We have applied these findings to improve the resolution of affinity chromatography used as an analytical tool (4 - 6).

REFERENCES

1. KASCHE, V. *Biochem. Biophys. Res. Comm.* *38*:875, 1970.
2. KASCHE, V. *Studia biophysica* *35*:45, 1973.
3. AMNEUS, H., GABEL, D., & KASCHE, V. *J. Chromatogr.* *120*:391, 1976.
4. KASCHE, V., AMNEUS, H., GABEL, D., & NASLUND, L. *Biochim. Biophys. Acta* *490*:1, 1977.
5. KASCHE, V. *Radioanal. Letters* *3*:51, 1970.
6. KASCHE, V. *Arch. Biochem. Biophys.* *173*:269, 1976.

USE OF AFFINITY CHROMATOGRAPHY FOR DETERMINATION OF DISSOCIATION CONSTANTS OF COMPLEXES OF TRYPSIN AND CHYMOTRYPSIN WITH THEIR FREE AND IMMOBILIZED INHIBITORS

Jaroslava Turkova

Institute of Organic Chemistry and Biochemistry
Czechoslovak Academy of Science
Prague, Czechoslovakia

Since affinity chromatography is based in principle on the formation of a specific complex, e.g. between an enzyme and its competitive inhibitor, this method has become a tool of choice also for studies on specific interactions. For example, an enzyme can be adsorbed on a column of the immobilized inhibitor and eluted by solutions of the free inhibitor of different concentrations. If the elution volumes of the enzyme are plotted versus the inhibitor concentrations used for elution, then the dissociation constants of the complex formed by the enzyme and the soluble inhibitor (K_I) and the enzyme and the immobilized inhibitor (K_L) can be determined. It is necessary that the same inhibitor be used both immobilized on a carrier and soluble as eluant; then the effect of the carrier and the attachment on the specific interaction between the enzyme and the immobilized inhibitor can be estimated from comparison of the K_I and K_L values. The determination of both constants can be done using zonal analysis (1) or frontal analysis (2) type experiments. Making use of our previous experience with the isolation of chymotrypsin and trypsin on specific adsorbents prepared by attachment of natural high molecular weight inhibitors and low molecular weight synthetic inhibitors to Spheron (3-5), we decided to study the effect of carrier and attachment on the specific interaction of the enzyme with its immobilized inhibitor by these two methods.

The K_I and K_L values of trypsin complexes determined by zonal and frontal arrangement of affinity chromatography were in good agreement with each other and with the K_I values obtained by kinetic measurements (6) or recorded in the literature. The plot of $1/(V_i-V_o)$ versus $1/K_I$ (determined by zonal analysis) or of V_i versus $K_I(V-V_i)$ (determined by frontal analysis) for various inhibitors of identical concentration was linear; this provided evidence that this quantitative approach could be used over the entire range of dissociation constants determined.

The comparison of K_I (1.9×10^{-5}M) and K_L (1.6×10^{-6}M), characterizing the complex of trypsin with p-aminobenzamidine, pointed to a certain degree of nonspecific adsorption in the binding of trypsin to p-aminobenzamidine-NH_2Spheron. The immobilization of the inhibitor was without effect on the formation of the complex between chymotrypsin and immobilized antilysine, a high molecular weight inhibitor ($K_I = 8.9 \times 10^{-6}$M; $K_L = 9.0 \times 10^{-6}$M). The difference in the K_L values characterizing the complex of chymotrypsin with a low molecular weight inhibitor, Z-Gly-Phe attached to NH_2-Spheron obviously was caused by differences in procedures used for the preparation of NH_2-Spheron. The adsorbent, which was prepared by attachment of hexamethylenediamine to BrCN-activated Spheron, gave a higher K_L value (4.5×10^{-6}M) than the carrier to which hexamethylenediamine was coupled through p-nitrophenol ester groups on the gel ($K_L = 0.2 \times 10^{-6}$M). This was because of the carboxyl groups newly formed during the bonding. The K_I and K_L values for chymotrypsin complexes could not be determined by frontal analysis, because of strong nonspecific adsorption in the presence of an excess of chymotrypsin.

REFERENCES

1. DUNN, B.M. & CHAIKEN, I.M. *Biochemistry* *14*:2343, 1975.
2. KASAI, K. & ISHII, S. *J. Biochem. (Tokyo)* *77*:261, 1975.
3. TURKOVA, J., HUBALKOVA, O., KRIVAKOVA, M. & COUPEK, J. *Biochim. Biophys. Acta* *322*:1, 1973.
4. TURKOVA, J., BLAHA, K., VALENTOVA, O., COUPEK, J., & SEIFERTOVA, A. *Biochim. Biophys. Acta* *427*:586, 1976.
5. TURKOVA, J. & SEIFERTOVA, A. *J. Chromatog.*, in press.
6. DIXON, M. *Biochem. J.* *55*:170, 1953.

PURIFICATION OF PYRUVATE DEHYDROGENASE BY AFFINITY CHROMATOGRAPHY

Jaap Visser and Marijke Strating

Department of Genetics
Agricultural University
Wageningen, The Netherlands

The pyruvate dehydrogenase complex of *Escherichia coli* consists of three enzymes: pyruvate decarboxylase (E_1) (EC 1.2.4.1), lipoate acetyltransferase (E_2) (EC 2.3. 1.12), and lipoamide dehydrogenase (E_3) (EC 1.6.4.3). Various purification procedures exist; but the one commonly used (1) involves protamine sulfate fractionation and isoelectric precipitations at pH 5.7 to remove the α-oxoglutarate dehydrogenase complex and at pH 5.0 to precipitate the pyruvate dehydrogenase complex. A further improvement was obtained by introducing several intermediate isoelectric precipitation steps and calcium phosphate gel-cellulose chromatography (2). Another method (3) is more rapid and uses chromatography on Biogel A-50-M followed by calcium phosphate gel-cellulose chromatography. We recently developed a method based on Biogel chromatography followed by affinity chromatography on thiamin pyrophosphate, but due to instability of the matrix this method is of limited value. In the study being reported a rapid and mild purification procedure is described using ethanolamine as a ligand.

E. coli Kl-1 LR 8-13, a constitutive pyruvate dehydrogenase mutant kindly provided by U. Henning, was used as the enzyme source. Sepharose 2B was activated (5) and SDS-polyacrylamide gel electrophoresis was performed (6) according to published methods.

Hydrophobic affinity chromatography of *E. coli* pyruvate dehydrogenase using alkyl- and ω-alkyl Sepharose matrices could not be realized due to very strong interactions, even with C_2 alkyl segments. Bisoxirane and CNBr activated Sepharose 2B were then reacted with ethanolamine. The pyruvate dehydrogenase complex remained completely bound to this affinity adsorbent and could be eluted by an increase in ionic strength. This became the basis of the following purification procedure, which was mild and rapid.

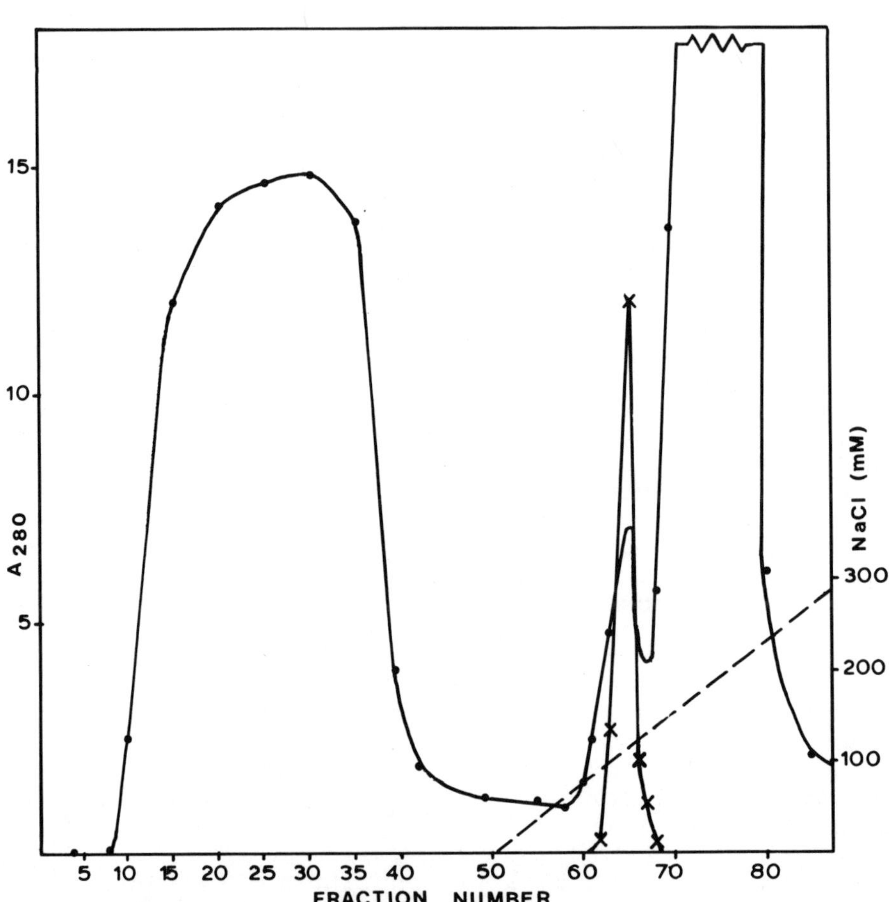

Fig. 1. Affinity chromatography of pyruvate dehydrogenase complex on ethanol-Sepharose 2B. Fractions of 15 ml were collected. A280 (o), enzyme activity (x), NaCl gradient concentration-(dashed line).

Cells (60-80 g) were grown (3) and broken with an LKB X-press. 250 ml of extraction buffer of the following composition was added: 0.1 M potassium phosphate pH 7.0, 3 mM magnesium chloride, 0.1 mM EDTA, 2 mM TPP, 5 mM β-mercaptoethanol, 50 μM PMSF, 5 μg/ml DNAase, and 10 μg/ml RNAase. The crude bacterial extract was dialyzed against a 50 mM phosphate buffer containing all components except DNAase, RNAase, and TPP and then applied to an affinity column (2.5 x 35 cm). The results are shown in Fig. 1. The adsorbed enzyme could be eluted by a salt gradient. Pyruvate dehydrogenase activity appeared before the large bulk of protein, which was adsorbed and collected by ammonium sulfate precipitation (60%). We did not observe a loss of enzyme activity. From 65-70 g of cells approximately 400 mg protein was obtained in the active fractions. The electrophoretic pattern of this pooled material demonstrated the large enrichment obtained. On the basis of activity the purification was 80-fold. Molecular sieving on Biogel A-50-M clearly showed a separation of intact 3-component pyruvate dehydrogenase, lower molecular weight contaminants, and transacetylase breakdown products. The same principle was used in the purification of other prokaryotes, as *Bacillus* species, *Pseudomonas fluorescens*, and *Azotobacter vinelandii*. Moreover in collaboration with J.G. Guest of the University of Sheffield, we were able to isolate inactive pyruvate dehydrogenase complexes from several mutants which map in the E_3 gene.

ACKNOWLEDGMENTS

The authors acknowledge the participation of W. van Dongen and D. van Lith. Discussions with C.J. Bos and H.W.J. van den Broek also are appreciated.

REFERENCES

1. REED, L.J. & MUKHERJEE, B.B. *Meth. in Enzymology 13*: 55, 1969.
2. SPECKHARD, D.C. & FREY, P.A., *Biochem. Biophys. Res. Commun. 62*:614, 1975.
3. VOGEL, O., BEIKIRCH, H., MULLER, H., & HENNING, U. *Eur. J. Biochem. 20*:169, 1972.
4. VISSER, J., STRATING, M., & VAN DONGEN, W., submitted.
5. SANDBERG, L. & PORATH, J. *J. Chromatog. 90*:87, 1974.
6. LAEMMLI, K.K. *Nature 227*:681, 1970.

STUDIES ON THE PURIFICATION OF *Aspergillus niger* ENDO-POLYGALACTURONASE BY AGAROSE GEL CHROMATOGRAPHY

J.F. Thibault and C. Mercier

Laboratoire De Biochimie Des Aliments
I.N.R.A.
Nantes, France

Endopolygalacturonase (endo PG) (poly (1,4-D-galacturonide) glycanohydrolase, E.C. 3.2.1.15), a pectolytic enzyme hydrolyzing randomly glycosidic linkages of pectic substances, can be purified by the usual methods. Nevertheless some original purification procedures have been developed using affinity chromatography on cross-linked pectic acid (1) or chromatography on agarose gel (2). On agarose it was observed that purified endo PG was delayed. In view of the resemblance between the anhydrogalactose residues of agarose and the anhydrogalacturonic ones of endo PG substrates, these workers suggested that biospecific interactions between agarose and endo PG were occurring. We used this suggestion to develop a method to purify endo PG from commercial Pectinase produced by *Aspergillus niger* and obtained from the Rapidase Company.

We found that endo PG from this crude preparation could be not only delayed but bound to the agarose gel (Sepharose 6B). This binding depended essentially upon three factors: dialysis treatment of the preparation, molarity, and pH of the elution buffer. Dialysis, which eliminated 40 to 50% of the 280 nm absorbing substances and led to 1.4 to 2 fold purification, seemed to be a necessary step before the chromatography for maximum binding of enzymic activity (90% instead of 46% with non-dialysed preparation). The molarity and pH of the elution acetate buffer strongly influenced the binding of endo PG to the gel. It was shown by viscosity and reducing group measurements that 90% of the endo PG activity was bound to the gel when the pH was between 4.30 and 4.42 and when the molarity was not greater than 20 mM. Beyond pH 4.42 the percentage decreased rapidly to zero at pH 5.20; in this latter case endo PG was not bound but simply delayed. However, about 7 to 12% of the bound proteins were not homogeneous; the endo PG was still contaminated by inert proteins. Therefore, some differential elution techniques were carried out. For instance, application of a solution of sodium polygalacturonate did not elute the enzyme. Poor results and poor separations were obtained with continuous or stepwise pH gradients. In contrast, the technique of elution with an NaCl gradient between 0 and 0.15 M gave satisfactory results. The 5 to 6% proteins still bound to the column were removed by elution with 20 mM acetate buffer containing 2 M NaCl. In this way endo PG was purified 62-fold in terms of viscosity units. This corresponded to 52% recovery. The low recovery could be explained by instability of the enzyme.

This purification method is fast, with good results; and an acceptable degree of purification is obtained in one operation. In addition it is simple, reproducible, and does not need a special gel other than Sepharose 6B.

The hypothesis of English *et al.* (2) of biospecific interactions between agarose and endo PG is strongly weakened by the fact that the enzyme was not eluted by its substrate. In contrast, the study of the influence of pH and molarity of the elution buffer upon the binding of endo PG showed ionic interactions between agarose and endo PG. These could be explained by the fact that commercial agarose is contaminated by agaropectin, giving to Sepharose 6B slight properties of a cation exchanger (3).

REFERENCES

1. REXOVA-BENKOVA, L. & TIBENSKY, V. *Biochim. Biophys. Acta 268*:187, 1972.
2. ENGLISH, P.D., MAGLOTHIN, A. & KEEGSTRA, K. *Plant Physiol. 49*:293, 1972.
3. CRONE, H.D. *J. Chromatog. 92*:127, 1974.

AFFINITY CHROMATOGRAPHY: RECENT ADVANCES

Peter D. G. Dean

Biochemistry Department
The University of Liverpool
Liverpool, U.K.

The chromatography of enzymes on group-specific affinity adsorbents, such as nucleotides, has been used to optimize the design of biospecific adsorbents (1). Despite the widespread application of these columns to the purification of many nucleotide-dependent enzymes, one drawback is apparent. The operational capacities of these columns are frequently low; 0.1 to 0.5% of the total ligand is common. It has been suggested that this phenomenon relates to the association constant (2). Some other small ligands show increased operational capacities when compared with nucleotides, for example, immobilized Cibacron Blue (3). Cibacron Blue-Sepharose in our hands is more specific for NAD^+-dependent enzymes than is suggested by the literature. On the other hand we have found that another triazine dye, Procion Red HE3B, also shows high operational capacities and appears to be specific for $NADP^+$-dependent dehydrogenases.

The use of agarose gels having decreasing pore sizes by interdigitating polyacrylamides (Ultrogels) has potential interest in widening the scope of applications of nucleotide matrices (3,4). However, using CNBr activation, such gels show anomalous behaviour and in some cases, like albumin binding to Cibacron Blue-Ultrogel, show quite different properties to their agarose counterparts. Both Ultrogel and agarose possess many diffusional limitations, which can be overcome using pellicular supports, such as Matrex (Amicon) beads. The capacities of $NADP^+$-Matrex for three enzymes changes by 30%

on increasing the flow rate from 1 ml/hr to 30 ml/hr. The capacities of $NADP^{+}$-Sepharose for the same enzymes fall below 50% at 15 ml/hr and go as low as 10% at 25 ml/hr.

Takeo and Nakamura (5) have described the use of electrophoresis to determine kinetic parameters in affinity chromatography. We have applied this technique to the desorption of immuno-adsorbents and other affinity interactions. Using low voltages and currents (typically 100V at 2 mA/cm^2) and a suitable cell, we have desorbed competent anti-steroid antibodies from immobilized steroid columns in high yields (6). Ferritin can also be recovered in this fashion from anti-ferritin-Sepharose-4B (7). Many other types of affinity matrices can be desorbed using this technique.

REFERENCES

1. HARVEY, M.J., CRAVEN, D.B., LOWE, C.R., & DEAN, P.D.G. *Meth. in Enzymology 34*:242, 1974.
2. DEAN, P.D.G. & WATSON, D.H. in "Proceedings of Symposium on Affinity Chromatography, Vienna" Pergamon Press, in press.
3. ANGAL, S. & DEAN, P.D.G. *Biochem. J. 167*:301, 1977.
4. DOLEY, S.G., HARVEY, M.J., & DEAN, P.D.G. *FEBS Lett. 65*:87, 1976.
5. TAKEO, K. & NAKAMURA, S. *Arch. Biochem. Biophys. 153*:1, 1972.
6. MORGAN, M., BROWN, P.J., LEYLAND, M., & DEAN, P.D.G. *FEBS LETT.*, in press.
7. BROWN, P.J., LEYLAND, M.J., KEENAN, J.P., & DEAN, P.D.G. *FEBS Lett. 83*:26, 1977.

APPENDIX

GUIDELINES FOR THE CHARACTERIZATION OF IMMOBILIZED ENZYMES: A PROPOSAL

Editor's Note: An *ad hoc committee* from the Federal Republic of Germany has proposed a detailed list of parameters, characteristics, and methods that should be specified or utilized in the characterization of immobilized enzymes, cells, and supports. Their proposal is included in this volume to stimulate discussion on this important topic. The inclusion in no way connotes official acceptance of the proposal by the conference participants or the conference organizers.

SUMMARY OF PROPOSAL

The aim of the proposal is (a) to define a rather complete set of data for characterizing an immobilized biocatalyst with respect to its potential in analytical or technical applications and (b) to develop standardized methods for determination of these data.

The characterization of an immobilized biocatalyst must take into account:

(a): the chemical basis and chemical behavior of the matrix (monomers, functional groups, method of synthesis, wet volume, and chemical stability).

(b): the mechanical characteristics of the matrix (shape and size of particles or membrane, mechanical stability, flow resistance, application range in fixed and fluidized beds).

(c): characterization and economics of immobilization procedure (binding method, content of total protein, yield).

TABLE 1

LIST OF CHARACTERISTICS

Item	Specify and Units
Method of Synthesis	Recipe
Chemical Composition	% by analysis or from reaction mix
Functional Groups	$-OH, -NH_3^+, -COO^-$, etc. μmole/mg
Water Regain at Specified pH	ml/g
Pore Volume	%
Stability	at pH and °C
Wet Mean Particle Diameter (Particle Diameter S.D.)	μm
Flow Resistance in Fixed Bed Applications	pressure drop, height, flow rate, viscosity
Sedimentation Velocity	cm/sec
Abrassiveness in Stirred Tanks	--
Native Enzyme E.C. Number	--
Native Enzyme Activity	kat, kat/mg
Method of Immobilization	concs., pH, I, temp., time
Amount of Bound Protein	mg/g
Yield of Bound Protein	%
Yield of Active Protein	%
Maximal Activity	apparent V_{max}, U/g matrix
Activation Energy	J/mole
Effectiveness (at spec. react. cond. for appl.)	%
Stability: Affect of °C, pH, medium	half life, loss in activity, %/day
Storage Stability	same as above
Operational Stability Continuous and Intermittant Use	same as above, number of times used

(d): the catalyst potential for practical use as immobilized enzymes or cells (maximal activity, efficiency under specified conditions).

(e): stability of the catalyst (under storage and specified operational conditions).

These general considerations are listed in more detail in Table 1.

A more detailed description of the proposed guidelines can be obtained from K. Buchholz.

CONFERENCE PARTICIPANTS

AND ATTENDEES

ANDRESEN, OTTO
Novo Industri A/S, Novo Alle 1, DK-2800 Bagsvaerd, Denmark.

ANTRIM, RICHARD L.
Clinton Corn Processing Co., Clinton, IA 52732, USA.

ARMBURSTER, FRED C.
CPC International Inc., Moffett Technical Center, Box 345, Argo, IL 60501, USA.

ASENJO, JUAN A.
Dept. Chemical and Biochemical Engineering, University College London, Torrington Place London WC 1, England, UK

BALLESTEROS, ANTONIO*
Instituto Catalisis y Petroleoquimica, CSIC, Serrano 119, Madrid 6, Spain.

BARICOS, WILLIAM H.*+
Dept. Biochemistry, Tulane University Medical School, 1430 Tulane Ave., New Orleans, LA 70112, USA

BARTHELEMY, P.
Fermentation Research, Roussel-Uclaf, 111 Rue de Noisy, F-93230 Romainville, France.

BARTOLI, FRANCESCO +
Snamprogetti SpA., Via E. Ramarini 32, I-00015 Monterotondo (Roma), Italy.

BERGKVIST, ROLF
Swedish Sugar Corp., Box 6, S-23200 Arlov, Sweden.

Footnote: * Invited Papers, Session Chairman, Member Executive Committee.
+ Contributed Paper.
= Workshop Chairman.

BEREZIN, ILIA V.*
Dept. Chemistry, Moscow State University, Lenin Hills, Moscow 117234, USSR.

BILLERBECK, MIGUEL A. DE
Velez Sarsfield 299, 5000 Cordoba, Argentina.

BONSE, DIRK
Miles Kali-Chemie AG, Hans Bockler Allee 20, D-3000 Hannover, F.R. Germany.

BORCHERT, AXEL
DECHEMA Institut, Theodor Heuss Allee 25, D-6000 Frankfurt (Main), F.R. Germany.

BRANNER-JORGENSEN, SVEN +
Novo Industri A/S, Novo Allee 1, DK-2880 Bagsvaerd, Denmark.

BRODELIUS, PETER E. +
Dept. Chemistry 0-058, University of California at San Diego, La Jolla, CA 92093, USA.

BROUN, GEORGES*+
Dept. Genie Biologique, Universite de Technologie Compiegne, 25 Rue Eugene-Jacquet, F-60206 Compiegne, France.

BRUMMER, WOLFGANG
Biochem. Forschung, E. Merck, Postfach 4119, D-6100 Darmstadt, F.R. Germany.

BRUNNER, GORIG*
Medizinische Hochschule Hannover, Oststadtkrankenhaus, Podbielskistrasse 380, D-3000 Hannover, F.R. Germany.

BUCHHOLZ, KLAUS*+=
DECHEMA, Theodor Heuss Allee 25, D-6000 Frankfurt (Main), F.R. Germany.

BUCKMANN, ANDREAS +
Gesellschaft fur Biotechnologische Forschung, Mascheroder Weg 1, D-3300 Braunschweig Stockheim, F.R. Germany.

CAMPBELL, JOHN
Dow Diagnostics R & D, Dow Chemical USA, PO Box 68511, Indianapolis, IN 46268, USA.

CARASIK, W.
Product Applications, Novo Laboratories, 59 Danbury Road, Wilton, CT 06897, USA.

CARRICO, ROBERT J.*
Ames R & D Labs., Miles Laboratories Inc., Elkhart, IN 46514, USA.

CHAMBERS, ROBERT P. +
Chemical Engineering Dept., Auburn University, Auburn, AL 36830, USA.

CHANG, THOMAS M. S.*
Dept. Physiology, McGill University, McIntyre Medical Science Building, Montreal, Quebec, Canada H3G 1Y6.

CHIBATA, ICHIRO*+=
Research Lab. Applied Biochemistry, Tanabe Seiyaku Co. Ltd., 16-89 Kashima-3-chome, Yodogawa-ku, Osaka, 532 Japan.

CHUN, MOONJIN
Dept. Agric. Chem., Korea University, Sungkuk-ku, Seoul 132, Rep. Korea.

COLE, SANDFORD S.*
Engineering Foundation Conferences, 345 East 47th St., New York, NY 10017, USA.

COLTON, CLARK K.*+
Dept. Chemical Engineering, Massachusetts Institute of Technology, Cambridge, MA 02139, USA.

COPPENS, GUIILAUME
Solvay & Cie, 310 Rue de Ransbeck, B-1120 Bruxelles, Belgium.

COUGHLIN, ROBERT W.*+
Dept. Chemical Engineering, University of Connecticut, Storrs, CT 06260, USA.

COULET, P. +
Lab. Biochimie Dynamique, 43 Blvd. du 11 Novembre 1918, F-69621 Villeurbanne, France.

DAHLMANS, J.J.
DSM Central Labor, Postbox 18, Geleen, Netherlands.

DAVIS, FRANK F.*
Dept. Biochemistry, Rutgers University, New Brunswick, NJ 08903, USA.

DEAN, PETER D.G. +=
Biochemistry Dept., University of Liverpool, Liverpool L 69 3BX, England, UK.

DELLWEG, HANSWERNER
Institut fur Garungsgewerbe und Biotechnologie, Seestrasse 13, D-1000 Berlin 65, F.R. Germany.

DIETRICH, H.H.
Institut fur Holzchemie, Leuschnerstrasse 91, D-2050 Hamburg 80, F.R. Germany.

DINELLI, DINO*
Via Maralago 3, I-00041 Albano, Roma, Italy.

DOHAN, LUC
Dept. Genie Chimique, Corning Centre Recherche, BP 3, F-77210 Avon, France.

DOLABDKIAN, BARKEW +
Institut Chemie Weihenstephan, Technical Universitat Munchen, D-8050 Freising-Weihenstephan, F.R. Germany.

DRIOLI, ENRICO
Instituto di Principi Ing. Chim., University of Napoli, Piazzale Tecchio, I-80125 Napoli, Italy.

DUNNILL, PETER +=
Dept. Chemical and Biochemical Engineering, University College London, Torrington Place, London WC1E 7JE, England, UK.

DUPPEL, WILFRIED
Deutsche Pharmacia, Munzinger Strasse 9, D-7800 Freiburg/Brsg., F.R. Germany

DUARTE, J.M. CARDOSO
Dept. Chemical and Biochemical Engineering, University College London, Torrington Place, London WC1, England, UK.

EHRENTHAL, EINHART
Institut Organische Chemie, Freie Universitat Berlin, Thielallee 63-67, D-1000 Berlin 33, F.R. Germany.

ENGASSER, J.M. +
Lab. Sciences Genie Chimique, CNRS, 1 Rue Grandville, F-54042 Nancy, France.

EVANS, TIMOTHY W.
Upjohn Co., Kalamazoo, MI 49001, USA.

EVERSE, JOHANNES
Dept. Biochemistry, School of Medicine, Texas Tech University, Lubbock, TX 79409, USA.

FASOLD, H.*
Institut fur Biochemie der Universitat, Sandhofstrasse 1, D-6000 Frankfurt, F.R. Germany.

FIECHTER, A.
Mikrobiologisches Institut, Eidg. Techn. Hochschule, Weinbergstrasse 38, CH-8006 Zurich, Switzerland.

FILIPPUSSON, H.
Lab. Biochemie, Faculty of Medicine, University of Iceland, Armuli 30, Reykjavik, Iceland.

FINK, DAVID J. +
Battelle Columbus Labs., 505 King Ave., Columbus, OH 43201, USA.

FISCHER, E.A.*
Forschung Diagnostica, Hoffmann-La Roche AG, CH-4000 Basel, Switzerland.

FLASCHEL, E.*
Institut Genie Chimique, Ecole Polytechnique Federal de Lausanne, CH-1024 Ecublens, Switzerland.

FLORA, ROBERT M.
Worthington Biochemical Corp., Halls Mill Road, Freehold, NJ 07728, USA.

FONTANA, ANGELO
Institut of Organic Chemistry, Via Marzolo, I-35100 Padova, Italy.

FREEMAN, AMIHAI +
Dept. Biochemistry, Tel Aviv University, Tel Aviv, Israel.

FROST, G.M.
John & E. Sturge Ltd., Denison Road, Selby, North Yorkshire Y08 8EF, England, UK.

FUKUI, SABURO*
Lab. Industrial Biochemistry, Dept. Industrial Chemistry, Fac. Engineering, Kyoto University, Yosida, Sakyo-ku, Kyoto 606, Japan.

GADDY, JAMES L.*
School Engineering, University of Missouri, Rolla, MO 65401, USA.

GAINER, JOHN L.*
Dept. Chemical Engineering, University of Virginia, Charlottsville, VA 22901, USA.

GAMS, TH. C.
Miles Kali-Chemie GmbH & Co., Hans Bockler Allee 20, D-3000 Hannover-Kleefeld, F.R. Germany.

GARDNER, COLIN R. +
Centre Recherche Merrell Intern. 16 Rue d'Ankara, F-67084 Strasbourg, France.

GAUTHERON, DANIELE C.*+
Lab. Biochimie Technol. Membranes, CNRS, Universite Claude Bernard de Lyon, 43 Blvd. 11 Nov. 1918, F-69621 Villeurbanne, France.

GESTRELIUS, STINA
Novo Industri A/S, Novo Alle 1, DK-2880 Bagsvaerd, Denmark.

GEYER, HANS ULRICH
Miles Kali-Chemie GmbH & Co., Hans Bockler Allee 20, D-3000 Hannover, F.R. Germany.

GIACOBBE, F.*
Biochem Design SpA, Roma, Italy.

GIBIAN, THOMAS G.
Technical Guidance International, 475 Steamboar Road, Greenwich, CT 06830, USA.

GIOVENCO, SILVIO
Snamprogetti SpA, Via E. Ramarini 32,
I-00015 Monterotondo, Roma, Italy.
GLOGER, MANFRED
Boehringer Mannheim GmbH, Bahnhofstrasse 5,
D-8132 Tutzing, F.R. Germany.
GODELMANN, BERNHARD
DECHEMA Institut, Theodoe Heuss Allee 25,
D-6000 Frankfurt (Main), F.R. Germany.
GOLKER, CHRISTIAN
Bayer AG, Friedrich Ebert Strasse 217,
D-5600 Wuppertal, F.R. Germany.
GOFFEAU, ANDRE
Lab. Enzymologie UCL, CCE, Place Croix du Sud 1,
B-1348 Louvain La Neuve, Belgium.
GOLDBERG, EUGENE P.*
Dept. Materials Science & Eng., University
of Florida, Gainsville, FL 32611, USA.
GOLDSTEIN, LEON*+
Dept. Biochemistry, Tel Aviv University,
Ramat Aviv, Tel Aviv, Israel.
GRATZ, A.
Behringwerke AG, Postfach 1140, D-3550
Marburg/Lahn, F.R. Germany.
GREGOR, HARRY*
Dept. Chemical Engineering Applied Chemistry,
Columbia University, Terrace Bldg., New York,
NY 10027, USA.
GREGORIADIS, GREGORY*
Division Clinical Investigation, Clinical
Research Centre, Watford Road, Harrow,
Middlesex HA1 3UJ, England, UK.
GUTHORLEIN, GERHARD
Behringwerke AG, Postfach 1140, D-3550
Marburg/Lahn, F.R. Germany.
HAMM, R.
Landwirtschaftl Versuchsstation, BASF AG,
D-6703 Limburgerhof, F.R. Germany.
HAN, MOON H.
Applied Biochemistry Lab., Institute of
Science & Technology, PO Box 131, Seoul, Korea.
HAMSHER, JAMES J.
Central Research, Pfizer Inc., Eastern
Point Rd., Groton, CT 06340, USA.
HANNIBAL-FRIEDRICH, OTTO
Institut Biochemie, Universitat Giessen,
D-6300 Giessen, F.R. Germany.

HARTMANN, GUNTER
Miles Kali-Chemie, Grosse Drakenburger Strasse 103, D-3070 Nienburg, F.R. Germany.

HARTIG, H.P.
G.R. Amylum NV, Van Wambekekaai 13, B-9300 Aalst, Belgium.

HARTMEIER, WINFRED
C.H. Boehringer Sohn, Postfach 200, D-6507 Ingelheim, F.R. Germany.

HEPNER, LEO
Walnak Hepner & Associates, Tavistock House North, Tavistock Square, London WC1H 9HX, England, UK.

HOCHULI, ERICH
Hoffmann-La Roche & Co. AG, Dept. F/ZFE 62/206, CH-4002 Basel, Switzerland.

HORNBY, WILLIAM E. =
Miles Laboratories Ltd., Stoke Court, Stoke Poges, Slough, Bucks SL2 4LY, England, UK.

HORVATH, CSABA*
Dept. Engineering Applied Science, Mason Lab., 9 Hillhouse Ave., Yale University, New Haven, CT 06520, USA.

HORWATH, R. OTTO
Biochemistry, Central Research Labs., Standard Brands Inc., PO Box 931, Stamford, CT 06904, USA.

HOWELL, JOHN A.*
Dept. Chemical Engineering, University College of Swansea, Swansea SA2 8PP, Wales, UK.

HUITRON, CARLOS
Instituto de Investigaciones Biomedicas, National University, Mexico 20, D.F., Mexico.

HULSMANN, HANS LEO
Dynamit Nobel AG, Werk Witten, D-5810 Witten/Ruhr, F.R. Germany.

HUGHES, M.A.
Development Pilot Plant, Beecham Pharmaceuticals, Clarendon Road, Worthing, Sussex, England, UK.

HUSTEDT, HELMUT +
Gesellschaft fur Biotechnologische Forschung, Mascheroder Weg 1, D-3300 Braunschweig-Stockheim, F.R. Germany.

JENSEN, VILLY*
R & D, Novo Industri A/S, Novo Alle 1, DK-2880 Bagsvaerd, Denmark.

JANCSIK, VERONIKA +
Enzymology Dept., Institute Biochemistry, Hungarian Academy of Sciences, H-1502 Budapest, Hungary.

KANG, CHUNGHEE K.
Biochemistry, Swift & Co., 1919 Swift Drive, Oak Brock, IL 60521, USA.

KASCHE, VOLKER*+
Biology Dept., Universitat Bremen, D-2800 Bremen 33, F.R. Germany.

KAWASHIMA, KOJI +
Radiation Utilization Lab., National Food Research Institute, 1-4-12 Shiohama, Kotoku, Tokyo 135, Japan.

KAZANSKAYA, N.F. +
Chemistry Dept., Moscow State University, Lenin-Hills, Moscow 117234, USSR.

KELETI, THOMAS +
Enzymology Dept., Institute Biochemistry, Hungarian Academy of Sciences, H-1502 Budapest, Hungary.

KENNEDY, JOHN F.*
Dept. Chemistry, University of Birmingham, Birmingham B15 2TT, England, UK.

KIESLICH, KLAUS
Mikrobiologische Chemie, Schering AG, Sellerstrasse 6, D-1000 Berlin 65, F.R. Germany.

KIRCH, PETER +
Institut Chemie Weihenstephan, Technical Universitat Munchen, D-8050 Freising-Weihenstephan, F.R. Germany.

KLEFENZ, HEINRICH
BASF AG, D-6700 Ludwigshafen/Rhein, F.R. Germany.

KLEIN, JOACHIM +
Institut fur Chemische Technologie, Technical University of Braunschweig, D-3300 Braunschweig, F.R. Germany.

KNAUSEDER, FRANZ
Biochemie Ges. mbH, A-6250 Kundl-Tirol, Austria.

KNOPFEL, HANS PETER
Process Engineering Co., 415 Alte Landstrasse, CH-8708 Mannedorf, Switzerland.

KONECNY, JAN O.*
Ciba-Geigy AG, CH-4002 Basel, Switzerland.

KOMINEK, LEO A.
Upjohn Co., 7000 Portage Rd., Kalamazoo, MI 49001, USA.

KRABBE, ERIK
US Environmental Protection Agency, 12-56 117 Street, New York, NY 11356, USA.

KRAEMER, DIETER M. +
Research Lab., Roehm GmbH., Kirschenallee, D-6100 Darmstadt, F.R. Germany.

KRAMPITZ, LESTER O.*
Dept. Microbiology, Case Western Reserve University, Cleveland, OH 44106, USA.

KRASNOBAJEW, VICTOR
Givaudan Research Comp. Ltd., Uberlandstrasse 138, CH-8600 Dubendorf, Switzerland.

KUSTERS, WERNER
BASF AG, D-6700 Ludwigshafen/Rhein, F.R. Germany.

KULA, MARIA-REGINA*
Gesellschaft fur Biotechnologische Forschung, Mascheroder Weg 1, D-3300 Braunschweig-Stockheim, F.R. Germany.

KUNSTMANN, MARTIN
Fermentation R & D, Lederle Laboratories, Middletown Rd., Pearl River, NY 10965, USA.

LANTERNO JR., ORESTE J. +
Miles Laboratories Inc., 1127 Myrtle St., Elkhart, IN 46514, USA.

LARSSON, P.O.*
Biochemical Div., Chemical Center, University of Lund, PO Box 740, S-22007 Lund 7, Sweden.

LASKIN, ALLEN I.
Biosciences Research, Exxon Research & Eng. Co., Exxon Research Center, Linden, NJ 07036, USA.

LAURENCE, ROBERT L.*
Dept. Chemical Engineering, University of Massachusetts, Amherst, MA 01003, USA.

LEE, CHI-YU +
Lab. Environmental Mutagenesis, National Institute of Environmental Health Sciences, Research Triangle Park, NC 27709, USA.

LEE, YOON Y.
Dept. Chemical Engineering, Auburn University, Auburn, AL 36830, USA.

LEHKY, PAVEL
Lonza AG, CH-3930 Visp, Switzerland.

LIEBERMANN, E.R.*
Biotechniques Inc., PO Box 954, Sommerville, NJ 08876, USA.

LILLY, MALCOLM D. =
Dept. Chemical & Biochemical Engineering, University College London, Torrington Place, London WC1, England, UK.

LIMON-LASON, JORGE +
Instituto de Investigaciones Biomedicas, Apartado Postal 70228, Ciudad Universitaria, Mexico 20, D. F., Mexico.

LINKO, PEKKA +
Chemistry Dept., Helsinki University of Technology, SF-02150 Espoo 15, Finland.

LINKO, YU-YEN +
Academy of Finland, Chemistry Dept., Helsinki University of Technology, SF-02150 Espoo 15, Finland.

MANECKE, GEORG*+=
Institut fur Organische Chemie, Freie Universitat Berlin, Thielallee 63/67, D-1000 Berlin 33, F.R. Germany.

MARCONI, W.*
Lab. Biochemical Processes, Snamprogetti SpA., Via E. Ramarini 32, I-00015 Monterotondo (Roma), Italy,

MAKAROVA, TAMARA N. +
State Committee USSR Council of Ministers for Science and Technology, Gorky Street, Moscow 103009, USSR.

MANDRAND, BERNRAD
Bio Merieux, Marcy l'Etaile, Charbonnieres les Bains, France.

MARTENSSON, KAY
Research Labs., Swedish Sugar Corp., Box 6, S-23200 Aarlov, Sweden.

MARTINY, STEEN C.
Chr. Hansen's Laboratory, Masnedogade 22, DK-2100 Copenhagen, Denmark.

MASON, J.R. +
Shell Biosciences Lab., Shell Research Ltd., Sittingbourne, Kent ME9 8AG, England, UK.

MATTIASSON, BO +
Biochemistry 2, Chemical Center, University of Lund, PO Box 740, S-22007 Lund 7, Sweden.

MEULEN, B. PH. TER
Central Technical Institute, TNO, PO Box 342, Apeldoorn, Netherlands.

MEYER, BORIS
Miles Kali-Chemie AG, Postfach 220, D-3000 Hannover, F.R. Germany.

MICHAELS, ALAN S.
Dept. Chemical Engineering, Stanford University, Stanford, CA 94305, USA.

MITSUGI, KOJI*
Central Research, Ajinomoto Co. Inc., 1-1 Suzuki-cho, Kawasaki-ku, Kawassaki, Kanagawa, Japan.

MOSBACH, KLAUS*=
Biochemistry 2, Chemical Center, University of Lund, PO Box 740, S-22007 Lund 7, Sweden.

MUNNECKE, DOUGLAS M. +
Institut fur Bodenbiologie, Bundesforschungsanstalt, fur Landwirtschaft, Bundesallee 50, D-3300 Braunschweig, F.R. Germany.

NEHEN, U.
Bayer AG, 5090 Leverkusen, Bayerwerk, F.R. Germany.

NIELSEN, HELWIIG
Novo Industri A/S, Novo Alle 1, DK-2880 Bagsvaerd, Denmark.

NYNS, ED. JACQUES
Lab. Enzymologie Appliquee, Universite de Louvain, Place Croix du Sud 1, B-1348 Louvain la Neuve, Belgium.

OKADA, HIROSUKE +
Dept. Fermentation Technology, Fac. Engineering, Osaka University, Yamada-kami, Suita-shi, Okaka 565, Japan.

OPPERMANN, ADOLF
Hoechst AG, Pharma-Production, D-6230 Frankfurt (Main) 80, F.R. Germany.

ORIEL, PATRICK J.
Dow Chemical Co., Bldg. 1701, Midland, MI 48640, USA.

OSHIMA, TAIRO*
Mitsubishi Kasei Institute of Life Sciences, 11 Minamiooya, Machida-shi, Tokyo, Japan.

OTTO, P. PH. H. L.
Central Lab., TNO, PO Box 217, Delft, Netherlands.

PACHALY, ROBERT W.
Technical Guidance International, 475 Steamboat Rd., Greenwich, CT 06830, USA.

PAICE, M.G. +
Pulp & Paper Research Institute of Canada, 570 St. Jean Blvd., Pointe Claire, Quebec, Canada.

PAPE, MARTIN
Basf AG, D-6700 Ludwigshafen/Rhein, F.R. Germany.

PITCHER JR., WAYNE H.*
Corning Glass Works, Corning, NY 14830, USA.

POULSEN, POUL B. +
Novo Industri A/S, Novo Alle 1, DK-2880, Bagsvaerd, Denmark.

POWELL, L.W.
Research Pharmaceuticals Ltd., Claredon Rd., Worthing, Sussex, England, UK.

PRENOSIL, J.E. +
Technisch-chemisches Lab., Federal Institute of Technology, CH-8092 Zurich, Switzerland.

PYE, E. KENDALL =
Dept. Biochemistry, School of Medicine, University of Pennsylvania, Philadelphia, PA 19104, USA.

RADOLA, B.J.
Institut fur Lebensmittel-technologie und Allgem. Chemie, Technische Universitat, D-8050 Freising-Weihenstephan, F.R. Germany.

RAMBACH, ALAIN*
Institut Pasteur, 28 Rue du Deroux, F-75724 Paris, Cedex 15, France.

RAPP, PETER
Gesellschaft fur Biotechnologische Forschung, Mascheroder Weg 1, D-3300 Braunschweig, F.R. Germany.

RAWLINS, C.J.
IPC Science & Technology Press Ltd., IPC House, 32 High Street, Guildford, Surrey GUI 3EW, England, UK.

REILLY, PETER J.*
Dept. Chemical Engineering & Nuclear Engineering, Iowa State University, Ames, IA 50011, USA.

REIMERDES, ERNST H. +
Institut Chemie, Bundesanstalt fur Milchforschung, Hermann Weigmann Strasse 3/11, D-2300 Kiel, F.R. Germany.

REINER, ROLAND R.*
Battelle Institut e. V., Am Romerhof 35, D-6000 Frankfurt (Main), F.R. Germany.

REINHARDT, GERD
Bayer AG, Friedrich Ebert Strasse, D-5600 Wuppertal, F.R. Germany.

REEN, DONALD W.
Marine Colloids Inc., PO Box 308, Rockland, ME 04841, USA.

ROSEN, CARL GUSTAF
Alfa-Laval AB, S-14700 Tumba, Sweden.

ROSENGREN, JAN
Pharmacia Fine Chemicals, PO Box 175, S-75104 Uppsala 1, Sweden.

ROESEVEAR, A. +
Biochemistry Group, AERE Harwell,
Oxon 11, England, UK.
ROYER, GARFIELD P.*+
Dept. Biochemistry, Ohio State University,
484 West 12th Ave., Columbus, OH 43210,
USA.
RUDKIN, GEORGE O.
ICI United States Inc., Wilmington, DE 19897,
USA.
RYU, DEWY +
Korea Advanced Institute of Science,
PO Box 150, Chung Ryang RI, Seoul, Korea.
SAMEJIMA, HIROTOSHI*
Tokuo Research Lab., Kyowa Hakko Kogyo Co. Ltd.,
3-6-6 Asahimachi, Machida-shi, Tokyo, 194, Japan.
SAUBER, KLAUS
Hoechst AG, D-6230 Frankfurt (Main) 80,
F.R. Germany.
SCHLUNSEN, JURGEN
Institut fur Organische Chemie, Freie
Universitat Berlin, Thielallee 63/67,
D-1000 Berlin 33, F.R. Germany.
SCHMID, ROLF D.
Biotechnology Dept., Henkel & Cie. GmbH.,
Postfach 1100, D-4000 Dusselforf 1, F.R. Germany.
SCHMIDT, HANNS LUDWIG +
Institut Chemie Weihenstephan, Technical
Universitat Munchen, D-8050
Freising-Weihenstephan, F.R. Germany.
SCHMIDT-KASTNER, GUNTER
Bayer AG, Dept. VE Biochemie, Friedrich
Ebert Strasse 217, D-5600 Wuppertal 1,
F.R. Germany.
SCOUTEN, WILLIAM H. +
Dept. Chemistry, Bucknell University,
Lewisburg, PA 17837, USA.
SEGARD, E.
Institut Technologie Surfaces Actives,
Universite de Technologie Compiegne,
F-60206 Compiegne, France.
SEQUEIRA, R.M.
Corn Products Research, Amstar Corp., PO Box
240, Woodland, CA 95695, USA.
SERNETZ, MANFRED +
Institut Biochemie, Universitat Giessen,
D-6300 Giessen, F.R. Germany.

SFAT, MICHAEL R.
Bio-Technical Resources Inc., 7th & Marshall St., Manitowoc, WI 54220, USA.

SICARD, P.J.
Societe Roquette Freres, F-62136 Lestrem, France.

SICSIC, SAM +
Centre d'Etudes Recherches de Chimie Organique Appliquee, CNRS, BP 28, F-94320 Thiais, France.

SIMON, HELMUT +
Institut Organic Chemische, Technical Universitat, Arcisstrasse 16, D-8000 Munchen, F.R. Germany.

SKOT, GEORG*
Novo Industri A/S, Novo Alle 1, DK-2880, Bagsvaerd, Denmark.

SMITH, JERRY H.
ICI United States Inc., Wilmington, DE 19897, USA.

SOLOMON, BARRY A. +
Dept. Chemical Engineering, Massachusetts Institute of Technology, Cambridge, MA 02139, USA.

SOMERVILLE, H.J. +
Shell Biosciences Lab., Shell Research Ltd., Sittingbourne, Kent ME9 8AG, England, UK.

SPROSSLER, BRUNO
Rohm GmbH., Postfach 42, D-6100 Darmstadt 1, F.R. Germany.

STEPANOV, V.M. +
Institute of Genetics & Selection of Industrial Microorganisms, Moscow, USSR.

SUKATSCH, D.A.
Hoechst AG, D-6230 Frankfurt (Main) 80, F.R. Germany.

SUNDARAM, P.V.*
Dept. Chemie, Max Planck Institut fur Experimentelle Medizin, Hermann Rein Strasse 3, D-3400 Gottingen, F.R. Germany.

SUZUKI, SHUICHI*
Research Lab. Resources Utilization, Tokyo Institute of Technology, Ookayama, Meguro-ku, Tokyo, Japan.

SWEIGART, R. DALE*
Biological Systems, Corning Glass Works, Corning, N.Y. 14830, USA.

SUNDBERG, LARS
Vitrum AB, Box 12 170, S-10224 Stockholm, Sweden.
SZENTIRMAI, ATTILA +
Dept. Microbiology, Research Institute Pharmaceutical Chemistry, Szabadsagharcos 47, HU-1045 Budapest, Hungary.
TAVES, MILTON A.
Hercules Inc., Research Center, Wilmington, DE 19899, USA.
THEVENOT, DANIEL +
EBAM, Lab. Energetique Biochimique, Universite Paris-Val de Marne, Ave. General de Gaulle, F-94010 Creteil, France.
THIBAULT, JEAN FRANCOIS +
Lab. Biochemie Aliments, INRA, Chemin de la Geraudiere, F-44072 Nantes, France.
THOMAS, DANIEL*
Lab. Technologie Enzymatique, Universite de Technologie Compiegne, 25 Rue Eugene Jacquet, F-60206 Compiegne, France.
TURKOVA, JAROSLAVA +
Dept. Protein Chemistry, Institute of Organic Chemistry & Biochemistry, Czechoslovak Academy of Science, 16610 Prague, Czechoslovakia.
UNDEN, AKE
AB Fermenta, S-15200 Strangnas, Sweden.
VAN BEYNUM, G. M. A.
Gist Brocades NV, PO Box 1, Delft, Netherlands.
VENKATASUBRAMANIAN, K. +
Corporate Research, H. J. Heinz Co., PO Box 57, Pittsburgh, PA 15230, USA.
VIETH, WOLF R.*
Dept. Chemical & Biochemical Engineering, Rutgers University, New Brunswick, NJ. 08903, USA.
VISSER, JAAP +
Dept. Genetics, Agricultural University, Gen. Foulkesweg 53, Wageningen, Netherlands.
VOGT, HANS-GUNTER +
Institut for Organische Chemie, Freie Universitat Berlin, Thielallee 63/67, D-1000 Berlin 33, F.R. Germany.
VOSS, H. FRED
Abbott Laboratories, Dept. 90T AP-9, North Chicago, IL 60064, USA.

WAGNER, FRITZ*+
Institut fur Biochemie Biotechnologie, Technical Universitat Braunschweig, Mascheroder Weg 1, D-3300 Braunschweig, F.R. Germany.

WAISSBLUTH, MARIO +
Consejo Nacional de Ciencia y Tecnologia, AP. Post. 20-033, Insurgentes Sur No. 1677, Mexico 20, D. F., Mexico.

WANDREY, CHRISTIAN*
Institut Technische Chemie, Technischen Universitat Hannover, Callinstrasse 46, D-3000 Hannover 1, F.R. Germany.

WARSCHAUER, K.A.
Div. Technology in Society, Organization for Industrial Research, TNO, PO Box 342, Apeldoorn, Netherlands.

WEAVER, JAMES C. +
Massachusetts Institute of Technology, Rm 26-461, Cambridge, MA 02139, USA.

WEETALL, HOWARD H.*
Corning Glass Works, Corning, NY 14830, USA.

WEIBEL, MICHAEL K.
R & D, Novo Laboratories Inc., 59 Danbury rd., Wilton, CT 06897, USA.

WELCH, G. RICKEY*
Dept. Biological Sciences, University of New Orleans, Lake Front, New Orleans, LA 70122, USA.

WESTLAKE, J.D.
British Petroleum Co. Ltd., BP Research Centre, Chertsey Road, Sanbury-on-Thames, Middx. TW16 7LN England, UK.

WIDMER, FRANCOIS +
Nestle Products Technical Assistance Co. Ltd., Research Dept., CH-1814 La Tour-de-Peilz, Switzerland.

WILCHEK, MEIR*+
Dept. Biophysics, Weizmann Institute of Science, Rehovot, Israel.

WINDISH, W.W.
Miles Kali-Chemie, Grosse Drakenburger Strasse 103, D-3070 Nienburg, F.R. Germany.

WINGARD JR., LEMUEL B.*+=
Dept. Pharmacology, School Medicine, 620 Scaife Hall, University of Pittsburgh, Pittsburgh, PA 15261, USA.

WINTER, OLAF
Process Planning, The Lummus Co., 1515
Broad St., Bloomfield, NJ 07003, USA.

YUGARI, Y.Y. +
Life Sciences Lab., Ajinomoto Co. Inc.,
1-1 Suzuki-cho, Kawasaki-ku, Kawasaki,
Kanagawa, Japan.

ZITTAN, LENA
Novo Industri A/S, Novo Alle 1, DK-2880
Bagsvaerd, Denmark.

SUBJECT INDEX

C.P.C. ABERYSTWYTH
LLYFRGELL